Bessel Functions and their Applications

ANALYTICAL METHODS AND SPECIAL FUNCTIONS
An International Series of Monographs in Mathematics

FOUNDING EDITOR: A.P. Prudnikov (Russia)
SERIES EDITORS: C.F. Dunkl (USA), H.-J. Glaeske (Germany) and M. Saigo (Japan)

This book is part of a series. The publisher will accept continuation orders which may be cancelled at any time and which provide for automatic billing and shipping of each title in the series upon publication. Please write for details.

Bessel Functions and their Applications

B.G. Korenev

CRC Press
Taylor & Francis Group
Boca Raton London New York

CRC Press is an imprint of the
Taylor & Francis Group, an **informa** business
A TAYLOR & FRANCIS BOOK

First published 2002 by Taylor & Francis

Published 2019 by CRC Press
Taylor & Francis Group
6000 Broken Sound Parkway NW, Suite 300
Boca Raton, FL 33487-2742

© 2002 by Taylor & Francis Group, LLC
CRC Press is an imprint of Taylor & Francis Group, an Informa business

First issued in paperback 2019

No claim to original U.S. Government works

ISBN-13: 978-0-367-45485-2 (pbk)
ISBN-13: 978-0-415-28130-0 (hbk)

Visit the Taylor & Francis Web site at
http://www.taylorandfrancis.com

and the CRC Press Web site at
http://www.crcpress.com

Translated by E.V. Pankratiev

Typeset in 10/12 Garamond by
Newgen Imaging Systems (P) Ltd, Chennai, India

British Library Cataloguing in Publication Data
A catalogue record for this book is available from the British Library

Library of Congress Cataloging in Publication Data
A catalogue record has been requested

Contents

Preface

A large number of diverse problems concerning practically all the most important areas of mathematical physics and various technical problems is connected with applications of Bessel functions. References where these functions are present are actually immense.

Different parts of Bessel function theory are widely used when solving problems of acoustics, radio physics, hydrodynamics, atomic and nuclear physics and so on.

Applications of Bessel functions to heat conduction theory, including dynamical and linked problems are very numerous.

In elasticity theory the solutions in Bessel functions are effective for all spatial problems, which are solved in spherical or cylindrical coordinates; also for different problems concerning the oscillations of plates and the equilibrium of plates on an elastic foundation; for a series of questions of the theory of shells; for problems on the concentration of the stresses near cracks and others.

In each of these areas there is a wide range of different applications of Bessel functions.

Bessel functions are well studied and there are a large number of reference books which contain the formulae and different tables.

The study of the theory of the Bessel function from the viewpoint of its applications has to be of interest for a large number of engineers and researchers. Special parts of different courses of mathematical physics are devoted to the presentation of the theory of Bessel functions (see the courses of R. Courant and D. Hilbert, P.M. Morse and H. Feshbach, A.N. Tikhonov and A.A. Samarski and others). At the same time, monographs devoted to the special functions also put an emphasis on the study of the Bessel functions. It concerns, for instance, the second part of "A course of modern analysis" by E.T. Whittaker and G.N. Watson, the book "Special functions and their applications" by N.N. Lebedev, the book "Transcendental functions" by A.N. Kratzer and V. Franz, the monograph by N.Ya. Vilenkin "Special functions and theory of group representations" and others.

There are several books concerning the theory of Bessel functions in Russian. We mention two editions of the book by R.O. Kuz'min "Bessel functions", the translation of the book by E. Grey and G. Matyuse "Bessel functions and their applications in physics and mechanics" as well as the translation of the most detailed course in the theory of Bessel functions, namely, the book by G.N. Watson "A treatise on the theory of Bessel functions".

This book contains a brief presentation of the elements of the theory of Bessel functions which is accompanied by examples of the applications of the theory to the solution of various physical problems.

The book is intended for the reader who is, basically, interested in the applications of the theory, not the theory itself. This fact determined the choice of the topics as well as the style of the presentation. The author's intention was to give the reader the opportunities to find the most necessary information so that afterwards the reader can consciously use the reference literature and can, all by himself or herself, apply the Bessel functions to the solution of applied problems.

Applying the methods of the theory of Bessel functions, an engineer-researcher must have a general notion of the foundation of the theory and clearly represent the mutual connections between its different areas. This is the reason why the deduction of the formulae, given in this book in a large number, are important for the reader who is interested in applications. For the same reason, when choosing the schemes of the deduction, we prefer those which use some properties of the Bessel functions in order to clarify other properties. The mutual connections between different parts of the theory can be seen more clearly by this approach.

Some short remarks or introductory sections of an heuristic nature are also intended to clarify the connections of the theory with physical applications. This is the viewpoint that should be used when considering some conclusions related, for example, with the addition formulae, the Bessel integral, improper integrals and others, as well as the applications of the Bessel functions considered in the second part of the book.

The book consists of two parts.

The first part of the book contains the presentation of the foundation of the theory of Bessel functions, the second part contains the applications. For the sake of convenience of reading, the first part is divided into two chapters.

In the first chapter, which contains comparatively simple material, we first consider the main properties of the solutions of the homogeneous Bessel equation, which are based on the representation of the solutions in the form of series in increasing powers of the argument. Then we give the Bessel integral, the Poisson integral and, in Section 19, some their generalizations (which can be skipped in the first reading because of their complexity). The Neumann addition theorems and the foundation of the theory of products of Bessel functions are considered. Questions concerning differential equations which can be reduced to the Bessel equation and the inhomogeneous Bessel equations are considered in detail. In this connection we consider here, in particular, the functions contiguous to Bessel functions. Moreover, we consider some integrals here whose deduction immediately follows from the properties of Bessel functions which are investigated in this chapter.

The second chapter can be skipped in the first reading. In this case one should return to this chapter only as the case may require. It contains the theory of definite and improper integrals as well as the elements of the theory of dual integral equations. Then the representation of the functions by the series of Fourier–Bessel, Dini and Schlömilch is presented. Before this group of sections we consider the roots of Bessel functions. Then we consider problems concerning the inhomogeneous Bessel equations which are more complicated than those considered in the first chapter, which lead to Lommel functions in two variables. From the same viewpoint we briefly discuss some facts concerning the partial cylindrical functions. In the conclusion of the chapter, we consider very briefly the asymptotic expansions of Bessel functions.

The second part of the book devoted to the applications also consists of two chapters. In the first one we consider problems concerning plates and shells of rotation. Among them some problems on the oscillations of a circular plate and on the equilibrium of a plate on a Winkler-type foundation are considered, as well as problems which can be reduced to the use of the method of compensating loadings. The method of initial parameters is considered in detail in the context of the problem of the axially symmetrical deformation of a circular conic shell.

In the second chapter we collected comparatively different problems concerning questions of the theory of oscillations, hydrodynamics, heat theory and others. Section 11 of this chapter stands apart. In this section we consider the boundary problems whose solution can be reduced to singular equations. Although we formally consider plates and membranes here, the solution has a more general nature and is less connected with concrete applications. If the reader wishes, he can read the second part of the book independently of the first one. In this case the reader can treat the second part as a continuation of the introduction, and look in the first part only to make inquires.

The author conceives it his pleasant duty to express his deep gratitude to Professor S.G. Mikhlin, who has attentively read the manuscript, for his very valuable notes and, not least measure, for the great attention he dave this work and advice given when discussing it.

The author sincerely thanks his disciples, A.I. Tseitlin for the notes produced after reading the manuscript, L.M. Reznikov and M.Ya. Volotski for help during the preparation of the manuscript for publication and verifying the numerous formulae.

B. Korenev

Introduction

In many problems of mathematical physics, whose solution is connected with the application of cylindrical and spherical coordinates, the process of separation of the variables results in the differential equation

$$z^2 \frac{d^2 u}{dz^2} + z \frac{du}{dz} + (z^2 - \nu^2)u = 0, \tag{0.1}$$

which is called the Bessel equation, and its solutions are called cylindrical or Bessel functions.

First, we shall show in some examples what kind of problems results in the Bessel equation, and then we shall consider different properties of Bessel functions and their applications.

Consider small free oscillations of a thin homogeneous circular membrane. The differential equation of the membrane oscillations has the form

$$\frac{1}{c^2} \frac{\partial^2 w}{\partial t^2} = \frac{\partial^2 w}{\partial r^2} + \frac{1}{r} \frac{\partial w}{\partial r} + \frac{1}{r^2} \frac{\partial^2 w}{\partial \varphi^2}, \tag{0.2}$$

where w is the membrane deflection; t is the time; r and φ are polar coordinates; $1/c^2 = \rho/N$ where ρ is the membrane mass relative to the unit of the area; N is the membrane tension.

We apply the method of separation of the variables, setting

$$w = u(r)T(t)\Phi(\varphi); \tag{0.3}$$

then from 0.2 we immediately obtain

$$T(t) = \sin(\alpha t + A), \quad \Phi(\varphi) = \sin(n\varphi + B),$$

where α and A are the circular frequency and initial phase of the oscillations, respectively; n is the number of nodal diameters: B is a constant which can easily be expressed via the angle between the first nodal diameter and the polar axis; $u(r)$ is a function of the variable r to be determined.

After substituting (0.3) into (0.2), we obtain the following equation with respect to u:

$$\frac{d^2 u}{dr^2} + \frac{1}{r} \frac{du}{dr} + \left(\frac{\alpha^2}{c^2} - \frac{n^2}{r^2} \right) u = 0. \tag{0.4}$$

If we introduce a new variable $\xi = \alpha r/c$, then equation (0.4) takes the form

$$\frac{d^2 u}{d\xi^2} + \frac{1}{\xi} \frac{du}{dr} + \left(1 - \frac{n^2}{\xi^2} \right) u = 0. \tag{0.5}$$

1

In the introduction we denote the function $u = u_n(\xi)$ and call ξ the argument and n the index. In the problem under consideration ξ is a real number and the index n is a natural number; non-integer indices will be denoted by ν.

Further, we consider the problem of the oscillations of a membrane which has the form of a circular sector with a central angle γ under the assumption that the membrane is inflexibly fixed along the whole contour.

We shall seek the solution of equation (0.2) in the form

$$w = u_\nu(r)\sin(\alpha t + A)\sin\frac{\pi(k+1)}{\gamma}\varphi, \quad k = 0, 1, 2, \ldots,$$

where k is the number of nodal radii lying inside the sector. If $\nu = \pi(k+1)/\gamma$ is a non-integer, then the problem reduces to solving the Bessel equation

$$\frac{d^2 u_\nu}{d\xi^2} + \frac{1}{\xi}\frac{du_\nu}{d\xi} + \left(1 - \frac{\nu^2}{\xi^2}\right)u_\nu = 0 \qquad (0.6)$$

with a non-integer index ν.

As a next example of an application of the Bessel equation we consider the problem of integration of the Laplace equation in cylindrical coordinates. Such problems occur very often in mathematical physics. For instance, considering the problem of heat conduction theory in the case when the temperature field is stationary and there are no heat sources, the temperature $T(r, \varphi, z)$ satisfies the equation

$$\frac{\partial^2 T}{\partial z^2} + \frac{\partial^2 T}{\partial r^2} + \frac{1}{r}\frac{\partial T}{\partial r} + \frac{1}{r^2}\frac{\partial^2 T}{\partial \varphi^2} = 0. \qquad (0.7)$$

If we seek a partial solution of equation (0.7) in the form

$$T_n = e^{-\alpha z}\sin(n\varphi + B)u_n(r), \qquad (0.8)$$

then, substituting (0.8) into (0.7), we obtain for the function $u_n(r)$ the equation

$$\frac{d^2 u_n}{dr^2} + \frac{1}{r}\frac{du_n}{dr} + \left(\alpha^2 - \frac{n^2}{r^2}\right)u_n = 0.$$

Let us consider the problem of the determination of the displacement $w(r, \varphi)$ of a membrane which lies on an elastic Winkler-type foundation (this means that along with the loading, the membrane is subjected to some reactive forces which are proportional with a coefficient k_0 to its deflections).

The differential equation of the elastic surface of the unloaded part of the membrane has the form

$$\nabla^2 w - \lambda^2 w = 0, \quad \lambda = \sqrt{k_0/N}, \qquad (0.9)$$

$$\left(\nabla^2 = \frac{\partial^2}{\partial r^2} + \frac{1}{r}\frac{\partial}{\partial r} + \frac{1}{r^2}\frac{\partial^2}{\partial \varphi^2}\right),$$

where N is the tension of the membrane; k_0 is called the *bedding constant of the foundation material*. If we write equation (0.9) in the polar coordinate system and assume that the membrane occupies a circular domain, then, seeking a partial solution in the form $w_n = u_n(\xi)\sin(n\varphi + B)$, we obtain for the function u_n the equation

$$\frac{d^2 u_n}{d\xi^2} + \frac{1}{\xi}\frac{du_n}{d\xi} + \left(-\frac{n^2}{\xi^2} - 1\right)u_n = 0, \qquad (0.10)$$

where $\xi = \lambda r$.

If we introduce a new variable $x = i\xi$, then equation (0.10) will be transformed into the Bessel equation with an imaginary argument.

The problem of equilibrium of a floating plate which has been considered by Hertz, can be reduced to the solution of the equation

$$\nabla^2 \nabla^2 w + \lambda^4 w = 0, \quad \lambda = \sqrt[4]{\gamma/D}, \tag{0.11}$$

where γ is the volumetric weight of the liquid, D is the *flexural rigidity* of the plate.

If we seek a partial solution of equation (0.11) in the form

$$w_n = u_n(\xi) \sin(n\varphi + B_n),$$

then, as can be easily verified, the function $u_n(\xi)$ can be represented in the form

$$u_n(\xi) = u_{1,n}(\xi) + u_{2,n}(\xi),$$

where one of the functions in the right-hand side satisfies the first of equations (0.12) and the second function satisfies the second equation[1]

$$\frac{d^2 u_n}{d\xi^2} + \frac{1}{\xi} \frac{du_n}{d\xi} - \left(\frac{n^2}{\xi^2} \pm i \right) u_n = 0. \tag{0.12}$$

Thus, the solution of this problem results in Bessel functions of a complex argument $\xi \sqrt{i}$.

One can find many other different applied problems whose solution connected with integration of the Laplace equation, the wave equation, the heat equation, a system of wave equations and so on, in cylindrical, spherical, and conic coordinates results after the separation of variables in the Bessel equation. Some differential equations of second order with variable coefficients can also be reduced to this equation. In the cases when there are sources distributed over the volume, the problems can often be reduced to integration of the inhomogeneous Bessel equations. A series of such problems will be considered in the second part of this book.

---

[1]This problem in detail will be considered in p. 155.

Part 1

Foundation of the theory of Bessel functions

CHAPTER 1

The Bessel equation. Properties of Bessel functions

1. The Bessel differential equation. Application of power series. Cylindrical functions of the first kind

Consider the *Bessel differential equation* with the index ν

$$z^2 \frac{d^2u}{dz^2} + z\frac{du}{dz} + (z^2 - \nu^2)u = 0. \tag{1.1}$$

Equation (1.1) is a linear differential equation of second order; hence, its general integral can be expressed in the form

$$u(z) = C_1 u_1(z) + C_2 u_2(z),$$

where $u_1(z)$ and $u_2(z)$ are linearly independent partial solutions of equation (1.1). Suppose that z and ν can admit any complex values. In the cases when the index ν is an integer, we shall denote it by the letter n. If the argument z is a real number, then we shall denote it by the letter x. We introduce the *Bessel operator* of index ν

$$\nabla_\nu \equiv z^2 \frac{d^2}{dz^2} + z\frac{d}{dz} + z^2 - \nu^2 \tag{1.2}$$

and we rewrite (1.1) in the following form: $\nabla_\nu u = 0$.

We shall seek a solution of equation (1.1) in the form of a generalized power series in increasing powers of the argument z

$$u(z) = \sum_{m=0}^{\infty} a_m z^{m+\alpha}, \tag{1.3}$$

where $a_0 \neq 0$.

Let us determine α and the coefficients a_m of series (1.3) here. For this purpose we find the first and the second derivatives of (1.3)

$$u'(z) = \sum_{m=0}^{\infty} a_m (m + \alpha) z^{m+\alpha-1},$$

$$u''(z) = \sum_{m=0}^{\infty} a_m (m + \alpha)(m + \alpha - 1) z^{m+\alpha-2}$$

and substitute the series obtained into the left-hand side of equation (1.1) instead of the function $u(z)$.

Collecting the terms containing equal powers of z, we obtain the following series:

$$\nabla_\nu u = [\alpha(\alpha - 1) + \alpha - \nu^2]a_0 z^\alpha + [(\alpha + 1)\alpha + (\alpha + 1) - \nu^2]a_1 z^{\alpha+1}$$
$$+ \{[(\alpha + 2)(\alpha + 1) + (\alpha + 2) - \nu^2]a_2 + a_0\}z^{\alpha+2} + \dots$$
$$= a_0(\alpha^2 - \nu^2)z^\alpha + a_1[(\alpha + 1)^2 - \nu^2]z^{\alpha+1}$$
$$+ \sum_{m=2}^{\infty} \{a_m[(\alpha + m)^2 - \nu^2] + a_{m-2}\}z^{\alpha+m}. \tag{1.4}$$

Since $\nabla_\nu u = 0$, it is evident that all of the coefficients of $z^{\alpha+m}$ in (1.4) should be equal to zero.

This condition gives us an infinite system

$$a_0(\alpha^2 - \nu^2) = 0, \quad [(\alpha + 1)^2 - \nu^2]a_1 = 0,$$
$$a_m[(\alpha + m)^2 - \nu^2] + a_{m-2} = 0, \quad m = 2, 3, 4, \dots .$$

For even and odd values of m we obtain, respectively, the following two systems of equation:

$$(\alpha^2 - \nu^2)a_0 = 0, \quad [(\alpha + 2)^2 - \nu^2]a_2 + a_0 = 0,$$
$$[(\alpha + 4)^2 - \nu^2]a_4 + a_2 = 0, \dots, \tag{1.5}$$
$$[(\alpha + 1)^2 - \nu^2]a_1 = 0, \quad [(\alpha + 3)^2 - \nu^2]a_3 + a_1 = 0, \dots \tag{1.6}$$

Since $a_0 \neq 0$, the first equation of system (1.5) implies the equation $\alpha^2 - \nu^2 = 0$, which is called *determining*; this equation implies that $\alpha = \pm\nu$.

The second and next equations of system (1.5) allow one to express the coefficients a_m with even subscripts via a_0 and result in the recurrence relations

$$a_2 = -\frac{a_0}{4(\alpha + 1)}, \quad a_4 = -\frac{a_2}{8(\alpha + 2)} = \frac{a_0}{4 \cdot 8 \cdot (\alpha + 1)(\alpha + 2)}, \quad \dots .$$

Equations (1.6) are satisfied, if we set $a_1 = a_3 = a_5 = \dots = 0$.

Setting $\alpha = \nu$, we formally obtain the first partial solution of equation (1.1) in the form

$$u_1(z) = a_0 z^\nu \left[1 - \frac{z^2}{4 \cdot 1 \cdot (\nu + 1)} + \frac{z^4}{4^2 \cdot 2!(\nu + 1)(\nu + 2)} \right.$$
$$\left. - \frac{z^6}{4^3 \cdot 3!(\nu + 1)(\nu + 2)(\nu + 3)} + \dots \right];$$

for $\alpha = -\nu$ we obtain the second partial solution

$$u_2(z) = a_0' z^{-\nu} \left[1 - \frac{z^2}{4 \cdot 1 \cdot (-\nu + 1)} + \frac{z^4}{4^2 \cdot 2!(-\nu + 1)(-\nu + 2)} \right.$$
$$\left. - \frac{z^6}{4^3 \cdot 3!(-\nu + 1)(-\nu + 2)(-\nu + 3)} + \dots \right].$$

Usually, the constants a_0 and a_0' are assigned the following values:

$$a_0 = \frac{1}{2^\nu \Gamma(\nu + 1)}, \quad a_0' = \frac{1}{2^{-\nu}\Gamma(-\nu + 1)}.$$

Using property (3) of gamma functions, one can rewrite the series obtained in a more compact form[1].

The first series $u_1(z)$ defines a function which is called the *Bessel* or *cylindrical function of the first kind* of the *Bessel function of index*[2] ν of the argument z

$$J_\nu(z) = \left(\frac{z}{2}\right)^\nu \sum_{m=0}^\infty \frac{(-1)^m (z/2)^{2m}}{m!\,\Gamma(\nu+m+1)};\qquad (1.7)$$

the second series defines the Bessel function of the negative index $-\nu$

$$J_{-\nu}(z) = \left(\frac{z}{2}\right)^{-\nu} \sum_{m=0}^\infty \frac{(-1)^m (z/2)^{2m}}{m!\,\Gamma(-\nu+m+1)}.\qquad (1.8)$$

It can easily be shown that the power series obtained converge on the whole plane of the complex variable z (except, may be, the point $z = 0$) and admit term by term differentiation.

Actually, the series for $J_\nu(z)$ converges absolutely and uniformly in any bounded domain if variation of the index ν and in any closed domain of variation of z (if $\Re\nu < 0$, then at $z = 0$ the function has a singularity of the type $z^{\Re\nu}$; therefore the origin is not included in the domain). The validity of this assertion follows from the fact that for $|\nu| \le N$ and $|z| < d$ the absolute value of the ratio of the next term of the series to the previous one is less than one

$$\left|\frac{-z^2/4}{m(\nu+m)}\right| \le \frac{d^2/4}{m(m-N)} < 1,$$

if m is greater than the positive root of the equation $m^2 - mN - d^2/4 = 0$, which does not depend on ν and z.

The Weierstrass criterion proves the indicated property of the power series. Hence, the Bessel function $J_\nu(z)$ is an analytical function for all values of ν in a neighbourhood of any value of z, except, may be, $z = 0$. This, in particular, implies that the series obtained can be term by term differentiated and integrated.

Obviously, for a non-integer index the functions $J_\nu(z)$ and $J_{-\nu}(z)$ are linearly independent. If $\nu = n$ is an integer, then $\Gamma(n) = (n-1)!$ and the function $J_n(z)$ can be written in the following form:

$$J_n(z) = \sum_{m=0}^\infty \frac{(-1)^m z^{2m+n}}{2^{2m+n} m!\,(n+m)!}.\qquad (1.9)$$

Let us show that in this case there exists the following dependence between the functions $J_n(z)$ and $J_{-n}(z)$:

$$J_{-n}(z) = (-1)^n J_n(z).\qquad (1.10)$$

Indeed, if k is a positive integer or zero, then, as it is indicated in the Appendix (265), $\Gamma(-k) = \infty$. Therefore, for the positive integer $\nu = n$ each of the first n terms of series (1.8) vanishes; rewriting (1.8) starting from the $(n+1)$th term, we

---

[1] Information on gamma functions is given in the Appendix; all references on the formulae of the Appendix contain a number consisting of one integer; more detailed information concerning gamma functions can be found, for instance, in [4], [53].

[2] Sometimes, one uses instead of the term "index" the equivalent term "order".

obtain

$$J_{-n}(z) = \frac{(-1)^n (z/2)^{-n+2n}}{n!\Gamma(-n+n+1)} + \frac{(-1)^{n+1}(z/2)^{-n+2n+2}}{(n+1)!\Gamma(-n+n+2)} + \cdots$$

$$= (-1)^n \left[\frac{(z/2)^n}{0!n!} - \frac{(z/2)^{n+2}}{1!(n+1)!} + \frac{(z/2)^{n+4}}{2!(n+2)!} - \cdots \right]$$

$$= (-1)^n J_n(z). \tag{1.11}$$

2. Cylindrical functions of the second kind (Neumann functions)

In Section 1 we showed that for a non-integer index the general solution of the Bessel equation can be written in the form

$$u = B_1 J_\nu(z) + B_2 J_{-\nu}(z).$$

Obviously, in this case the function

$$u = B_3 J_\nu(z) + B_4 Z_\nu(z),$$

where

$$Z_\nu(z) = C_1 J_\nu(z) + C_2 J_{-\nu}(z),$$

is also a solution of this equation. Here B_1, B_2, B_3, B_4, C_1 and C_2 do not depend on the argument z; $C_2 \neq 0$.

If we set $C_1 = \cot \nu\pi$ and $C_2 = -\csc \nu\pi$, then we obtain a function which has been introduced by Weber and is denoted by $Y_\nu(z)$:

$$Y_\nu(z) = \frac{J_\nu(z) \cos \nu\pi - J_{-\nu}(z)}{\sin \nu\pi}. \tag{2.1}$$

In the literature this function is often called the *Neumann function* and is sometimes denoted by $N_\nu(z)$. The function $Y_\nu(z)$ is also called the *Bessel* or *cylindrical function of the second kind* of the index ν of the argument z.

For an integer value $\nu = n$ the right-hand side of (2.1) is an indeterminacy of the type $\frac{0}{0}$. In order to find $Y_(z)$ we remove the indeterminacy by the L'Hospital rule and we define

$$Y_n(z) = \lim_{\nu \to n} \frac{\frac{\partial}{\partial \nu}[J_\nu(z)\cos \nu\pi - J_{-\nu}(z)]}{\frac{\partial}{\partial \nu} \sin \nu\pi}. \tag{2.2}$$

As a result, we obtain the formula

$$Y_n(z) = \frac{2}{\pi} J_n(z) \left(\ln \frac{z}{2} + C \right) - \frac{1}{\pi} \sum_{m=0}^{n-1} \frac{(n-m-1)!}{m!} \left(\frac{z}{2} \right)^{-n+2m}$$

$$- \frac{1}{\pi} \sum_{m=0}^{\infty} \frac{(-1)^m (z/2)^{n+2m}}{m!(m+n)!} \left\{ \sum_{k=1}^{n+m} \frac{1}{k} + \sum_{k=1}^{m} \frac{1}{k} \right\}, \tag{2.3}$$

where C is the Euler constant; it is approximately equal to 0.5772157.

The Neumann function $Y_\nu(z)$ and the function $J_\nu(z)$ form a fundamental system of solutions to the Bessel equation for any, including integer, value of the index.

Let us deduce formula (2.3). It is sufficient to obtain this formula for the special case $n = 0$ (then, using the recurrence relations considered below in Section 6, one can obtain the result for any n).

From (2.2) we have

$$Y_0(z) = \lim_{\nu \to 0} \frac{\frac{\partial}{\partial \nu}[J_\nu(z)\cos\nu\pi - J_{-\nu}(z)]}{\frac{\partial}{\partial \nu}\sin\nu\pi} = \frac{2}{\pi}\left[\frac{\partial J_\nu(z)}{\partial \nu}\right]_{\nu=0};$$

differentiating (1.7) term by term with respect to ν and using formulae (9), (11), we obtain

$$Y_0(z) = \frac{2}{\pi}J_0(z)\left(\ln\frac{z}{2} + C\right)$$
$$- \frac{2}{\pi}\sum_{k=1}^{\infty}\frac{(-1)^k}{(k!)^2}\left(\frac{z}{2}\right)^{2k}\left[1 + \frac{1}{2} + \cdots + \frac{1}{k}\right]. \tag{2.4}$$

3. Cylindrical functions of the third kind (Hankel functions)

Any linear combination of the solutions obtained in Sections 1, 2, is also an integral of the Bessel equation.

Consider the functions

$$H_\nu^{(1)}(z) = J_\nu(z) + iY_\nu(z),$$
$$H_\nu^{(2)}(z) = J_\nu(z) - iY_\nu(z), \tag{3.1}$$

which are called *cylindrical functions of the third kind*. They are also called *Hankel functions* of the first and second kind, respectively.

One can see from (3.1) that there exist dependencies between Hankel, Bessel, and Neumann functions which are analogous to the connection between the exponential function of an imaginary argument, cosine, and sine.

The Bessel function of a real argument in the problem of integration of the wave equation is an image of a standing wave, and the Hankel functions give an image of a spreading wave. Therefore, it is obvious, what important role is played by Hankel functions in applications, especially, when studying wave processes in unbounded domains.

4. Cylindrical functions of a pure imaginary argument

In this section we consider the Bessel functions of an imaginary argument $z = ix$. After changing the variable in (1.1) by the formula $z = ix$, we obtain the equation

$$x^2\frac{d^2u}{dx^2} + x\frac{du}{dx} - (x^2 + \nu^2)u = 0. \tag{4.1}$$

One of solutions of equation (4.1) is the Bessel function of an imaginary argument $J_\nu(ix)$ defined by expression (1.7) after replacing z by ix in this expression. The function $J_n(ix)$ is real for any even n and takes imaginary values for odd n. In order to get rid of this inconvenience one introduces, usually, in consideration the *modified Bessel function*

$$I_\nu(x) = e^{-\nu\pi i/2}J_\nu(ix), \tag{4.2}$$

which is real for any ν. The expansion of the function $I_\nu(x)$ into a power series has the following form

$$I_\nu(x) = \sum_{m=0}^{\infty}\frac{(x/2)^{2m+\nu}}{m!\Gamma(\nu + m + 1)}. \tag{4.3}$$

As the second integral of equation (4.1), usually, the following function is taken

$$K_\nu(x) = \frac{1}{2}\pi i e^{\nu\pi i/2} H_\nu^{(1)}(ix),\tag{4.4}$$

which is real for any real value of the index.

This function is called the *Macdonald function*; for integer $\nu = n$ the power series expansion of this function has the form

$$K_n(x) = (-1)^{n+1} I_n(x)\ln\frac{x}{2} + \frac{1}{2}\sum_{m=0}^{n-1}(-1)^m \left(\frac{x}{2}\right)^{-n+2m}\frac{(n-m-1)!}{m!}$$

$$+ (-1)^{n+1}\frac{1}{2}\sum_{m=0}^{\infty}\frac{(x/2)^{n+2m}}{m!(n+m)!}\left[2C - \sum_{k=1}^{m+n}\frac{1}{k} - \sum_{k=1}^{m}\frac{1}{k}\right],\tag{4.5}$$

where, as before, C is the Euler constant. This formula can easily be obtained, if we introduce $J_\nu(z)$ and $Y_\nu(z)$ into (4.4) with the help of the first of formulae (3.1) and afterwards change the argument z by ix in formulae (1.9) and (2.3) and express $J_n(ix)$ via $I_n(x)$ using formula (4.2).

Note that the formulae given above for the modified functions can be applied, if we replace x by a complex argument z; however, in this case the modified function $I_\nu(z)$ is defined in the following way:

$$I_\nu(z) = e^{-\nu\pi i/2} J_\nu(ze^{\pi i/2}), \qquad -\pi < \arg z \le \frac{\pi}{2},$$

$$I_\nu(z) = e^{3\nu\pi i/2} J_\nu(ze^{-3\pi i/2}), \qquad \frac{\pi}{2} < \arg z \le \pi.$$

For large values of the argument x the functions $I_n(x)$ and $K_n(x)$ behave similarly to the exponential function of a real positive and real negative argument, respectively. Therefore, sometimes the functions $e^x K_n(x)$ and $e^{-x} I_n(x)$ are tabulated.

5. Cylindrical functions of a complex argument

If we set $\nu = 0$ and $z = x\sqrt{-i}$ in (1.1), then this equation takes the form

$$\frac{d^2u}{dx^2} + \frac{1}{x}\frac{du}{dx} - iu = 0.\tag{5.1}$$

One of the two solutions of this equation is the function of the complex argument $I_0(x\sqrt{i})$, whose power series expansion consists of alternating real and imaginary terms.

The real and imaginary parts of this equation are denoted by ber x and bei x and are called, usually, the *Thomson functions*.

Replacing the argument x by $x\sqrt{i}$ and setting $\nu = 0$ in (4.3), we obtain

$$I_0(x\sqrt{i}) = \text{ber } x + i\,\text{bei } x,\tag{5.2}$$

where

$$\text{ber } x = \sum_{m=0}^{\infty}\frac{(-1)^m (x/2)^{4m}}{(2m!)^2},\tag{5.3}$$

$$\text{bei } x = \sum_{m=1}^{\infty}\frac{(-1)^{m-1}(x/2)^{4m-2}}{[(2m-1)!]^2}.\tag{5.4}$$

Similarly, the function

$$I_0(x\sqrt{-i}) = \mathrm{ber}\, x - i\,\mathrm{bei}\, x \tag{5.5}$$

is one of the solutions of the equation

$$\frac{d^2u}{dx^2} + \frac{1}{x}\frac{du}{dx} + iu = 0. \tag{5.6}$$

Comparing (4.3) with (1.9), we obtain

$$I_0(x\sqrt{\pm i}) = J_0(x\sqrt{\mp i}).$$

Thus, the functions $J_0(x\sqrt{-i})$ and $J_0(x\sqrt{i})$ are solutions of equations (5.1) and (5.6), respectively.

For the sake of brevity, we introduce the notation

$$u_0(x) = \mathrm{ber}\, x, \quad v_0(x) = -\,\mathrm{bei}\, x,$$

which is used in many papers.

As the second solution of equation (5.6) we can take the following function of the complex argument

$$H_0^{(1)}(x\sqrt{i}) = f_0(x) + ig_0(x), \tag{5.7}$$

where, by virtue of (1.9), (2.4), and (3.1) we have

$$f_0(x) = \frac{u_0(x)}{2} - \frac{2}{\pi}\left[R_1(x) + v_0(x)\ln\frac{\gamma x}{2}\right], \tag{5.8}$$

$$g_0(x) = \frac{v_0(x)}{2} + \frac{2}{\pi}\left[R_2(x) + u_0(x)\ln\frac{\gamma x}{2}\right], \tag{5.9}$$

and

$$R_1(x) = \sum_{k=0}^{\infty}(-1)^k[(2k+1)!]^{-2}\left(\frac{x}{2}\right)^{4k+2}\sum_{m=1}^{2k+1}\frac{1}{m},$$

$$R_2(x) = \sum_{k=0}^{\infty}(-1)^k[(2k+1)!]^{-2}\left(\frac{x}{2}\right)^{4k+4}\sum_{m=1}^{2k+2}\frac{1}{m},$$

$$\gamma = \exp C.$$

As the second solution of equation (5.1) we take the function

$$H_0^{(2)}(x\sqrt{-i}) = f_0 - ig_0(x). \tag{5.10}$$

The solutions of the more general equation

$$\left(\frac{d^2}{dx^2} + \frac{1}{x}\frac{d}{dx} - \frac{\nu^2}{x^2}\right)u \pm iu = 0$$

are, respectively, the following functions of a complex argument

$$J_\nu(x\sqrt{\pm i}) = u_\nu(x) \pm iv_\nu(x), \tag{5.11}$$

$$H_\nu^{(1,2)}(x\sqrt{\pm i}) = f_\nu(x) \pm ig_\nu(x). \tag{5.12}$$

For an integer index, the expansions of the functions u_n, v_n, f_n, and g_n, can be obtained as before with the help of expressions (1.9), (2.3), and (3.1), if we replace there z by $x\sqrt{\pm i}$ and then divide the real and imaginary parts.

In various problems of physics and mechanics, in particular, in problems on circular plates compressed or stretched by axial forces, which are considered in Part 2, there occur some functions of the argument $\rho e^{i\varphi} = a + bi$.

Let us introduce the notation

$$J_\nu(\rho e^{\pm i\varphi}) = \tilde{u}_\nu(\rho) \pm i\tilde{v}_\nu(\rho),$$ (5.13)

$$H_\nu^{(1,2)}(\rho e^{\pm i\varphi}) = \tilde{f}_\nu(\rho) \pm i\tilde{g}_\nu(\rho),$$ (5.14)

where φ is a constant, and $\tilde{u}_\nu$, $\tilde{v}_\nu$, $\tilde{f}_\nu$, and $\tilde{g}_\nu$ are defined, as before, with the help of formulae (1.9), (2.4), and (3.1) after substituting in them $\rho e^{\pm i\varphi}$ instead of z and dividing then the real and imaginary parts; some formulae concerning these functions are given in Section 6.

Let us note that as $\rho \to 0$ we have

$$\tilde{g}_0(\rho) \sim \frac{2}{\pi}\left(\ln\frac{\rho}{2} + C\right),$$ (5.15)

$$\tilde{f}_0(\rho) \sim \frac{1}{2\pi}\rho^2 \ln\rho \sin 2\varphi + 1 - \frac{2\varphi}{\pi}.$$ (5.16)

When $\varphi = \pi/4$, expressions (5.13) and (5.14) are transformed into (5.11) and (5.12).

The cylindrical functions of a complex argument whose imaginary part is small compared with the real one, i.e. when $|\tan\varphi| \ll 1$ are also of practical interest. This case often occurs in problems of oscillation theory and acoustics and will be considered in Section 6.

6. Formulae of differentiation, recurrence relations

Dividing (1.7) by z^ν, we have

$$\frac{J_\nu(z)}{z^\nu} = \frac{1}{2^\nu} \sum_{m=0}^\infty \frac{(-1)^m (z/2)^{2m}}{m!\Gamma(\nu+m+1)};$$

differentiating this equality with respect to the argument z, we obtain the relation

$$\frac{d}{dz}\frac{J_\nu(z)}{z^\nu} = \frac{1}{2^\nu} \sum_{m=1}^\infty \frac{(-1)^m (z/2)^{2m-1}}{(m-1)!\Gamma(\nu+m+1)} = -\frac{J_{\nu+1}(z)}{z^\nu},$$

which can be rewritten in the following form:

$$\frac{1}{z}\frac{d}{dz}\frac{J_\nu(z)}{z^\nu} = -\frac{J_{\nu+1}(z)}{z^{\nu+1}}.$$ (6.1)

Similarly, we can obtain the formula

$$\frac{d}{z\,dz}[z^\nu J_\nu(z)] = z^{\nu-1}J_{\nu-1}(z).$$ (6.2)

After differentiating the left-hand side of formulae (6.1) and (6.2) and simplificating the expressions obtained, we have the equalities

$$\frac{d}{dz}J_\nu(z) = -J_{\nu+1}(z) + \frac{\nu J_\nu(z)}{z},$$ (6.3)

$$\frac{d}{dz}J_\nu(z) = J_{\nu-1}(z) - \frac{\nu J_\nu(z)}{z},$$ (6.4)

which imply the following recurrence relations:

$$J_{\nu-1}(z) + J_{\nu+1}(z) = \frac{2\nu J_\nu(z)}{z},$$ (6.5)

$$J_{\nu-1}(z) - J_{\nu+1}(z) = 2\frac{d}{dz}J_\nu(z).$$ (6.6)

One can replace $J_\nu(z)$ in all these formulae by any of the functions: $Y_\nu(z)$, $H_\nu^{(1)}(z)$, $H_\nu^{(2)}(z)$. Repeatedly differentiating formulae (6.1) and (6.2), one can obtain

$$\left(\frac{d}{z\,dz}\right)^m [z^\nu J_\nu(z)] = z^{\nu-m} J_{\nu-m}(z), \tag{6.7}$$

$$\left(\frac{d}{z\,dz}\right)^m [z^{-\nu} J_\nu(z)] = (-1)^m z^{-\nu-m} J_{\nu+m}(z). \tag{6.8}$$

For the modified cylindrical functions we have the following formulae of differentiation, which are obtained as a result of the change of the argument z by ix and representation of the functions $J_\nu(z)$ and $H_\nu^{(1)}(z)$ via the functions $I_\nu(x)$ and $K_\nu(x)$:

$$\frac{d}{dx} I_\nu(z) = \frac{1}{2}[I_{\nu-1}(x) + I_{\nu+1}(x)], \tag{6.9}$$

$$\frac{d}{dx} K_\nu(z) = -\frac{1}{2}[K_{\nu-1}(x) + K_{\nu+1}(x)], \tag{6.10}$$

The corresponding recurrence relations have the form

$$I_{\nu-1}(x) - I_{\nu+1}(x) = \frac{2\nu}{x} I_\nu(x), \tag{6.11}$$

$$K_{\nu-1}(x) - K_{\nu+1}(x) = -\frac{2\nu}{x} K_\nu(x). \tag{6.12}$$

For the complex argument $z = x\sqrt{i}$, the formulae of differentiation of the functions u, v, f, and g with respect to the absolute value of x can be found, replacing the argument z by $x\sqrt{i}$ in expressions (6.3) and (6.4) and dividing the real and imaginary parts in the expressions obtained:

$$u_\nu'(x) = \frac{1}{\sqrt{2}}[u_{\nu-1}(x) - v_{\nu-1}(x)] - \frac{\nu}{x} u_\nu(x), \tag{6.13}$$

$$v_\nu'(x) = \frac{1}{\sqrt{2}}[u_{\nu-1}(x) + v_{\nu-1}(x)] - \frac{\nu}{x} v_\nu(x), \tag{6.14}$$

$$f_\nu'(x) = \frac{1}{\sqrt{2}}[f_{\nu-1}(x) - g_{\nu-1}(x)] - \frac{\nu}{x} f_\nu(x), \tag{6.15}$$

$$g_\nu'(x) = \frac{1}{\sqrt{2}}[f_{\nu-1}(x) + g_{\nu-1}(x)] - \frac{\nu}{x} g_\nu(x). \tag{6.16}$$

From the practical standpoint the following functions are of interest, which can be obtained from the Bessel differential equation by the change of the argument z by $x\sqrt{i}$:

$$\Delta_\nu u_\nu = v_\nu, \quad \Delta_\nu v_\nu = -u_\nu, \quad \Delta_\nu f_\nu = g_\nu, \quad \Delta_\nu g_\nu = -f_\nu, \tag{6.17}$$

where

$$\Delta_\nu = \frac{d^2}{dx^2} + \frac{1}{x}\frac{d}{dx} - \frac{\nu^2}{x^2}. \tag{6.18}$$

For the functions of the complex argument $z = \rho\, e^{i\varphi}$ the formulae of differentiation with respect to the modulus ρ have the form

$$\tilde{u}'_\nu(\rho) = -\frac{\nu}{\rho}\tilde{u}_\nu(\rho) + \tilde{u}_{\nu-1}(\rho)\cos\varphi - \tilde{v}_{\nu-1}(\rho)\sin\varphi, \tag{6.19}$$

$$\tilde{v}'_\nu(\rho) = -\frac{\nu}{\rho}\tilde{v}_\nu(\rho) + \tilde{u}_{\nu-1}(\rho)\sin\varphi + \tilde{v}_{\nu-1}(\rho)\cos\varphi, \tag{6.20}$$

$$\tilde{f}'_\nu(\rho) = -\frac{\nu}{\rho}\tilde{f}_\nu(\rho) + \tilde{f}_{\nu-1}(\rho)\cos\varphi - \tilde{g}_{\nu-1}(\rho)\sin\varphi, \tag{6.21}$$

$$\tilde{g}'_\nu(\rho) = -\frac{\nu}{\rho}\tilde{g}_\nu(\rho) + \tilde{f}_{\nu-1}(\rho)\sin\varphi + \tilde{g}_{\nu-1}(\rho)\cos\varphi. \tag{6.22}$$

For $\nu = 0$ we have

$$\tilde{u}'_0(\rho) = -\tilde{u}_1(\rho)\cos\varphi + \tilde{v}_1(\rho)\sin\varphi, \tag{6.23}$$

$$\tilde{v}'_0(\rho) = -\tilde{u}_1(\rho)\sin\varphi - \tilde{v}_1(\rho)\cos\varphi. \tag{6.24}$$

We also give the formulae similar to (6.17) for $\nu = 0$

$$\left.\begin{aligned}\Delta_0\tilde{u}_0 &= -\tilde{u}_0\cos 2\varphi + \tilde{v}_0\sin 2\varphi,\\ \Delta_0\tilde{v}_0 &= -\tilde{u}_0\sin 2\varphi - \tilde{v}_0\cos 2\varphi.\end{aligned}\right\} \tag{6.25}$$

In (6.23)–(6.25) one can replace $\tilde{u}$ and $\tilde{v}$ by $\tilde{f}$ and $\tilde{g}$, respectively.

Consider functions of the complex argument $z = x(1 + \kappa i)$, where $|\kappa| \ll 1$. Let us introduce the following notation:

$$Z_\nu[x(1 + \kappa i)] = u_\nu^{(1)}(x) + i v_\nu^{(1)}(x), \tag{6.26}$$

where Z_ν is the cylindrical function of index ν.

If after changing the argument we only keep terms containing powers of x no higher than one in the appropriate power series, then as a result, we obtain the following approximate formulae:

$$u_\nu^{(1)}(x) \approx Z_\nu(x), \quad v_\nu^{(1)}(x) \approx \kappa x\frac{dZ_\nu(x)}{dx}, \tag{6.27}$$

which can be used in practice, however, only for relatively small values of $|x|$.

7. Cylindrical functions with a half-integer index

Setting the index $\nu = 1/2$ in the expansion for $J_\nu(z)$ and replacing the gamma-functions by their values according to formula (15), after elementary simplifications we obtain

$$J_{1/2}(z) = \sqrt{\frac{2}{\pi z}}\sum_{k=0}^{\infty}\frac{(-1)^k z^{2k+1}}{(2k+1)!} = \sqrt{\frac{2}{\pi z}}\sin z. \tag{7.1}$$

Differentiating (7.1), we obtain

$$J'_{1/2}(z) = \sqrt{\frac{2}{\pi z}}\cos z - \frac{1}{z}\sqrt{\frac{1}{2\pi z}}\sin z. \tag{7.2}$$

The we use formula (6.4); setting $\nu = 1/2$, one can easily obtain

$$zJ'_{1/2}(z) + \frac{1}{2}J_{1/2}(z) = zJ_{-1/2}(z),$$

then we find that

$$J_{-1/2}(z) = \sqrt{\frac{2}{\pi z}} \cos z. \tag{7.3}$$

Using the recurrence relations given in Section 6, one can find the Bessel function for any index of the form $n + 1/2$, where n is an integer, and prove that for any positive integer n the following formulae hold:

$$\left. \begin{aligned} J_{n+\frac{1}{2}}(z) &= \frac{(-1)^n (2z)^{n+1/2}}{\sqrt{\pi}} \frac{d^n}{(dz^2)^n} \left(\frac{\sin z}{z} \right), \\ J_{-n-\frac{1}{2}}(z) &= \frac{(-1)^n (2z)^{n+1/2}}{\sqrt{\pi}} \frac{d^n}{(dz^2)^n} \left(\frac{\cos z}{z} \right). \end{aligned} \right\} \tag{7.4}$$

In the same way one can obtain formulae similar to (7.1)–(7.4) for the modified functions; in particular,

$$I_{1/2}(z) = \sqrt{\frac{2}{\pi z}} \sinh z, \quad K_{1/2}(z) = \sqrt{\frac{2}{\pi z}} e^{-z}. \tag{7.5}$$

With the help of recurrence relations one can also easily obtain formulae for other values of the half-integer index.

8. Some notation for functions of half-integer and fractional index. Airy integral

Functions of half-integer index occur often in applications, in particular, in problems of mathematical physics solved in spherical coordinates. The method of separation of variables in these problems gives, as a result, solutions which contain as a term the product of the cylindrical function and the power function of the same argument. In this connection, different authors gave different names to various products of this kind. A review of these results can be found in [7]. We give there the notation of Sommerfeld

$$\psi_n(z) = \left(\frac{1}{2} \pi z \right)^{1/2} J_{n+1/2}(z),$$

$$\zeta_n(z) = \left(\frac{1}{2} \pi z \right)^{1/2} [J_{n+1/2}(z) + (-1)^n i J_{-n-1/2}(z)].$$

In many papers the functions

$$\sqrt{\frac{\pi}{2z}} J_{n+1/2}(z), \quad \sqrt{\frac{\pi}{2z}} Y_{n+1/2}(z), \quad \sqrt{\frac{\pi}{2z}} H^{(1)}_{n+1/2}(z)$$

are called the *spherical Bessel functions* of the first, second, and third kind, respectively.

Let us now consider the following differential equation

$$U''(s) + s^\alpha U(s) = 0. \tag{8.1}$$

The solutions of equation (8.1) can be expressed[3] in terms of so-called *generalized Airy functions* $U_1(s, \alpha)$ and $U_2(s, \alpha)$

$$U = C_1 U_1(s, \alpha) + C_2 U_2(s, \alpha). \tag{8.2}$$

---

[3] See also p. 32

These functions and their derivatives with respect to the argument s are connected with the Bessel functions by the following relations:

$$
\left.\begin{array}{l}
U_1(s,\alpha) = (\alpha+2)^{-1/(\alpha+2)}\Gamma\left(\dfrac{\alpha+1}{\alpha+2}\right)\sqrt{s}\,J_{-1/(\alpha+2)}\left(\dfrac{2}{\alpha+2}s^{(\alpha+2)/2}\right), \\[3mm]
U_2(s,\alpha) = (\alpha+2)^{1/(\alpha+2)}\Gamma\left(\dfrac{\alpha+3}{\alpha+2}\right)\sqrt{s}\,J_{1/(\alpha+2)}\left(\dfrac{2}{\alpha+2}s^{(\alpha+2)/2}\right), \\[3mm]
U_1'(s,\alpha) = -(\alpha+2)^{-1/(\alpha+2)}\Gamma\left(\dfrac{\alpha+1}{\alpha+2}\right)s^{(\alpha+1)/2}J_{(\alpha+1)/(\alpha+2)}\left(\dfrac{2}{\alpha+2}s^{(\alpha+2)/2}\right), \\[3mm]
U_2'(s,\alpha) = (\alpha+2)^{1/(\alpha+2)}\Gamma\left(\dfrac{\alpha+3}{\alpha+2}\right)s^{(\alpha+1)/2}J_{-(\alpha+1)/(\alpha+2)}\left(\dfrac{2}{\alpha+2}s^{(\alpha+2)/2}\right).
\end{array}\right\}
$$

$$(8.3)$$

For $\alpha = 1$ we obtain the functions which are called the *Airy functions*:

$$
\left.\begin{array}{l}
U_1(s,1) = \dfrac{\Gamma(2/3)}{\sqrt[3]{3}}\sqrt{s}\,J_{-1/3}\left(\dfrac{2}{3}s^{3/2}\right), \\[3mm]
U_2(s,1) = \sqrt[3]{3}\,\Gamma\left(\dfrac{4}{3}\right)\sqrt{s}\,J_{1/3}\left(\dfrac{2}{3}s^{3/2}\right), \\[3mm]
U_1'(s,1) = -\dfrac{\Gamma(2/3)}{\sqrt[3]{3}}s\,J_{2/3}\left(\dfrac{2}{3}s^{3/2}\right), \\[3mm]
U_2'(s,1) = \sqrt[3]{3}\,\Gamma\left(\dfrac{4}{3}\right)s\,J_{-2/3}\left(\dfrac{2}{3}s^{3/2}\right).
\end{array}\right\}
$$

$$(8.4)$$

The name of the functions U_1 and U_2 is stipulated by the fact that they are closely connected with the *Airy integral*

$$
\int_0^\infty \cos(t^3 \pm xt)\,dt, \tag{8.5}
$$

namely

$$
\int_0^\infty \cos(t^3 + xt)\,dt = \frac{1}{3}\pi\sqrt{\frac{x}{3}}\left[I_{-1/3}\left(\frac{2x\sqrt{x}}{3\sqrt{3}}\right) - I_{1/3}\left(\frac{2x\sqrt{x}}{3\sqrt{3}}\right)\right]
$$

$$
= \frac{\sqrt{x}}{3}K_{1/3}\left(\frac{2x\sqrt{x}}{3\sqrt{3}}\right), \tag{8.6}
$$

$$
\int_0^\infty \cos(t^3 - xt)\,dt = \frac{1}{3}\pi\sqrt{\frac{x}{3}}\left[J_{-1/3}\left(\frac{2x\sqrt{x}}{3\sqrt{3}}\right) + J_{1/3}\left(\frac{2x\sqrt{x}}{3\sqrt{3}}\right)\right]. \tag{8.7}
$$

In order to prove the validity of these formulae, we have to differentiate the integral (8.5) twice with respect to the parameter x and remark that this integral satisfies the following differential equation

$$
\frac{d^2v}{dx^2} \pm \frac{1}{3}xv = 0,
$$

which is a special case of equation (8.1).

Let us give the following notation for the Airy integral

$$\frac{1}{\pi} \int\limits_0^\infty \cos\left(tz + \frac{1}{3}t^3\right) dt = \text{Ai}(z) \tag{8.8}$$

and the *adjoint function*

$$\frac{1}{\pi} \int\limits_0^\infty \left[\exp\left(tz - \frac{1}{3}t^3\right) + \sin\left(tz + \frac{1}{3}t^3\right)\right] dt = \text{Bi}(z). \tag{8.9}$$

9. Wronski determinant

If $u_1(z)$ and $u_2(z)$ are linearly independent solutions of the Bessel equation, then they satisfy the relation

$$u_1 u_2' - u_1' u_2 = C_1/z, \tag{9.1}$$

where C_1 is a constant.

Actually, introducing u_1 in the Bessel equation (1.1), then u_2 and multiplying the first equation by u_2 and the second one by u_1, then subtracting the first equation from the second one, we have after a simple transformation

$$u_1 \frac{d}{dz}(zu_2') - u_2 \frac{d}{dz}(zu_1') = 0$$

or

$$\frac{d}{dz}[z(u_1 u_2' - u_1' u_2)] = 0,$$

which immediately implies (9.1).

Note that the left-hand side of equation(9.1) which represents the determinant

$$\begin{vmatrix} u_1, & u_2 \\ u_1' & u_2' \end{vmatrix}, \tag{9.2}$$

is usually denoted by $\mathfrak{W}(u_1, u_2)$ and is called the *Wronski determinant* or the *Wronskian of the Bessel equation*.

The value of the constant C_1 can easily be determined, if we pass to the limit as $z \to 0$ in formula (9.1) and use the expansions of the Bessel functions obtained in Sections 1–3.

Suppose that the equation has an non-integer index ν and let us find the Wronskian

$$\mathfrak{W}(J_\nu(z), J_{-\nu}(z)) = C_1/z. \tag{9.3}$$

Note that if ν is non-integer, then

$$J_\nu = \frac{(z/2)^\nu}{\Gamma(\nu+1)}[1 + O(z^2)],$$

$$J_\nu' = \frac{(z/2)^{\nu-1}}{2\Gamma(\nu)}[1 + O(z^2)],$$

as $z \to 0$, where $O(z^2)$ denotes a quantity, whose ratio to z^2 is bounded as $z \to 0$.

Introducing these expressions into (9.1), we obtain

$$J_\nu(z)J'_{-\nu}(z) - J'_\nu(z)J_{-\nu}(z) = \frac{1}{z}\left[\frac{1}{\Gamma(\nu+1)\Gamma(-\nu)} - \frac{1}{\Gamma(\nu)\Gamma(-\nu+1)}\right] + O(z)$$

$$= \frac{1}{z}\frac{-2}{\Gamma(\nu)\Gamma(1-\nu)} + O(z)$$

$$= -\frac{2\sin\pi z}{\pi z} + O(z); \qquad (9.4)$$

when deducing this formula, we used formula (3).

Multiplying (9.3) by z and using (9.4), after passing to the limit as $z \to 0$, we obtain

$$C_1 = -\frac{2\sin\pi\nu}{\pi},$$

hence,

$$J_\nu(z)J'_{-\nu}(z) - J'_\nu(z)J_{-\nu}(z) = (-2\sin\pi\nu)/(\pi z). \qquad (9.5)$$

Just in the same way, using relations (2.1) between the Bessel functions of the first and second kind, one can obtain

$$J_\nu(z)Y'_\nu(z) - J'_\nu(z)Y_\nu(z) = 2/(\pi z). \qquad (9.6)$$

The Wronski determinant for the functions $I_\nu(z)$ and $K_\nu(z)$ is equal to

$$I_\nu(z)K'_\nu(z) - I'_\nu(z)K_\nu(z) = -1/z, \qquad (9.7)$$

and for $J_\nu(z)$, $H_\nu^{(1)}(z)$ and $H_\nu^{(1)}(z)$, $H_\nu^{(2)}(z)$ we have, respectively,

$$J_\nu(z)\frac{dH_\nu^{(1)}(z)}{dz} - H_\nu^{(1)}(z)\frac{dJ_\nu(z)}{dz} = 2\frac{i}{\pi z}, \qquad (9.8)$$

$$H_\nu^{(1)}(z)\frac{dH_\nu^{(2)}(z)}{dz} - H_\nu^{(2)}(z)\frac{dH_\nu^{(1)}(z)}{dz} = -\frac{4i}{\pi z}. \qquad (9.9)$$

Setting $z = x\sqrt{i}$ in (9.8) and separating the real and imaginary parts, for the functions $u_\nu(x)$, $v_\nu(x)$, $f_\nu(x)$, and $g_\nu(x)$ we obtain

$$u_\nu(x)f'_\nu(x) - v_\nu(x)g'_\nu(x) = u'_\nu(x)f_\nu(x) - v'_\nu(x)g_\nu(x), \qquad (9.10)$$

$$v_\nu(x)f'_\nu(x) + u_\nu(x)g'_\nu(x) = u'_\nu(x)g_\nu(x) + v'_\nu(x)f_\nu(x) + 2/(\pi x), \qquad (9.11)$$

where the prime denotes differentiation with respect to the variable x.

In many problems the *Basset formulae* are also very convenient. In order to obtain these formulae we need to use the Wronskian and the Bessel differential equation:

$$J_\nu(z)Y_\nu''(z) - Y_\nu(z)J_\nu''(z) = -\frac{2}{\pi z^2}, \tag{9.12}$$

$$J_\nu'(z)Y_\nu''(z) - Y_\nu'(z)J_\nu''(z) = \frac{2}{\pi z}\left(1 - \frac{\nu}{z^2}\right), \tag{9.13}$$

$$J_\nu(z)Y_\nu'''(z) - Y_\nu(z)J_\nu'''(z) = \frac{2}{\pi z}\left(\frac{\nu^2 + 2}{z^2} - 1\right), \tag{9.14}$$

$$J_\nu'(z)Y_\nu'''(z) - Y_\nu'(z)J_\nu'''(z) = \frac{2}{\pi z^2}\left(\frac{3\nu^2}{z^2} - 1\right), \tag{9.15}$$

$$J_\nu''(z)Y_\nu'''(z) - Y_\nu''(z)J_\nu'''(z) = \frac{2}{\pi z}\left(1 - \frac{2\nu^2 + 1}{z^2} + \frac{\nu^4 - \nu^2}{z^4}\right), \tag{9.16}$$

$$J_\nu(z)Y_\nu^{(IV)}(z) - Y_\nu(z)J_\nu^{(IV)}(z) = \frac{4}{\pi z^2}\left(1 - \frac{3\nu^2 + 3}{z^2}\right), \tag{9.17}$$

$$J_\nu'(z)Y_\nu^{(IV)}(z) - Y_\nu'(z)J_\nu^{(IV)}(z) = -\frac{2}{\pi z}\left(\frac{\nu^4 + 11\nu^2}{z^4} - \frac{2\nu^2 + 3}{z^2} + 1\right). \tag{9.18}$$

A formula, which establishes a relationship between cylindrical functions whose indices differ by one, has been obtained by Lommel and Hankel; this formula has the form

$$J_\nu(z)Y_{\nu+1}(z) - J_{\nu+1}(z)Y_\nu(z) = -2/(\pi z). \tag{9.19}$$

Using the Wronski determinant and recurrence relations, one can immediately obtain similar relations for cylindrical functions whose indices differ by an integer; we shall call these formulae the *Lommel–Hankel formulae*. For example,

$$J_\nu(z)Y_{\nu+2}(z) - J_{\nu+2}(z)Y_\nu(z) = -\frac{4(\nu + 1)}{\pi z^2}, \tag{9.20}$$

$$J_\nu(z)Y_{\nu+3}(z) - J_{\nu+3}(z)Y_\nu(z) = -\frac{8(\nu + 1)(\nu + 2)}{\pi z^3} + \frac{2}{\pi z}, \tag{9.21}$$

$$J_\nu(z)Y_{\nu+4}(z) - J_{\nu+4}(z)Y_\nu(z) = -\frac{16(\nu + 1)(\nu + 2)(\nu + 3)}{\pi z^4} + \frac{8(\nu + 2)}{\pi z^2}. \tag{9.22}$$

Denoting

$$J_\nu(z)Y_{\nu+m}(z) - J_{\nu+m}(z)Y_\nu(z) = B(\nu, m; z), \tag{9.23}$$
$$m = 1, 2, 3, \ldots,$$

we can immediately establish the following recurrence relation

$$B(\nu, m + 1; z) = \frac{2(\nu + m)}{z}B(\nu, m; z) - B(\nu, m - 1; z). \tag{9.24}$$

With the help of (9.24) one can easily establish that the right-hand sides of the Lommel–Hankel formulae are polynomials in $1/z$, which can be represented in the form

$$B(\nu, m; z) = \frac{1}{\pi} \left\{ -\left(\frac{z}{2}\right)^{-m} \prod_{k=1}^{m-1} (\nu + k) + (m - 2) \left(\frac{z}{2}\right)^{-m+2} \prod_{k=2}^{m-2} (\nu + k) + \ldots \right\}$$
$$- \frac{2}{\pi z} \sin \frac{m\pi}{2}$$
$$= \frac{1}{\pi} \sum_{i=1}^{r} (-1)^i \prod_{k=i}^{m-i} (\nu + k) \left(\frac{z}{2}\right)^{-m+(i-1)2} C_{m-i}^{i-1} - \frac{2}{\pi z} \sin \frac{m\pi}{2}, \qquad (9.25)$$

where

$$C_{m-i}^{i-1} = \frac{(m - i)!}{(m - 2i + 1)!(i - 1)!},$$

$$r = \begin{cases} (m - 1)/2, & \text{if } m \text{ is odd}, \\ m/2, & \text{if } m \text{ is even}. \end{cases}$$

Similar formulae can be obtained for the modified functions

$$I_\nu(z)K_{\nu+1}(x) + I_{\nu+1}(x)K_\nu(z) = \frac{1}{x}, \qquad (9.26)$$

$$I_\nu(z)K_{\nu+2}(x) - I_{\nu+2}(x)K_\nu(z) = \frac{2(\nu + 1)}{x^2}, \qquad (9.27)$$

$$I_\nu(z)K_{\nu+3}(x) + I_{\nu+3}(x)K_\nu(z) = \frac{4(\nu + 1)(\nu + 2)}{x^3} + \frac{1}{x}, \qquad (9.28)$$

$$I_\nu(z)K_{\nu+4}(x) - I_{\nu+4}(x)K_\nu(z) = \frac{8(\nu + 1)(\nu + 2)(\nu + 3)}{x^4} + \frac{4(\nu + 2)}{x^2}. \qquad (9.29)$$

Furthermore, the *functional equations of the Wronski determinant type, which contain functions with different indices* are of interest; they can also be easily obtained with the help of recurrence relations. So, for instance,

$$J_\nu'(z)Y_{\nu+1}(z) - J_{\nu+1}(z)Y_\nu'(z) = -\frac{2}{\pi z^2}, \qquad (9.30)$$

$$J_\nu''(z)Y_{\nu+1}(z) - J_{\nu+1}(z)Y_\nu''(z) = \frac{2\nu(1 - \nu)}{\pi z^3} + \frac{2}{\pi z}, \qquad (9.31)$$

$$J_\nu'(z)Y_{\nu+2}(z) - J_{\nu+2}(z)Y_\nu'(z) = -\frac{4\nu(\nu + 1)}{\pi z^3} + \frac{2}{\pi z}, \qquad (9.32)$$

$$J_\nu''(z)Y_{\nu+2}(z) - J_{\nu+2}(z)Y_\nu''(z) = \frac{4\nu(1 - \nu^2)}{\pi z^4} + \frac{4\nu + 2}{\pi z^2}. \qquad (9.33)$$

One can easily obtain similar dependencies for the modified functions, for instance,

$$I_\nu'(z)K_{\nu+1}(x) + I_{\nu+1}(x)K_n'(x) = \frac{\nu}{x^2}. \qquad (9.34)$$

10. Bessel integral and Jacobi expansion

Consider the power series expansions of the functions $\exp(zt/2)$ and $\exp\left(-\frac{z}{2t}\right)$. Since

$$\exp z = \sum_{k=0}^{\infty} \frac{z^k}{k!},$$

the product of these expansions gives

$$\exp\left[\frac{z}{2}\left(t - \frac{1}{t}\right)\right] = \sum_{r=0}^{\infty}\sum_{s=0}^{\infty} \frac{(-1)^s z^{r+s} t^{r-s}}{r!s!2^{r+s}}. \tag{10.1}$$

Setting $s = m$, $r = m+n$ and recalling that $\frac{1}{\Gamma(m+n+1)} = 0$ for $n < -m$, we rewrite (10.1) in the following form:

$$\exp\left[\frac{z}{2}\left(t - \frac{1}{t}\right)\right] = \sum_{m=0}^{\infty}\sum_{n=-m}^{\infty} \frac{(-1)^m z^{n+2m} t^n}{\Gamma(m+n+1)\Gamma(m+1)2^{n+2m}}$$

$$= \sum_{m=0}^{\infty}\sum_{n=-\infty}^{\infty} \frac{(-1)^m z^{n+2m} t^n}{\Gamma(m+n+1)\Gamma(m+1)2^{n+2m}}$$

and by virtue of (1.7) we immediately obtain

$$\exp\left[\frac{z}{2}\left(t - \frac{1}{t}\right)\right] = \sum_{n=-\infty}^{\infty} J_n(z) t^n. \tag{10.2}$$

Using relation (1.10), we represent (10.2) in the form

$$\exp\left[\frac{z}{2}\left(t - \frac{1}{t}\right)\right] = J_0(z) + \sum_{n=1}^{\infty}[t^n + (-1)^n t^{-n}]J_n(z).$$

Setting $t = \pm \exp i\theta$ here, we obtain

$$\exp(\pm iz \sin\theta) = J_0(z) + 2\sum_{n=1}^{\infty} J_{2n}(z)\cos 2n\theta$$

$$\pm 2i\sum_{n=0}^{\infty} J_{2n+1}(z)\sin(2n+1)\theta. \tag{10.3}$$

This equality implies

$$\cos(z\sin\theta) = J_0(z) + 2\sum_{n=1}^{\infty} J_{2n}(z)\cos 2n\theta, \tag{10.4}$$

$$\sin(z\sin\theta) = 2\sum_{n=0}^{\infty} J_{2n+1}(z)\sin(2n+1)\theta. \tag{10.5}$$

If we replace θ by $\pi/2 - \eta$ in (10.4) and (10.5), then we obtain

$$\cos(z\cos\eta) = J_0(z) + 2\sum_{n=1}^{\infty}(-1)^n J_{2n}(z)\cos 2n\eta, \tag{10.6}$$

$$\sin(z\cos\eta) = 2\sum_{n=0}^{\infty}(-1)^n J_{2n+1}(z)\cos(2n+1)\eta. \tag{10.7}$$

These *expansions* have been obtained by *Jacobi* and are called by his name. They give a representation of a plane wave in cylindrical coordinates.

Replacing θ by φ in (10.4), multiplying the left- and right-hand sides by $\cos n\varphi$ and integrating with respect to φ from 0 to π, we obtain

$$\int_0^{\pi} \cos(z\sin\varphi)\cos n\varphi\, d\varphi = \begin{cases} \pi J_n(z) & \text{for even } n, \\ 0 & \text{for odd } n. \end{cases} \tag{10.8}$$

Similarly, from (10.5) we obtain

$$\int_0^\pi \sin(z \sin \varphi) \sin n\varphi \, d\varphi = \begin{cases} 0 & \text{for even } n, \\ \pi J_n(z) & \text{for odd } n. \end{cases} \tag{10.9}$$

Adding (10.8) to (10.9), we obtain that for any integer n

$$\frac{1}{\pi} \int_0^\pi \cos(z \sin \varphi - n\varphi) \, d\varphi = J_n(z). \tag{10.10}$$

The integral in the left-hand side of (10.10) is called the *Bessel integral*; Bessel took this equality (more precisely, slightly different) as a definition of the function $J_n(z)$. Note that for a non-integer index this integral does not give the Bessel function and is denoted by

$$\frac{1}{\pi} \int_0^\pi \cos(\nu\theta - z \sin \theta) \, d\theta = \mathbf{J}_\nu(z). \tag{10.11}$$

where $\mathbf{J}_\nu(z)$ is a function which is usually called the *Anguer function*; the properties of this function will be discussed in Section 16.

For $n = 0$ integral (10.10) is called the *Parseval integral*

$$\frac{1}{\pi} \int_0^\pi \cos(z \sin \theta) \, d\theta = J_0(z). \tag{10.12}$$

The deduction of the formulae given above illustrates in a sense the character of applications of the Bessel integral. It arises naturally in cases when one should pass from a solution of the Helmholtz equation in Cartesian coordinates to a solution in polar coordinates. Below we shall show that integrals of the Bessel integral type can also be obtained when passing from other coordinate systems to the polar system. In applications the Bessel integral also appears when passing from polar coordinates to Cartesian ones. One of typical problems associated with application of the Bessel integral is given in the second part (see page 248).

Consider once again the Parseval integral. Let us take a partial solution of the Helmholtz equation

$$\frac{\partial^2 u}{\partial \xi^2} + \frac{\partial^2 u}{\partial \eta^2} + \lambda^2 u = 0$$

in the form $u = \cos \lambda \xi = \cos(\lambda R \cos \theta)$; here R and θ are the polar coordinates of the point with Cartesian coordinates ξ and η.

Let us form the superposition of such solutions for $0 \leq \theta \leq 2\pi$, assuming that the ξ-axis rotates around the origin

$$u^*(R) = \int_0^{2\pi} \cos(\lambda R \cos \theta) d\theta.$$

Obviously, $u^*(R)$ is a solution of the axially symmetric problem for the Helmholtz equation without a singularity at the origin; hence, $u^*(R) = B J_0(\lambda R)$. Let us set

$R = 0$. Since $J_0(0) = 1$, we have $B = \int_0^{2\pi} d\theta = 2\pi$ and

$$J_0(\lambda R) = \frac{1}{2\pi} \int_0^{2\pi} \cos(\lambda R \cos \theta) d\theta. \qquad (10.13)$$

In the case when $u = \sin \lambda \xi$, we have

$$u^*(R) = \int_0^{2\pi} \sin(\lambda R \cos \theta) d\theta = B J_0(\lambda R).$$

Setting here $R = 0$, we obtain $B = 0$.

One can also easily obtain a modified Parseval integral which only slightly differs from integral (10.12).

Setting $u = \cos \alpha \xi \cos \beta \eta$, where $\alpha^2 + \beta^2 = \lambda^2$, we have

$$B J_0(\lambda R) = \int_0^{2\pi} \cos(\alpha R \cos \theta) \cos(\beta R \sin \theta) d\theta.$$

Obviously, $B = 2\pi$. Hence,

$$J_0(\sqrt{\alpha^2 + \beta^2} R) = \frac{1}{2\pi} \int_0^{2\pi} \cos(\alpha R \cos \theta) \cos(\beta R \sin \theta) d\theta. \qquad (10.14)$$

If we replace the function $u(\xi, \eta)$ by an arbitrary function which is a solution of the Helmholtz equation and has no singularities in the finite domain under consideration, then rotating this function we also obtain up to a constant factor the Bessel function of zero index.

So, if we use an elliptic coordinate system and set $u(\xi, \eta) = ce_0(\xi) Ce_0(\eta)$ (see [37]), then, obviously, we have

$$B J_0(R) = \int_0^{2\pi} ce_0(R \cos \theta) Ce_0(R \sin \theta) d\theta. \qquad (10.15)$$

One can continue the generalizations connected with the fact that one can interchange the arguments of the functions ce_0 and Ce_0 as well as introduce other indices different from zero.

Thus, the Parseval integral (10.13) can be considered as the superposition of plane waves and integrals (10.14) and (10.15) can be considered as the superposition of waves of a more complicated structure.

It should be noted that, using the Parseval integral, one can expand the Bessel function of zero index into a power series. For this purpose it is sufficient to represent the cosine in (10.13) in the form of a power series, denote $\lambda R = r$ and change the order of integration and summation.

The Parseval integral can easily be generalized to the case when $\lambda = i\gamma$ is a pure imaginary number.

Taking a partial solution of the equation

$$\Delta^2 u - \gamma^2 u = 0 \qquad (10.16)$$

in the form

$$u_1(\gamma r) = A \cosh \alpha x \cosh \beta y = A \cosh[\alpha r \cos \varphi] \cosh(\beta r \sin \varphi),$$

where $r = \sqrt{x^2 + y^2}$, $y/x = \tan\varphi$, $\alpha^2 + \beta^2 = \gamma^2$, α and β are parameters with real values, we immediately obtain

$$I_0(\gamma r) = \frac{1}{2\pi} \int_0^{2\pi} \cosh(\alpha r \cos\varphi) \cosh(\beta r \sin\varphi) d\varphi. \qquad (10.17)$$

Now we go from the Parseval integral to the Bessel integral. In what follows the restriction that the index n is an integer is essential.

Consider the integral

$$\int_0^{2\pi} A \cos n(\theta - \varphi) \cos(\alpha r \cos\varphi) \cos(\beta r \sin\varphi) d\varphi$$

$$= A \cos n\theta \int_0^{2\pi} \cos n\varphi \cos(\alpha r \cos\varphi) \cos(\beta r \sin\varphi) d\varphi = u_n(\gamma r) \cos n\theta.$$

The function $u_n(\gamma r) \cos(n\theta)$ satisfies the equation $\nabla^2 u + \gamma^2 u = 0$. Hence, up to a constant factor, $u_n(\gamma r)$ is the Bessel function of the index n and the argument γr. Thus, for an appropriate choice of the constant A, the following equality holds

$$A \int_0^{2\pi} \cos n\varphi \cos(\alpha r \cos\varphi) \cos(\beta r \sin\varphi) d\varphi = J_n(\gamma r).$$

The constant A can be obtained as a result of simple calculations and is equal to

$$A = 1/(2\pi \cos n\delta); \quad \delta = \arcsin(\alpha/\gamma).$$

It is convenient to mention another superposition of plane waves here which can be used in order to represent the Bessel functions of non-integer index; its distinction from the previous one consists in the assumption that the axis along which the plane wave is moving rotates around the top of a circular cone and is always situated on the surface of this cone. This result is given in the monograph [50].

Another heuristic method is due to Weyrich [84], who considered a cylindrical wave and its mathematical image — the appropriate Bessel function — as a superposition of spherical waves symmetrical with respect to the center, if the center moves along a straight line. The method of Weyrich admits obvious generalizations such as, for instance, consideration of the problem on the movement of a dipole or a source with abundance changing according to a sinusoidal law.

11. Addition theorems

The addition theorem for the Bessel function of zero index is defied as the equality

$$J_0(\sqrt{x^2 + y^2 - 2xy\cos\varphi}) = 2\sum_{m=0}^{\infty}{}' J_m(x)J_m(y)\cos m\varphi, \qquad (11.1)$$

where the symbol $\displaystyle\sum_{m=0}^{\infty}{}'$ differs from $\displaystyle\sum_{m=0}^{\infty}$ by an additional factor equal to $1/2$ for $m = 0$.

Equality (11.1) is called the *Neumann addition the-orem*. We give its geometrical interpretation, assuming that x, y, and φ are real numbers. The vertices O_1, O, and C of the triangle O_1OC (Fig. 1) are called the first pole, the second pole, and the observation point, respectively. The lengths of the sides O_1O, OC, and O_1C will be denoted x, y, and z, respectively. The angles of the vertices O_1 and O will be denoted ψ and φ, respectively. Obviously, $z = \sqrt{x^2 + y^2 - 2xy\cos\varphi}$. The left-side of (11.1) represents a standing cylindrical wave with the pole O_1; each term in the right-hand side describes a cylindrical wave with the second pole. Thus, formula (11.1) allows one to solve the problem of

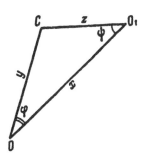

Figure 1

decomposition of a cylindrical wave into cylindrical waves with another pole. A proof of the theorem can easily be obtained with the help of the Parseval integral. We shall give a proof of a more general equality

$$\cos n\psi J_n(\sqrt{x^2 + y^2 - 2xy\cos\varphi}) = 2\sum_{m=0}^{\infty}{}' J_{m+n}(x)J_m(y)\cos m\varphi, \qquad (11.2)$$

where n is an integer. Obviously, this equality can be interpreted just in the same way as (11.1).

Let us represent the Bessel function with the help of the Bessel integral

$$J_n(z) = \frac{1}{\pi}\int_0^\pi \cos(z\sin\theta - n\theta)d\theta$$

$$= \frac{1}{\pi}\int_0^\pi e^{i(z\sin\theta - n\theta)}d\theta$$

$$= \frac{e^{-n\alpha i}}{2\pi}\int_{-\pi}^\pi e^{i[z\sin(\theta+\alpha) - n\theta]}d\theta.$$

This formula holds for arbitrary values of α. Let us take α equal to $\angle CO_1O$, then $z\sin\psi = y\sin\varphi$, $z\cos\psi = x - y\cos\varphi$, hence, $z\sin(\theta + \psi) = x\sin\theta + y\sin(\varphi - \theta)$.

Using the Jacobi formulae (10.6) and (10.7), making elementary transformations and once again using the Bessel integral, we obtain

$$J_n(z) = \frac{1}{2\pi}e^{-\psi ni}\int_{-\pi}^\pi e^{ix\sin\theta}e^{iy\sin(\varphi-\theta)}e^{-in\theta}d\theta$$

$$= \frac{e^{-in\psi}}{2\pi}\sum_{m=-\infty}^{\infty}\int_{-\pi}^\pi J_m(y)e^{im\sin(\varphi-\theta)}e^{-in\theta}e^{ix\sin\theta}d\theta$$

$$= e^{-in\psi}\sum_{m=-\infty}^{\infty} J_m(y)J_{m+n}(x)e^{im\varphi}.$$

This immediately implies (11.2).

Let us note that for a non-integer index the Bessel integral on which the deduction is based transforms into integral representation of the Anguer function (see 16). Hence, for a non-integer index the above reasoning gives an addition formula for the Anguer function, in the right-hand side of which the functions $J_{m+n}(x)$ are replaced by the Anguer functions $\mathbf{J}_{m+\nu}(x)$.

The addition theorems for the Bessel functions with an arbitrary index can be deduced in a similar way. However, in this case we use instead of the Bessel integral other integral representations (a proof of the addition theorems for the Bessel functions with an arbitrary index is given in [7], [4]).

Now we formulate without proof the addition theorems for Bessel functions, when the index ν is an arbitrary number. The result which is due to Graaf, can be written in the following way:

$$e^{i\nu\psi}J_\nu(z) = \sum_{m=-\infty}^{\infty} J_{\nu+m}(x)J_m(y)e^{mi\varphi}, \tag{11.3}$$

where $\psi = \text{arccotan}\,\frac{x-y\cos\varphi}{y\sin\varphi}$.

Changing the signs of φ and ψ in this formula, we obtain

$$e^{-i\nu\psi}J_\nu(z) = \sum_{m=-\infty}^{\infty} J_{\nu+m}(x)J_m(y)e^{-mi\varphi}. \tag{11.4}$$

Formulae (11.3) and (11.4) imply that

$$J_\nu(z){\sin^{\cos}}\nu\psi = \sum_{m=-\infty}^{\infty} J_{m+\nu}(x)J_m(y){\sin^{\cos}}m\varphi. \tag{11.5}$$

If we change signs of ν and m in (11.5) and recall the relations between the cylindrical functions of the first, second, and third kinds, then we can obtain

$$Z_\nu(z){\sin^{\cos}}\nu\psi = \sum_{m=-\infty}^{\infty} Z_{\nu+m}(x)J_m(y){\sin^{\cos}}m\varphi, \tag{11.6}$$

where $Z_\nu(z)$ denotes a cylindrical function of the first, second or third kind. Replacing z, x, and y in (11.5) and (11.6) by iz, ix, iy, respectively, one can easily obtain the following formulae:

$$I_\nu(z){\sin^{\cos}}\nu\psi = \sum_{m=-\infty}^{\infty} (-1)^m I_{\nu+m}(x)I_m(y){\sin^{\cos}}m\varphi, \tag{11.7}$$

$$K_\nu(z){\sin^{\cos}}\nu\psi = \sum_{m=-\infty}^{\infty} K_{\nu+m}(x)I_m(y){\sin^{\cos}}m\varphi. \tag{11.8}$$

Formulae (11.3)–(11.8) are valid under the condition that $|ye^{\pm i\varphi}| < |x|$.

One can easily deduce the addition formulae for the cylindrical functions of a complex argument. Replacing z, x, and y in (11.6) by $z\sqrt{i}$, $x\sqrt{i}$, and $y\sqrt{i}$, respectively, setting $\nu = 0$ and $Z_\nu(z) = H_0^{(1)}(z)$ and separating the real and imaginary parts, we obtain

$$\Re H_0^{(1)}(z\sqrt{i}) = f_0(z) = \begin{cases} 2\sum\limits_{m=0}^{\infty}{}'[f_m(y)u_m(x) - g_m(y)v_m(x)]\cos m\varphi & \text{for } x \le y, \\ 2\sum\limits_{m=0}^{\infty}{}'[f_m(x)u_m(y) - g_m(x)v_m(y)]\cos m\varphi & \text{for } x \ge y; \end{cases}$$

(11.9)

$$\Im H_0^{(1)}(z\sqrt{i}) = g_0(z) = \begin{cases} 2\sum\limits_{m=0}^{\infty}{}'[v_m(y)f_m(x) + u_m(y)g_m(x)]\cos m\varphi & \text{for } x \le y, \\ 2\sum\limits_{m=0}^{\infty}{}'[v_m(x)f_m(y) + u_m(x)g_m(y)]\cos m\varphi & \text{for } x \ge y. \end{cases}$$

(11.10)

At the beginning of this section we indicated that formula (11.1) gives the expansion of a solution of the Helmholtz equation in polar coordinates into a series in solutions of the same equation but for another pole. The physical meaning of the addition formulae consists in rebuilding the solution in polar coordinates under the necessity of changing the location of the pole.

This point of view on the addition formulae can be well illustrated by a number of applications in which an axially non-symmetric solution of the Helmholtz equation for a circular domain is considered (see §§1, 3 of Part II).

In order that the reasoning below is clearer from the physical standpoint, let us consider as an example one such application, namely, the problem on a circular membrane on an elastic foundation. Taking into account what was said above, we give the deduction of the addition formulae based on using the expansion of a point force into a series in trigonometric functions and on applying the Wronski determinant of the Bessel equation.

Consider the test Green function of the equation $\nabla^2 u - u = 0$, which represents a solution of this equation bounded at infinity and with a singularity at the origin of the point force type, i.e. this function satisfies the condition

$$-\int\limits_0^{2\pi} r\frac{\partial u}{\partial r}d\varphi\bigg|_{r\to 0} \to 1.$$

Obviously, the function considered has the form $\frac{1}{2\pi}K_0(r)$.

Let us pass to another polar system and consider the observation point (R, θ). Let

$$z = \sqrt{r^2 + R^2 - 2Rr\cos(\theta - \varphi)}$$

denote the distance between this point and the point (r, φ) at which the force is applied. We represent the function $K_0(z)$ in a form more convenient for problems associated with application of the second coordinate system. Consider the circumference of the radius r with the center at the origin of the second polar coordinate system. The unit point force is applied at the point (r, φ) of this circumference.

We expand the force into a trigonometric series

$$q(\theta - \varphi) = \frac{1}{\pi r} \sum_{m=1}^{\infty} \left(\frac{1}{2} + \cos[m(\theta - \varphi)] \right)$$

$$= \frac{1}{\pi r} \sideset{}{'}\sum_{m=0}^{\infty} \cos[m(\theta - \varphi)].$$

Accordingly, we represent the Green function in the form

$$\frac{1}{2\pi} K_0(z) = \sideset{}{'}\sum_{m=0}^{\infty} w_m(R) \cos[m(\theta - \varphi)].$$

Obviously, the function $w_m(R)$ is a solution of the Bessel equation of an imaginary argument with the index m. It should be analytic at the origin and tend to zero at infinity. At $R = r$ the derivative of this function should have a discontinuity such that

$$\frac{dw_m}{dR}\bigg|_{r-0} - \frac{dw_m}{dR}\bigg|_{r+0} = \frac{1}{\pi r}. \tag{11.11}$$

We shall seek the solution n the form

$$w_m(R) = \begin{cases} A_m I_m(R), & R \leq r; \\ B_m K_m(R), & R \geq r. \end{cases}$$

The condition of continuity of the function $w_m(R)$ implies

$$A_m I_m(r) - B_m K_m(r) = 0.$$

Moreover, from (11.11) we have

$$A_m I'_m(r) - B_m K'_m(r) = \frac{1}{\pi r}.$$

Solving this system and using formula (9.7) for simplification, we obtain

$$A_m = \frac{1}{\pi} K_m(r), \quad B_m = \frac{1}{\pi} I_m(r),$$

hence,

$$K_0(z) = \begin{cases} 2 \sideset{}{'}\sum_{m=0}^{\infty} I_m(R) K_m(r) \cos[m(\theta - \varphi)], & R \leq r, \\ 2 \sideset{}{'}\sum_{m=0}^{\infty} I_m(r) K_m(R) \cos[m(\theta - \varphi)], & R \geq r. \end{cases} \tag{11.12}$$

Differentiating (11.12) with respect to the variable r, one can obtain a solution of the problem on a membrane to which a point moment of force is applied.

At the conclusion of this section we give a formula which is called the *Gegenbauer addition theorem*, which can be obtained from (11.1) by differentiating this equality n times with respect to $\cos\varphi$:

$$\frac{J_n(z)}{z^n} = 2 \sideset{}{'}\sum_{m=0}^{\infty} \frac{J_{m+n}(x)}{x^n} \frac{J_{m+n}(y)}{y^n} \frac{d^n \cos(m+n)\varphi}{[d(\cos\varphi)]^n}; \tag{11.13}$$

there are no restrictions here on the ratio x/y.

The Gegenbauer theorem for the cylindrical function $Z_\nu(z)$ has the form

$$\frac{Z_\nu(z)}{z^\nu} = 2^\nu \Gamma(\nu) \sum_{m=0}^{\infty}{}' \frac{Z_{m+\nu}(x)}{x^\nu} \frac{J_{m+\nu}(y)}{y^\nu} C_m^\nu(\cos\varphi) \qquad (11.14)$$

for $|ye^{\pm i\varphi}| < |x|$. Here $C_m^\nu(\cos\varphi)$ is the coefficient of the term which contains β^m in the expansion of $(1 - 2\beta\cos\varphi + \beta^2)^{-\nu}$.

12. Lommel expansion

Let us expand the function $(z+h)^{-\nu/2} J_\nu(\sqrt{z+h})$, which is an analytic function of the argument $(z + h)$ for any its value, into a series in the powers of h, using formulae (6.8) when computing the derivatives. We obtain

$$(z + h)^{-\nu/2} J_\nu(\sqrt{z+h}) = \sum_{m=0}^{\infty} \frac{h^m}{m!} \frac{d^m}{dz^m} \left\{ z^{-\nu/2} J_\nu\left(\sqrt{z}\right) \right\}$$

$$= \sum_{m=0}^{\infty} \frac{(-h/2)^m}{m!} z^{-(\nu+m)/2} J_{\nu+m}\left(\sqrt{z}\right). \qquad (12.1)$$

The function $(z + h)^{\nu/2} J_\nu(\sqrt{z+h})$ is analytic everywhere, except the point $h = -z$. If $|h| < |z|$, then

$$(z + h)^{\nu/2} J_\nu(\sqrt{z+h}) = \sum_{m=0}^{\infty} \frac{h^m}{m!} \frac{d^m}{dz^m} \left\{ z^{\nu/2} J_\nu\left(\sqrt{z}\right) \right\}$$

$$= \sum_{m=0}^{\infty} \frac{(h/2)^m}{m!} z^{(\nu-m)/2} J_{\nu-m}\left(\sqrt{z}\right). \qquad (12.2)$$

Setting $\nu = 1/2$ and $\nu = -1/2$ in expansion (12.1), we, respectively, obtain

$$\left(\frac{2}{\pi z}\right)^{1/2} \cos\left(\sqrt{z^2 - 2zt}\right) = \sum_{m=0}^{\infty} \frac{t^m}{m!} J_{m-1/2}(z), \qquad (12.3)$$

$$\left(\frac{2}{\pi z}\right)^{1/2} \sin\left(\sqrt{z^2 + 2zt}\right) = \sum_{m=0}^{\infty} \frac{t^m}{m!} J_{1/2-m}(z), \quad |t| < \frac{1}{2}|z|. \qquad (12.4)$$

For $|h| < |z|$ we can replace the functions of the first kind in formulae (12.1) and (12.2) by the functions of the second or third kind. Both formulae are also valid in the case when the index is integer. Some additional results concerning the Lommel expansion are given in 14 and 16.

13. Differential equations reducible to the Bessel equation

In this section we consider the problem to which differential equations can be reduced to the Bessel equation and how the necessary transformations should be performed. This problem is treated from the practical standpoint and the presentation which follows the settled tradition can hardly be called systematic. This representation is more likely a collection of the most successful and interesting results ordered in increasing of their complexity. The most part of these results is due to Lommel.

Many differential equations can be reduced to the Bessel equation using transformations of dependent and independent variables. We begin with the simplest special case.

Consider the equation

$$\frac{d^2u}{dz^2} - c^2u = \frac{p(p+1)}{z^2}u. \tag{13.1}$$

Let us set $v = z^{-1/2}u$. With respect to this new variable v the equation takes the form

$$z^2\frac{d^2v}{dz^2} + z\frac{dv}{dz} - \left[c^2z^2 + \left(p + \frac{1}{2}\right)^2\right]v = 0.$$

Denoting $p + 1/2 = \nu$ and $icz = x$, we reduce this equation to the form

$$x^2\frac{d^2v}{dx^2} + x\frac{dv}{dx} + (x^2 - \nu^2)v = 0,$$

i.e. we obtain the ordinary Bessel equation. Hence, $v = Z_\nu(x)$, where $Z_\nu(x)$ is the cylindrical function of index ν. As a result of the inverse substitution, we obtain

$$u = z^{1/2}Z_{p+1/2}(icz).$$

Setting $u = wz^{-p}$ in (13.1), we obtain as a result of the substitution the equation

$$\frac{d^2w}{dz^2} - \frac{2p}{z}\frac{dw}{dz} - c^2w = 0, \tag{13.2}$$

whose solution has the form

$$w = z^pu = z^{p+1/2}Z_{p+1/2}(icz). \tag{13.3}$$

Setting $z = \zeta^q/q$ and $q = 1/(2p+1)$ in (13.2), we obtain the equation

$$\frac{d^2w}{d\zeta^2} - c^2\zeta^{2q-2}w = 0. \tag{13.4}$$

We find the solution of this equation replacing the argument z by ζ^q/q in (13.3). Thus,

$$w = \zeta^{1/2}(1/q)^{1/(2q)}Z_{1/(2q)}(ic\zeta^q/q).$$

Equations of a larger class can also be reduced by a direct transformation to the Bessel equation. In particular, Lommel has shown that the general solution of the equation

$$z^2\frac{d^2u}{dz^2} + (2\alpha - 2\beta\nu + 1)z\frac{du}{dz} + [\beta^2\gamma^2z^{2\beta} + \alpha(\alpha - 2\beta\nu)]u = 0 \tag{13.5}$$

is

$$u = z^{\beta\nu - \alpha}Z_\nu(\gamma z^\beta). \tag{13.6}$$

In order to obtain (13.6) we need to introduce new variables $w = uz^{\alpha - \beta\nu}$ and $t = \gamma z^\beta$. It can easily be shown that the function $w(t)$ satisfies the Bessel equation with index ν.

Let us give some special cases of equation (13.5):

$$\frac{d^2u}{dz^2} \pm zu = 0, \qquad u = z^{1/2} Z_{1/3}\left(\frac{2}{3}z^{3/2}\right), \qquad u = z^{1/2} Z_{1/3}\left(\frac{2}{3}iz^{3/2}\right);$$
$$(13.7)$$

$$\frac{d^2u}{dz^2} \pm z^{-1/2}u = 0, \qquad u = z^{1/2} Z_{2/3}\left(\frac{4}{3}z^{3/4}\right), \qquad u = z^{1/2} Z_{2/3}\left(\frac{4}{3}iz^{3/4}\right);$$
$$(13.8)$$

$$z\frac{d^2u}{dz^2} + (1-\nu)\frac{du}{dz} - \frac{1}{4}u = 0, \qquad u = z^{\nu/2}Z_\nu(i\sqrt{z}); \qquad (13.9)$$

$$z\frac{d^2u}{dz^2} + \frac{du}{dz} + \frac{1}{4}\left(1 - \frac{\nu^2}{z}\right)u = 0, \qquad u = Z_\nu(\sqrt{z}); \qquad (13.10)$$

$$\frac{d^2u}{dz^2} + \frac{1}{z}\frac{du}{dz} + 4\left(z^2 - \frac{\nu^2}{z^2}\right)u = 0, \qquad u = Z_\nu(z^2); \qquad (13.11)$$

$$\frac{d^2y}{dz^2} + \frac{1-2k}{z}\frac{dy}{dz} + \beta^2 z^{\nu-2}y = 0, \qquad y = z^k Z_{2k/\nu}\left(\frac{2\beta}{\nu}z^{\nu/2}\right); \qquad (13.12)$$

$$\frac{d^2y}{dz^2} + \frac{1-2k}{z}\frac{dy}{dz} + \left(\beta^2 + \frac{c}{z^2}\right)y = 0, \qquad y = z^k Z_{\sqrt{k^2-c}}(\beta z). \qquad (13.13)$$

Now we shall deduce a more general result which is also due to Lommel. For this purpose we consider equation (13.2). Replacing in this equation w by $y/\chi(z)$, z by $\psi(z)$ and $2p$ by $2\nu - 1$ and setting $c = i$, we obtain

$$\frac{d^2y}{dz^2} - \left[\frac{\psi''(z)}{\psi'(z)} + (2\nu-1)\frac{\psi'(z)}{\psi(z)} + 2\frac{\chi'(z)}{\chi(z)}\right]\frac{dy}{dz}$$
$$+ \left\{\left[\frac{\psi''(z)}{\psi'(z)} + (2\nu-1)\frac{\psi'(z)}{\psi(z)} + 2\frac{\chi'(z)}{\chi(z)}\right]\frac{\chi'(z)}{\chi(z)} - \frac{\chi''(z)}{\chi(z)} + [\psi'(z)]^2\right\}y = 0. \quad (13.14)$$

The solution of equation (13.14) is

$$y = \chi(z)[\psi(z)]^\nu Z_\nu[\psi(z)].$$

Introducing into the consideration a new function $\varphi(z)$ which is defined by the equation

$$\frac{\varphi'(z)}{\varphi(z)} = \frac{\psi''(z)}{\psi'(z)} + (2\nu-1)\frac{\psi'(z)}{\psi(z)} + 2\frac{\chi'(z)}{\chi(z)}, \qquad (13.15)$$

and eliminating the function $\chi(z)$ from (13.14), we obtain

$$\frac{d^2y}{dz^2} - \frac{\varphi'(z)}{\varphi(z)}\frac{dy}{dz} + \left\{\frac{3}{4}\left[\frac{\varphi'(z)}{\varphi(z)}\right]^2 - \frac{1}{2}\frac{\varphi''(z)}{\varphi(z)} - \frac{3}{4}\left[\frac{\psi''(z)}{\psi'(z)}\right]^2 \right.$$
$$\left. + \frac{1}{2}\frac{\psi'''(z)}{\psi'(z)} + \left[\psi^2(z) - \nu^2 + \frac{1}{4}\right]\left[\frac{\psi'(z)}{\psi(z)}\right]^2\right\}y = 0. \quad (13.16)$$

The solution of equation (13.16) has the form

$$y = \sqrt{\frac{\varphi(z)\psi(z)}{\psi'(z)}} Z_\nu[\psi(z)].$$

In the special case, when $\varphi(z) \equiv 1$,

$$\frac{d^2y}{dz^2} + \left\{ \frac{1}{2}\frac{\psi'''(z)}{\psi'(z)} - \frac{3}{4}\left[\frac{\psi''(z)}{\psi'(z)}\right]^2 + \left[\psi^2(z) - \nu^2 + \frac{1}{4}\right]\left[\frac{\psi'(z)}{\psi(z)}\right]^2 \right\}y = 0. \quad (13.17)$$

The solution of equation (13.17) is

$$y = \sqrt{\frac{\psi(z)}{\psi'(z)}} Z_\nu[\psi(z)].$$

Coming back to the initial equation (13.2) and considering the case $\chi(z) \equiv [\psi(z)]^{\mu-\nu}$, we find that the solution of the equation

$$\frac{d^2y}{dz^2} - \left[\frac{\psi''(z)}{\psi'(z)} + (2\mu - 1)\frac{\psi'(z)}{\psi(z)}\right]\frac{dy}{dz} + [\mu^2 - \nu^2 + \psi^2(z)]\left[\frac{\psi'(z)}{\psi(z)}\right]^2 y = 0 \quad (13.18)$$

has the form

$$y = [\psi(z)]^\mu Z_\nu[\psi(z)].$$

A special case of (13.18) is the equation

$$\frac{d^2y}{dz^2} + (e^{2z} - \nu^2)y = 0, \quad (13.19)$$

whose solution is the function

$$y = Z_\nu(e^z). \quad (13.20)$$

The following equations can be reduced to the Bessel equation with the help of the Lommel transformations[4]:

$$\frac{d^2y}{dz^2} + \left[b^2\exp(2\beta i) - \frac{4p^2 - 1}{z^2}\right]y = 0, \quad y = z^{1/2}Z_p(bze^{\beta i});$$

$$\frac{d^2y}{dz^2} + \frac{a}{z}\frac{dy}{dz} + \left[b^2e^{2cz} + \frac{a(a-2)}{4z^2} - d^2\right]y = 0, \quad y = z^{-a/2}Z_{d/c}\left(\frac{b}{c}e^{cz}\right);$$

$$\frac{d^2y}{dz^2} + \frac{a}{z}\frac{dy}{dz} + \left[\frac{a(a-2)}{4z^2} + \frac{1}{z^4}(b^2e^{2c/z} - p^2)\right]y = 0, \quad y = z^{1-a/2}Z_{p/c}\left(\frac{b}{c}e^{c/z}\right);$$

$$\frac{d^2y}{dz^2} + \left(\frac{1}{z} + h\right)\frac{dy}{dz} + \left(\frac{h^2}{4} + \frac{h+2b^2}{2z} - \frac{p^2}{4z^2}\right)y = 0, \quad y = \exp[-hz/2]\cdot Z_p(2b\sqrt{z});$$

$$\frac{d^2y}{dz^2} + \left(\frac{a}{z} + h\right)\frac{dy}{dz} + \left[b^2z^m + \frac{h}{2}\left(\frac{a}{z} + \frac{h}{2}\right) + \frac{(1-a)2}{4z^2}\right]y = 0,$$

$$y = \exp[-hz/2]z^{(1-a)/2}Z_0\left(\frac{2b}{m+2}z^{m/2+1}\right);$$

$$\frac{d^2y}{dz^2} + \frac{2}{\sinh 2z}\frac{dy}{dz} + \cosh^2 z\left(1 - \frac{p^2}{\sinh^2 z}\right)y = 0, \quad y = Z_p(\sinh z);$$

$$\frac{d^2y}{dz^2} + \frac{2b}{2bz+c}\frac{dy}{dz} + \frac{b^2}{2bz+c}\left(1 - \frac{p^2}{2bz+c}\right)y = 0, \quad y = Z_p(\sqrt{2bz+c}).$$

---

[4]These equations are taken from [80], where Appendix contains a large and useful list of similar equations.

Now we are going to consideration of equations of the forth order. Note that in many problems of mathematical physics the method of separation of variables gives, as a result, ordinary differential equations of the following form:

$$\Delta_\nu^2 u + a\Delta_\nu u + bu = 0, \quad \Delta_\nu = \frac{d^2}{dz^2} + \frac{1}{z}\frac{d}{dz} - \frac{\nu^2}{z^2}.$$

If the equation has the form

$$\Delta_\nu^2 u - 2b_1\Delta_\nu u + \lambda^4 u = 0, \tag{13.21}$$

then, passing to the variable $\xi = \lambda z$ and keeping the same notation for the operator Δ_ν $\left(\Delta_\nu = \frac{d^2}{d\xi^2} + \frac{1}{\xi}\frac{d}{d\xi} - \frac{\nu^2}{\xi^2}\right)$, we obtain

$$\Delta_\nu\Delta_\nu u - 2b_0\Delta_\nu u + u = 0, \quad b_0 = b_1/\lambda^2. \tag{13.22}$$

Let us note that $\Delta_\nu = \frac{1}{z^2}\nabla_\nu - 1$, where ∇_ν is the Bessel operator. We shall seek a solution of the form

$$u = AZ_\nu(\xi\sqrt{s})$$

and we introduce this expression into equation (13.22). As a result, we obtain the characteristic equation $s^2 + 2b_0 s + 1 = 0$, whose roots are $s_{1,2} = -b_0 \pm \sqrt{b_0^2 - 1}$.

If b_0 is equal to zero, then $s_{1,2} = \pm i$ and the case occurs which is considered in detail on page 162.

If b_0 is non-zero but small with respect to one, then we can approximately assume that $s_{1,2} \approx -b_0 \pm i$. In the more general case, for $0 < b_0 < 1$, we have

$$s_{1,2} = a \pm ib,$$

where $a = -b_0$, $b = \sqrt{1 - b_0^2}$.

Setting $\xi\sqrt{a \pm ib} = \rho e^{\pm i\varphi}$ and squaring this equation, we obtain

$$\xi^2(a \pm ib) = \rho^2(\cos 2\varphi \pm \sin 2\varphi).$$

Since $|a| \le 1$ and $|b| \le 1$, setting

$$\xi = \rho, \quad a = \cos 2\varphi, \quad b = \sin 2\varphi,$$

where $\varphi = \frac{1}{2} \operatorname{arccotan} \frac{a}{b}$, we can write the solution in the following form

$$u = A_1^* J_\nu(\rho e^{i\varphi}) + A_2^* H_\nu^{(1)}(\rho e^{i\varphi}) + A_3^* J_\nu(\rho e^{-i\varphi}) + A_4^* H_\nu^{(2)}(\rho e^{-i\varphi}),$$

or

$$u = A_1\bar{u}_\nu(\rho) + A_2\tilde{v}_\nu(\rho) + A_3\tilde{f}_\nu(\rho) + A_4\tilde{g}_\nu(\rho),$$

where

$$A_1 = A_1^* + A_3^*, \qquad\qquad A_2 = i(A_1^* - A_3^*),$$
$$A_3 = A_2^* + A_4^*, \qquad\qquad A_4 = i(A_2^* - A_4^*).$$

If $b_0 = 1$, then the characteristic equation has multiple roots.

If $b_0^2 > 1$, then both roots of the characteristic equation are real and the solution of equation (13.21) can be written in the following form:

for $b_0 < -1$

$$u = A_1 J_\nu(\xi\sqrt{s_1}) + A_2 Y_\nu(\xi\sqrt{s_1}) + A_3 J_\nu(\xi\sqrt{s_2}) + A_4 Y_\nu(\xi\sqrt{s_2}),$$

for $b_0 > 1$

$$u = A_1 I_\nu(\xi\sqrt{|s_1|}) + A_2 K_\nu(\xi\sqrt{|s_1|}) + A_3 I_\nu(\xi\sqrt{|s_2|}) + A_4 K_\nu(\xi\sqrt{|s_2|}).$$

Now let us consider problems in which the transition to a system of Bessel equations is less obvious.

We give a result which has been obtained by Lommel and afterwards generalized by Watson.

Equation (13.5) can be written in the form

$$(D + a)(D + a - 2\beta\nu)u = -\beta^2\gamma^2 z^{2\beta}u, \qquad D = z\frac{d}{dz}. \qquad (13.23)$$

Let us write the equation

$$\prod_{r=0}^{n-1}(D + a - 2r\beta)(D + a - 2\beta\nu - 2r\beta)u = (-1)^n\beta^{2n}c^{2n}z^{2n\beta}u, \qquad (13.24)$$

which for $n = 1$ coincides with (13.23). It can easily be verified that a solution of equation (13.24) has the form

$$u = z^{\beta\nu-u}Z_\nu(\gamma z^\beta),$$

where

$$\gamma = c\exp(r\pi i/n), \qquad r = 0, 1, \ldots, n - 1.$$

Assigning to r the values $0, 1, \ldots, n - 1$ successively, we obtain n solutions which form a fundamental system.

Setting $n = 2$ in (13.24), we obtain an equation of the fourth order

$$(D + a)(D + a - 2\beta)(D + a - 2\beta\nu)(D + a - 2\beta\nu - 2\beta)u = \beta^4c^4z^{4\beta}u. \quad (13.25)$$

Kirchhoff has considered the equation

$$\frac{d^2}{dz^2}\left(z^4\frac{d^2u}{dz^2}\right) = z^2u, \qquad (13.26)$$

which describes the transverse vibrations of a conic rod. Introducing a new independent variable $\beta = 2\sqrt{z}$ and denoting $u\beta^2 = \gamma$, we reduce equation (13.26) to a system of two equations

$$\frac{d^2\gamma}{d\beta^2} + \frac{1}{\beta}\frac{d\gamma}{d\beta} \pm \gamma - \frac{4}{\beta^2}\gamma = 0,$$

which implies

$$u = \frac{1}{z}\left[A_1 J_2(2\sqrt{z}) + A_2 Y_2(2\sqrt{z}) + A_3 I_2(2\sqrt{z}) + A_4 K_2(2\sqrt{z})\right]. \qquad (13.27)$$

The equation considered by Mononobe

$$\frac{d^2}{dz^2}\left(z^3\frac{d^2u}{dz^2}\right) = zu \qquad (13.28)$$

has a solution which can be obtained, if we set the indices in (13.27) to be equal to 1 and replace $1/z$ by $1/\sqrt{z}$. This equation can be encountered in problems on vibration of a wedge. A more general problem has been studied by A.N. Dinnik [12], who considered the differential equation of rod vibrations which has the form

$$\frac{d^2}{dz^2}\left(z^m\frac{d^2u}{dz^2}\right) - z^{m-2}u = 0. \qquad (13.29)$$

Obviously, the Mononobe and Kirchhoff problems can be obtained as special cases of Dinnik's problem. We write the solution of Dinnik's equation in the form

$$u = z^{1-m/2}\left\{A_1 J_{m-2}(2\sqrt{z}) + A_2 Y_{m-2}(2\sqrt{z}) + A_3 I_{m-2}(2\sqrt{z}) + A_4 K_{m-2}(2\sqrt{z})\right\}.$$

We also formulate a result recently published which is due to T. Lardner [32] who has shown that a solution of the differential equation

$$\frac{d^2}{dz^2}\left(z^m \frac{d^2 u}{dz^2}\right) - pz^n u = 0 \tag{13.30}$$

can be expressed via Bessel functions, if $m = P/Q$, $n = (P+2)/Q - 4$, where P and Q are natural numbers and $Q \neq 0$; moreover, at least one of the numbers $|P|$ and $|Q|$ is odd.

Lardner does not give the general solution in [32], emphasizing its complexity. He considers only the special case $P = 9$, $Q = 2$.

Setting $z_1 = \dfrac{pz^\theta}{\theta^4}$, $t = \ln(4z_1^{1/4})$, we can rewrite equation (13.30) in the form

$$[D_1(D_1 - 4)(D_1 + 6)(D_1 + 10) - e^{4t}]u(t) = 0, \tag{13.31}$$

where $D_1 = d/dt$.

The solution of equation (13.31) has the following form

$$u(t) = e^{-12t} D_1^2 (D_1 - 2)(D_1 - 4)^2 (D_1 - 8)^2 (D_1 - 12)\Phi(t),$$

where $\Phi(t)$ is the solution of the equation

$$[D_1^2 (D_1 - 2)^2 - e^{4t}]\Phi(t) = 0,$$

which can be expressed via the Bessel functions,

$$\Phi(t) = A_1 J_0(e^t) + A_2 Y_0(e^t) + A_3 I_0(e^t) + A_4 K_0(e^t).$$

The equations of the second and fourth orders considered are homogeneous. Using the Cauchy functions given in Section 17, one can easily obtain the solutions of the corresponding inhomogeneous equations.

14. The Poisson integral

Parallel with the Bessel integral in the theory of Bessel functions and its applications another integral representation plays an important role. This representation is due to Poisson who in his papers on heat conduction considered integrals of the type

$$\int_0^\pi \cos(z \cos \theta) \sin^{2n+1} \theta \, d\theta,$$

$$\int_0^\pi \cos(z \cos \theta) \sin^{2n} \theta \, d\theta,$$

where n is a positive integer or zero.

Later on, similar integrals were considered by Lommel.

Let us consider the *Poisson integral*

$$J_\nu(z) = \frac{(z/2)^\nu}{\Gamma(\nu + 1/2)\Gamma(1/2)} \int_0^\pi \cos(z \cos \theta) \sin^{2\nu} \theta \, d\theta. \tag{14.1}$$

In order to prove (14.1) we replace the first factor of the integrand in the right-hand by the series

$$\cos(z\cos\theta) = \sum_{m=0}^{\infty} (-1)^m \frac{(z\cos\theta)^{2m}}{(2m)!}, \tag{14.2}$$

then we change the order of summation and integration and use formulae (7), (8). As a result, we obtain

$$\frac{(z/2)^\nu}{\Gamma(\nu+1/2)\Gamma(1/2)} \int_0^\pi \cos(z\cos\theta)\sin^{2\nu}\theta\,d\theta$$

$$= \frac{(z/2)^\nu}{\Gamma(\nu+1/2)\Gamma(1/2)} \sum_{m=0}^{\infty} (-1)^m \frac{z^{2m}}{\Gamma(2m+1)} \int_0^\pi \cos^{2m}\theta\sin^{2\nu}\theta\,d\theta$$

$$= \sum_{m=0}^{\infty} \frac{(-1)^m (z/2)^{2m+\nu} 2^{2m}}{\Gamma(\nu+1/2)\Gamma(1/2)\Gamma(2m+1)} \frac{\Gamma(\nu+1/2)\Gamma(m+1/2)}{\Gamma(m+\nu+1)}$$

$$= \sum_{m=0}^{\infty} \frac{(-1)^m (z/2)^{2m+\nu}}{\Gamma(m+1)\Gamma(m+\nu+1)} = J_\nu(z), \tag{14.3}$$

that prove the validity of (14.1). In contrast to the Bessel integral, the result obtained holds not only for an integer index n but also for any real $\nu > -1/2$ as well as for complex ν provided that $\Re\nu > -1/2$.

Taking into account the equality

$$\int_0^\pi \sin(z\cos\theta)\sin^{2\nu}\theta\,d\theta = 0, \tag{14.4}$$

we can rewrite formula (14.1) in the following form:

$$J_\nu(z) = \frac{(z/2)^\nu}{\Gamma(\nu+1/2)\Gamma(1/2)} \int_0^\pi e^{iz\cos\theta}\sin^{2\nu}\theta\,d\theta. \tag{14.5}$$

After making the change of the variable $\cos\theta = t$, we can write the Poisson integral in the form

$$J_\nu(z) = \frac{(z/2)^\nu}{\Gamma(\nu+1/2)\Gamma(1/2)} \int_{-1}^{1} e^{izt}(1-t^2)^{\nu-1/2}dt. \tag{14.6}$$

In Section 19 we will consider some generalizations of this integral.

As an example of using the Poisson integral we give the deduction of the Lommel expansion considered in Section 12 for $\nu = 0$. Let us note that before obtaining the general solution given in Section 12, Lommel considered a special case of the expansion when $\nu = 1$ (see, for instance, [7]). The deduction given below is based on the obvious fact that the Poisson integral (10.12) does not change when the origin of the polar angle is shifted, hence, it can be written in the form

$$J_0(r) = \frac{1}{\pi} \int_0^\pi \cos[r\cos(\theta - \varphi)]\,d\varphi$$

$$= \frac{1}{\pi} \int_0^\pi \cos[r\cos\theta\cos\varphi + r\sin\theta\sin\varphi]\,d\varphi$$

$$= \frac{1}{\pi} \int_0^\pi [\cos(r\cos\theta\cos\varphi)\cos(r\sin\theta\sin\varphi) - \sin(r\cos\theta\cos\varphi)\sin(r\sin\theta\sin\varphi)]\,d\varphi$$

$$= \frac{1}{\pi} \int_0^\pi \cos(r\cos\theta\cos\varphi)\cos(r\sin\theta\sin\varphi)\,d\varphi, \quad 2\pi \geq \theta \geq 0.$$

Let us note that a similar formula was obtained before, in Section 10 by another method (see (10.14)). Expanding the second factor of the integrand into a power series, changing the order of integration and summation and using the Poisson integral, we obtain

$$J_0(r) = \frac{1}{\pi} \int_0^\pi \cos(r\cos\theta\cos\varphi) \sum_{m=0}^\infty (-1)^m \frac{r(\sin\theta\sin\varphi)^{2m}}{(2m)!}\,d\varphi$$

$$= \frac{1}{\pi} \sum_{m=0}^\infty \frac{(-1)^m(r\sin\theta)^{2m}}{(2m)!} \int_0^\pi \cos(r\cos\theta\cos\varphi)\sin^{2m}\varphi\,d\varphi$$

$$= \frac{1}{\pi} \sum_{m=0}^\infty \frac{(-1)^m(r\sin\theta)^{2m}}{(2m)!} \frac{\Gamma\left(m+\frac{1}{2}\right)\Gamma\left(\frac{1}{2}\right)}{\left(\frac{1}{2}\right)^m} \frac{J_m(r\cos\theta)}{(r\cos\theta)^m}$$

$$= \frac{1}{\pi} \sum_{m=0}^\infty \frac{(-1)^m(r\sin\theta)^{2m}}{m!\,2^m} \frac{J_m(r\cos\theta)}{(r\cos\theta)^m}. \tag{14.7}$$

Denoting $r\cos\theta = \sqrt{z}$, $r\sin\theta = \sqrt{h}$, $r = \sqrt{z+h}$, we find

$$J_0(\sqrt{z+h}) = \sum_{m=0}^\infty \frac{(-h/2)^m}{m!} z^{-m/2} J_m(\sqrt{z}). \tag{14.8}$$

From this formula, using formula (6.8) and the formula for differentiations

$$\frac{d^n}{d(z^2)^n} J_0(z) = \frac{(-1)^n J_n(z)}{2^n z^n},$$

which can be easily deduced from formulae (6.2) and (1.10), we obtain for integer values of n after simple transformations

$$(z+h)^{-n/2} J_n(\sqrt{z+h}) = \sum_{m=0}^\infty \frac{(-h/2)^m z^{-(n+m)/2}}{m!} J_{n+m}(\sqrt{z}). \tag{14.9}$$

Expansions of a similar type can also be constructed for the integral

$$\int_0^\pi \cos[z\cos(\theta - \varphi)]\sin^{2\nu}\varphi\,d\varphi, \tag{14.10}$$

which is an obvious generalization of the Poisson integral (14.1) and when $\nu = 0$ becomes the Parseval integral.

The Poisson integral multiplied by z^ν can be considered as a superposition of degree ν moments for the displacements of the plane waves type with respect to the axis coinciding with the fixed radius.

The series expansion of integral (14.10) similar to the Lommel expansion can be obtained in just the same way as we act when considering (14.1). Thus,

$$
\int_0^\pi \cos[r\cos(\theta - \varphi)] \sin^{2\nu}\varphi\, d\varphi
$$

$$
= \int_0^\pi \cos(\sqrt{z}\cos\varphi)\cos(\sqrt{h}\sin\varphi)\sin^{2\nu}\varphi\, d\varphi
$$

$$
= \int_0^\pi \cos(\sqrt{z}\cos\varphi) \sum_{m=0}^\infty \frac{(-1)^m (\sqrt{h}\sin\varphi)^{2m}}{(2m)!} \sin^{2\nu}\varphi\, d\varphi
$$

$$
= 2^\nu \Gamma\left(\frac{1}{2}\right) \sum_{m=0}^\infty \frac{(-1)^m 2^m h^m}{(2m)!} \frac{J_{\nu+m}(\sqrt{z})}{(\sqrt{z})^{m+\nu}} \Gamma\left(\nu + m + \frac{1}{2}\right).
$$

$$(14.11)$$

If $\theta = 0$, then, obviously, we obtain the Poisson integral because in this case $h = 0$ and only the first term remains in the expansion.

15. Some indefinite integrals

First, we consider some indefinite integrals which are consequences of the differentiation formulae and relations following from consideration of the Wronski determinant and the Basset formulae.

Recurrence relations and the differentiation formulae allow one to obtain the following indefinite integrals immediately:

$$
\int z^{\nu+1} Z_\nu(z)\, dz = z^{\nu+1} Z_{\nu+1}(z), \tag{15.1}
$$

$$
\int z^{-\nu+1} Z_\nu(z)\, dz = -z^{-\nu+1} Z_{\nu-1}(z), \tag{15.2}
$$

In the case when $z = \rho e^{-i\varphi}$, $\nu = 0$, we have

$$
\int \rho \tilde{f}_0(\rho)\, d\rho = \rho[\tilde{f}_1(\rho)\cos\varphi + \tilde{g}_1(\rho)\sin\varphi], \tag{15.3}
$$

$$
\int \rho \tilde{g}_0(\rho)\, d\rho = \rho[\tilde{g}_1(\rho)\cos\varphi - \tilde{f}_1(\rho)\sin\varphi]. \tag{15.4}
$$

One can replace $\tilde{f}$ and $\tilde{g}$ in these formulae by $\tilde{u}$ and $\tilde{v}$, respectively.

The following integrals containing two Bessel functions or squared Bessel functions are obtained by Lommel with the help of using the Wronskian of the Bessel equation:

$$\int \frac{dz}{z J_\nu^2(z)} = -\frac{\pi}{2 \sin \nu\pi} \cdot \frac{J_{-\nu}(z)}{J_\nu(z)}, \tag{15.5}$$

$$\int \frac{dz}{z J_\nu(z) J_{-\nu}(z)} = \frac{\pi}{2 \sin \nu\pi} \ln \frac{J_\nu(z)}{J_{-\nu}(z)}, \tag{15.6}$$

$$\int \frac{dz}{z J_\nu(z) Y_\nu(z)} = \frac{\pi}{2} \ln \frac{Y_\nu(z)}{J_\nu(z)}, \tag{15.7}$$

$$\int \frac{dz}{z J_\nu^2(z)} = \frac{\pi}{2} \frac{Y_\nu(z)}{J_\nu(z)}, \tag{15.8}$$

$$\int \frac{dz}{z Y_\nu^2(z)} = -\frac{\pi}{2} \frac{J_\nu(z)}{Y_\nu(z)}, \tag{15.9}$$

In order to prove the first of these formulae, we consider the derivative of the fraction

$$\frac{d}{dz}\left[\frac{J_{-\nu}(z)}{J_\nu(z)}\right] = \frac{J'_{-\nu}(z) J_\nu(z) - J_{-\nu}(z) J'_\nu(z)}{J_\nu^2(z)} = -\frac{2 \sin \nu\pi}{\pi z J_\nu^2(z)}.$$

(We have used formula (9.5) here.) Integrating with respect to z, we immediately obtain (15.5); in this case we assume that ν is non-integer. Let us note that we can obtain similar formulae for other cylindrical functions in the same way.

Let us now consider the derivative of the logarithm of the fraction $\frac{J_\nu(z)}{J_{-\nu}(z)}$.

$$\frac{d}{dz}\left[\ln \frac{J_\nu(z)}{J_{-\nu}(z)}\right] = \frac{J'_\nu(z) J_{-\nu}(z) - J'_{-\nu}(z) J_\nu(z)}{J_\nu(z) J_{-\nu}(z)} = \frac{2 \sin \nu\pi}{\pi z J_\nu(z) J_{-\nu}(z)}.$$

Integrating this relation with respect to z, we obtain for non-integer ν

$$\int \frac{dz}{z J_\nu(z) J_{-\nu}(z)} = \frac{\pi}{2 \sin \nu\pi} \ln \frac{J_\nu(z)}{J_{-\nu}(z)},$$

that prove (15.6).

In the same way we can also prove formulae (15.7)–(15.9), where ν is non-integer. Let us also notice that

$$\int \frac{dz}{z H_\nu^{(1)}(z) H_\nu^{(2)}(z)} = -\frac{\pi}{4i} \ln \frac{H_\nu^{(2)}(z)}{H_\nu^{(1)}(z)}, \tag{15.10}$$

where ν can be integer as well as non-integer.

In the same way, but with the help of the Basset formula (9.13) we can obtain indefinite integrals which contain derivatives of the cylindrical functions

$$\frac{2}{\pi} \int \frac{(\nu^2/z^2 - 1)dz}{z[Y'_\nu(z)]^2} = \frac{J'_\nu(z)}{Y'_\nu(z)}, \tag{15.11}$$

$$\frac{2}{\pi} \int \frac{(1 - \nu^2/z^2)dz}{z[J'_\nu(z)]^2} = \frac{Y'_\nu(z)}{J'_\nu(z)}, \tag{15.12}$$

$$\frac{2}{\pi} \int \frac{(\nu^2/z^2 - 1)dz}{z J'_\nu(z) Y'_\nu(z)} = \ln \frac{J'_\nu(z)}{Y'_\nu(z)}, \tag{15.13}$$

$$\frac{2}{\pi} \int \frac{(1 - \nu^2/z^2)dz}{z J'_\nu(z) Y'_\nu(z)} = \ln \frac{Y'_\nu(z)}{J'_\nu(z)}, \tag{15.14}$$

The product of the right-hand sides of formulae (15.8) and (15.11) can be represented with the help of the Basset formula (9.13) and the Wronski determinant in the following form:

$$\frac{2}{\pi} \int \frac{J_\nu'(z)Y_\nu'(z) - (1 - \nu^2/z^2)Y_\nu(z)J_\nu(z)}{z[J_\nu(z)Y_\nu'(z)]^2} dz = \frac{J_\nu'(z)Y_\nu(z)}{J_\nu(z)Y_\nu'(z)}. \tag{15.15}$$

Obviously, we can introduce the functions $H_\nu^{(1)}(z)$, $H_\nu^{(2)}(z)$, $J_{-\nu}(z)$ instead of the functions $J_\nu(z)$ and $Y_\nu(z)$ in these integrals without any efforts. One can easily obtain similar formulae for the modified functions $I_\nu(z)$ and $K_\nu(z)$; for instance,

$$\int \frac{(\nu^2/z^2 + 1)dz}{z[I_\nu'(z)]^2} = \frac{K_\nu'(z)}{I_\nu'(z)}, \tag{15.16}$$

$$\int \frac{-I_\nu'(z)K_\nu'(z) + (1 + \nu^2/z^2)I_\nu(z)K_\nu(z)}{z[I_\nu(z)K_\nu'(z)]^2} dz = \frac{I_\nu'(z)K_\nu(z)}{I_\nu(z)K_\nu'(z)}. \tag{15.17}$$

Let us consider the integral

$$\int z^n J_0(\alpha z)\, dz = \frac{1}{\alpha} \int z^{n-1} \frac{d}{dz}[zJ_1(\alpha z)]\, dz$$

$$= \frac{z^n}{\alpha} J_1(\alpha z) + \frac{n-1}{\alpha^2} z^{n-1} J_0(\alpha z) - \frac{(n-1)^2}{\alpha^2} \int z^{n-2} J_0(\alpha z)dz. \tag{15.18}$$

Obviously, repeatedly applying formulae (15.18) to the integral $\int z^n J_0(\alpha z)\, dz$, for odd n we can obtain an expression containing only the functions $J_0(\alpha z)$ and $J_1(\alpha z)$; if n is even, then the result contains one more summand containing $\int J_0(\alpha z)\, dz$.

Similar results can be obtained in the case when we have the functions $z^n Y_0(\alpha z)$ and $z^n H_0^{(1)}(\alpha z)$ under the integral sign.

A more general formula

$$\int z^{\mu+1} Z_\nu(z)\, dz = z^{\mu+1} Z_{\nu+1}(z) + (\mu - \nu)z^\mu Z_\nu(z) - (\mu^2 - \nu^2) \int z^{\mu-1} Z_\nu(z)\, dz$$

can easily be verified by differentiating and using formulae (6.3) and (6.4).

Let us consider very important indefinite integrals due to Lommel which can be obtained as a result of using the different transformations of the Bessel equation given in Section 13. If y and η are solutions of the equations

$$\frac{d^2y}{dz^2} + P(z)y = 0,$$

$$\frac{d^2\eta}{dz^2} + Q(z)\eta = 0,$$

then the following formula holds

$$\int (P - Q)y\eta\, dz = y\frac{d\eta}{dz} - \eta\frac{dy}{dz}. \tag{15.19}$$

Suppose that $P(z)$ is defined by (13.17) for $\psi(z) = kz$ and $\nu = \mu$; respectively, $Q(z)$ is obtained, by setting $\psi(z) = lz$, $\nu = \mu_2$, $k \neq l$ in (13.17). Then

$$\int \{(k^2 - l^2)z - (\mu_1^2 - \mu_2^2)/z\} Z_{\mu_1}(kz) Z_{\mu_2}^*(lz)\, dz$$

$$= z \left\{ Z_{\mu_1}(kz) \frac{dZ_{\mu_2}^*(lz)}{dz} - Z_{\mu_2}^*(lz) \frac{dZ_{\mu_1}(kz)}{dz} \right\}. \quad (15.20)$$

Assuming $\mu_1 = \mu_2 = \mu$, we obtain

$$\int z Z_\mu(kz) Z_\mu^*(lz)\, dz = \frac{z}{k^2 - l^2} \left\{ Z_\mu(kz) \frac{dZ_\mu^*(lz)}{dz} - Z_\mu^*(lz) \frac{dZ_\mu(kz)}{dz} \right\}$$

$$= \frac{z}{k^2 - l^2} \{ k Z_{\mu+1}(kz) Z_\mu^*(lz) - l Z_\mu(kz) Z_{\mu+1}^*(lz) \}, \quad (15.21)$$

$$\mu \neq l;$$

for $l = k$ we have the indefiniteness of the type $\dfrac{0}{0}$.

With the help of the L'Hospital rule we can obtain

$$\int z Z_\mu(kz) Z_\mu^*(kz)\, dz = \frac{z^2}{4} \{ 2 Z_\mu(kz) Z_\mu^*(kz) - Z_{\mu-1}(kz) Z_{\mu+1}^*(kz)$$

$$- Z_{\mu+1}(kz) Z_{\mu-1}^*(kz) \}. \quad (15.22)$$

If we set $P(z) = e^{2z - \nu^2}$ and $Q(z) = e^{2z + \nu^2}$, then formulae (13.19) and (13.20) imply that

$$\int (e^{2z - \nu^2} - e^{2z + \nu^2}) Z_\nu(e^z) Z_{i\nu}(e^z)\, dz = Z_\nu(e^z) \frac{dZ_{i\nu}(e^z)}{dz} - Z_{i\nu}(e^z) \frac{dZ_\nu(e^z)}{dz}.$$

$$(15.23)$$

One can easily obtain similar formulae for solutions of Inhomogeneous Bessel equations (see Section 16).

16. Functions contiguous to Bessel functions. Particular solutions of the inhomogeneous Bessel equation

In applications an essential role is played by functions which are particular solutions of the inhomogeneous Bessel equation $\nabla_\nu u = z^2 \varphi(z)$.

The simplest of these functions are those which can be obtained from the Bessel equation whose right-hand side is represented by a power or linear function of the argument and which are called *functions contiguous to Bessel functions*. These functions are considered in this section; in accordance with the type of the right-hand side and relations between the index of the equation and the exponent they are called the *Lommel, Struve, Anguer and Weber functions*. Below, in Section 27 we consider the *Lommel functions in two variables* and their particular cases: the *probability integral* and the *Fresnel integral*. Section 28 is devoted to a significantly more general and large class of functions — *incomplete cylindrical functions*; all these functions can be considered as particular solutions of the inhomogeneous Bessel equation whose right-hand side represents the product of a power function and an exponential one.

In Section 17 we consider two simple classes of particular solutions of the Bessel equation:

1) when the right-hand side represents a step-function;

2) when the right-hand side represents a Bessel function.

It is expedient, before investigating different particular solutions of the inhomogeneous Bessel equation separately, to consider some general properties of these solutions which can be established in a rather simple way. Subsection 1 of this section is devoted to this question.

1. We consider here the following questions:

1) functional equations containing the Wronskian;
2) indefinite integrals;
3) the Lommel expansion;
4) integral representations which can be obtained with the help of the Green functions of the Bessel equation, the fundamental Cauchy functions with the properties of unit matrix or by using the method of variations of arbitrary variables.

First, we consider the Wronskians of the inhomogeneous Bessel equations. Let u_1 and u_2, respectively, denote solutions of the homogeneous and inhomogeneous Bessel equations with index ν and the argument z

$$\frac{d^2u_1}{dz^2} + \frac{1}{z}\frac{du_1}{dz} + \left(1 - \frac{\nu^2}{z^2}\right)u_1 = 0,$$

$$\frac{d^2u_2}{dz^2} + \frac{1}{z}\frac{du_2}{dz} + \left(1 - \frac{\nu^2}{z^2}\right)u_2 = \varphi_1(z).$$

After multiplying the first of these equations by u_2 and the second one by u_1 and subtracting the second equation from the first one, we obtain the following equation:

$$\frac{d}{dz}[u_1'u_2 - u_1u_2'] + \frac{1}{z}[u_1'u_2 - u_1u_2'] = -u_1\varphi_1(z).$$

Denoting $t_1 = u_1'u_2 - u_1u_2'$, we seek t_1 in the form $t_1 = p(z)/z$. Then

$$p'(z)/z = -u_1\varphi_1(z), \qquad p(z) = -\int zu_1\varphi_1(z)\,dz,$$

hence,

$$(u_1'u_2 - u_1u_2')\big|_0^z = -\frac{1}{z}\int_0^z zu_1\varphi_1(z)\,dz.$$

If the function u_1 together with its first derivative vanish at $z = 0$, then we have

$$u_1'u_2 - u_1u_2' = -\frac{1}{z}\int_0^z zu_1\varphi_1(z)\,dz.$$

Using the recurrence relations for the functions u_1 and u_2, one can easily obtain different modifications of this equality so that the left-hand side contains functions of distinct indices which differ by an integer as well as to write the Basset and Lommel–Hankel formulae for the ingomogeneous equation (see Section 9). Especially simple results can be obtained for the special case when the right-hand side of the inhomogeneous equation represents a power function.

If both functions u_2 and u_3 are particular solutions of the inhomogeneous equations $\nabla_\nu u = z^2\varphi(z)$ whose right-hand sides are equal to $z^2\varphi_1(z)$ and $z^2\varphi_2(z)$, respectively, then we have

$$u_2' u_3 - u_3' u_2 \big|_0^z = \frac{1}{z} \int\limits_0^z z[u_3 \varphi_2(z) - u_2 \varphi_3(z)] \, dz.$$

In Section 15 we give some indefinite integrals whose integrand contains the Bessel functions. Now we are going to show how one can obtain integrals containing functions contiguous to the Bessel functions in a similar way.

Denoting

$$-\frac{1}{z} \int\limits_0^z z u_1 \varphi(z) \, dz = A_1(\varphi, z),$$

we have

$$\frac{d}{dz}\left(\frac{u_1}{u_2}\right) = \frac{u_1' u_2 - u_2' u_1}{u_2^2} = \frac{A_1(\varphi, z)}{u_2^2}.$$

Hence,

$$\frac{u_1}{u_2} = \int \frac{A_1(\varphi, z) \, dz}{u_2^2};$$

similarly,

$$\frac{u_2}{u_1} = -\int \frac{A_1(\varphi, z) \, dz}{u_1^2}.$$

Furthermore,

$$\frac{d}{dz}\left[\ln \frac{u_1}{u_2}\right] = \frac{u_1' u_2 - u_2' u_1}{u_2^2} \cdot \frac{u_2}{u_1};$$

hence,

$$\int \frac{A_1(\varphi, z) \, dz}{u_1 u_2} = \ln \frac{u_1}{u_2}.$$

Different integrals containing functions contiguous to the Bessel functions can be obtained with the help of Lommel's method.

Suppose that a function u_1 satisfies the homogeneous equation $u_1'' + P u_1 = 0$ and a function u_2 satisfies the inhomogeneous equation $u_2'' + Q u_2 = \varphi(z)$ which can be reduced to the Bessel equation. It can easily be verified that

$$\int (P - Q) u_1 u_2 \, dz = u_1 \frac{du_2}{dz} - u_2 \frac{du_1}{dz} - \int u_1 \varphi(z) \, dz.$$

From this equality one can easily obtain different integrals of products of cylindrical functions and functions contiguous to Bessel functions. Just in the same way, as the result of an obvious generalization one can obtain the corresponding integrals of two different contiguous functions.

Let us give without proof several definite integrals containing a function $s_{\mu,\nu}(z)$ which will be defined below by formulae (16.10)–(16.12). These integrals can be easily obtained by using Lommel's method which was applied by Lommel for cylindrical functions and is presented at the beginning of Section 15:

$$\int_0^z \frac{(\mu + \nu - 1)J_\nu(z)s_{\mu-1,\nu-1}(z) - J_{\nu-1}(z)s_{\mu,\nu}(z)}{J_\nu^2(z)}\, dz = \frac{s_{\mu,\nu}(z)}{J_\nu(z)}, \tag{16.1}$$

$$\int_0^z \frac{(\mu + \nu - 1)J_\nu(z)s_{\mu-1,\nu-1}(z) - J_{\nu-1}(z)s_{\mu,\nu}(z)}{s_{\mu,\nu}^2(z)}\, dz = -\frac{J_\nu(z)}{s_{\mu,\nu}(z)}, \tag{16.2}$$

$$\int_0^z \frac{(\mu + \nu - 1)J_\nu(z)s_{\mu-1,\nu-1}(z) - J_{\nu-1}(z)s_{\mu,\nu}(z)}{s_{\mu,\nu}(z)J_\nu(z)}\, dz = \ln\left[\frac{-J_\nu(z)}{s_{\mu,\nu}(z)}\right], \tag{16.3}$$

$$\int_0^z \frac{z^{\mu-1}J_\nu(z)\, dz}{(\mu + \nu - 1)J_\nu(z)s_{\mu-1,\nu-1}(z) - J_{\nu-1}(z)s_{\mu,\nu}(z)}$$
$$= \ln[(\mu + \nu - 1)zJ_\nu(z)s_{\mu-1,\nu-1}(z) - zJ_{\nu-1}(z)s_{\mu,\nu}(z)], \tag{16.4}$$

$$\int_0^z \left[(1 - \alpha^2) + \frac{\nu_2^2 - \nu_1^2}{z^2}\right] zJ_{\nu_1}(z)s_{\mu,\nu_2}(\alpha z)\, dz$$
$$= z[J_{\nu_1}(z)s_{\mu,\nu_2}'(\alpha z) - s_{\mu,\nu_2}(\alpha z)J_{\nu_1}'(z)]$$
$$- \alpha^{\mu+1}z[(\mu + \nu_1 - 1)J_{\nu_1}(z)s_{\mu-1,\nu_1-1}(z) - J_{\nu_1-1}(z)s_{\mu,\nu_1}(z)]. \tag{16.5}$$

An integral representation of the function $\Phi_\nu(z)$ which is a particular solution of the differential equation

$$\frac{d^2\Phi}{dz^2} + \frac{1}{z}\frac{d\Phi}{dz} + \left(1 - \frac{\nu^2}{z^2}\right)\Phi = \varphi(z),$$

according to the arbitrary variables variation method, can be sought in the form

$$\Phi_\nu(z) = A(z)J_\nu(z) + B(z)Y_\nu(z).$$

It can be easily shown that
for an non-integer index

$$\Phi_\nu(z) = \frac{-\pi}{2\sin\nu\pi}\int_{z_0}^z \xi[J_\nu(\xi)J_{-\nu}(z) - J_\nu(z)J_{-\nu}(\xi)]\varphi(\xi)\, d\xi, \tag{16.6}$$

$$z > z_0,$$

for an integer index

$$\Phi_n(z) = \frac{\pi}{2}\int_{z_0}^z \xi[J_n(\xi)Y_n(z) - J_n(z)Y_n(\xi)]\varphi(\xi)\, d\xi, \tag{16.7}$$

$$z > z_0.$$

In the cases when the right-hand side is a power function, the particular solution contains the integrals

$$A_{2\mu,\nu}(\sqrt{z+h}) = \int_0^z (z+h)^\mu J_\nu(\sqrt{z+h})\, d(\sqrt{z+h}),$$

which can be expressed in the form of a series being an analogue of the Lommel expansion:

$$A_{2\mu,\nu}(\sqrt{z+h}) = \sum_{m=0}^\infty \frac{(-h/2)^m}{m!} \sum_{k=0}^m \frac{h^k n!}{k!(n-k)!} A_{2\mu-m-2k,\nu-m}(\sqrt{z}). \qquad (16.8)$$

Let us consider a partial solution of the inhomogeneous Bessel equation in some cases.

In Subsection 2 a particular solution will be obtained in the case when the right-hand side represents a power function. This solution can be expressed via the Lommel functions. In Subsection 3 we consider a special case when the exponent is greater by one than the index of the Bessel operator and the solution is expressed via the Struve functions. In Subsection 4 the Anguer and Weber functions are considered. These functions allow one to find a particular solution in the case when the right-hand side represents a linear function of the argument z.

2. Consider a particular solution of the equation

$$\nabla_\nu u = kz^{\mu+1}, \qquad (16.9)$$

where k and μ are constants. One can easily see that the particular solution of this equation represented as a power series in increasing powers of the argument z, has the form

$$\begin{aligned}
y &= k\left[\frac{z^{\mu+1}}{(\mu+1)^2-\nu^2} - \frac{z^{\mu+3}}{\{(\mu+1)^2-\nu^2\}\{(\mu+3)^2-\nu^2\}} + \cdots\right] \\
&= kz^{\mu-1}\sum_{m=0}^\infty \frac{(-1)^m \left(\frac{1}{2}z\right)^{2m+2} \Gamma\left(\frac{1}{2}\mu-\frac{1}{2}\nu+\frac{1}{2}\right)\Gamma\left(\frac{1}{2}\mu+\frac{1}{2}\nu+\frac{1}{2}\right)}{\Gamma\left(\frac{1}{2}\mu-\frac{1}{2}\nu+m+\frac{3}{2}\right)\Gamma\left(\frac{1}{2}\mu+\frac{1}{2}\nu+m+\frac{3}{2}\right)} \\
&= ks_{\mu,\nu}(z). \qquad (16.10)
\end{aligned}$$

The function $s_{\mu,\nu}(z)$ is called the *Lommel function*.

The expansion (16.10) is, obviously, inelegible, if $\mu\pm\nu$ is a negative odd number.

In order to obtain another representation of the Lommel function, we apply the method of arbitrary variables variation to equation (16.9).

According to formulae (16.6) and (16.7) we define the Lommel function $s_{\mu,\nu}(z)$ in the following way:

$$s_{\mu,\nu}(z) = \frac{\pi}{2\sin\nu\pi}\left[J_\nu(z)\int_0^z z^\mu J_{-\nu}(z)\, dz - J_{-\nu}(z)\int_0^z z^\mu J_\nu(z)\, dz\right] \qquad (16.11)$$

for non-integer ν and

$$s_{\mu,\nu}(z) = \frac{\pi}{2}\left[Y_\nu(z)\int_0^z z^\mu J_\nu(z)\, dz - J_\nu(z)\int_0^z z^\mu Y_\nu(z)\, dz\right] \qquad (16.12)$$

for integer ν.

If $\mu \pm \nu$ is an odd positive number, then the particular solution of equation (16.9) for $k = 1$ can be represented in the form of a Lommel function $S_{\mu,\nu}(z)$, which for non-integer ν has the form

$$S_{\mu,\nu}(z) = s_{\mu,\nu}(z) + \frac{2^{\mu-1}\Gamma\left(\frac{1}{2}\mu - \frac{1}{2}\nu + \frac{1}{2}\right)\Gamma\left(\frac{1}{2}\mu + \frac{1}{2}\nu + \frac{1}{2}\right)}{\sin\nu\pi}$$

$$\times \left[\cos\left(\frac{1}{2}(\mu - \nu)\pi\right)J_{-\nu}(z) - \cos\left(\frac{1}{2}(\mu + \nu)\pi\right)J_\nu(z)\right]. \quad (16.13)$$

If ν is an integer, then we have

$$S_{\mu,\nu}(z) = s_{\mu,\nu}(z) + 2^{\mu-1}\Gamma\left(\frac{1}{2}\mu - \frac{1}{2}\nu + \frac{1}{2}\right)\Gamma\left(\frac{1}{2}\mu + \frac{1}{2}\nu + \frac{1}{2}\right)$$

$$\times \left[\sin\left(\frac{1}{2}(\mu - \nu)\pi\right)J_\nu(z) - \cos\left(\frac{1}{2}(\mu - \nu)\pi\right)Y_\nu(z)\right]. \quad (16.14)$$

From (16.11) and (16.12) we can immediately obtain the recurrence relations which are due to Lommel [7]:

$$s_{\mu+2,\nu}(z) = z^{\mu+1} - \{(\mu+1)^2 - \nu^2\}s_{\mu,\nu}(z), \quad (16.15)$$

$$s'_{\mu,\nu}(z) + \frac{\nu}{z}s_{\mu,\nu}(z) = (\mu + \nu - 1)s_{\mu-1,\nu-1}(z), \quad (16.16)$$

$$s'_{\mu,\nu}(z) - \frac{\nu}{z}s_{\mu,\nu}(z) = (\mu - \nu - 1)s_{\mu-1,\nu+1}(z). \quad (16.17)$$

These formulae immediately imply that

$$\frac{2\nu}{z}s_{\mu,\nu}(z) = (\mu + \nu - 1)s_{\mu-1,\nu-1}(z) - (\mu - \nu - 1)s_{\mu-1,\nu+1}(z), \quad (16.18)$$

$$2s'_{\mu,\nu}(z) = (\mu + \nu - 1)s_{\mu-1,\nu-1}(z) + (\mu - \nu - 1)s_{\mu-1,\nu+1}(z). \quad (16.19)$$

Notice that the functions $s_{\mu,\nu}(z)$ in (16.15)–(16.19) can be replaced by $S_{\mu,\nu}(z)$.

The case when $\mu \pm \nu$ is an odd negative integer, requires rather complicated reasoning which can be found in papers of Watson.

In the literature one can meet works where the integrals $\int \mathrm{Ai}(t)\, dt$ and $\int \mathrm{Bi}(t)\, dt$ (see Section 8) are considered which can be easily expressed via the Lommel functions.

From the practical viewpoint, the Lommel functions are very important special functions, because a great number of problems of mathematical physics can be reduced to the Bessel equation whose right-hand side represents a power function.

The Lommel functions allow one to obtain a particular solution in a convenient form in the cases when the right-hand side of the inhomogeneous Bessel equation is a piecewise continuous function representing a polynomial on every interval of continuity.

3. The Struve function with index ν is usually defined by the expression

$$\mathbf{H}_\nu(z) = \frac{2(z/2)^\nu}{\Gamma(\nu + 1/2)\Gamma(1/2)}\int_0^1 (1 - t^2)^{\nu-1/2}\sin(zt)\, dt$$

$$= \frac{2(z/2)^\nu}{\Gamma(\nu + 1/2)\Gamma(1/2)}\int_0^{\pi/2} \sin(z\cos\theta)\sin^{2\nu}\theta\, d\theta, \quad (16.20)$$

where $t = \cos\theta$.

Under the condition $\Re \nu > -1/2$ the function $\mathbf{H}_\nu(z)$ can be represented by a power series

$$\mathbf{H}_\nu(z) = \sum_{m=0}^{\infty} \frac{(-1)^m (z/2)^{\nu+2m+1}}{\Gamma(m+3/2)\Gamma(\nu+m+3/2)}. \tag{16.21}$$

We can easily prove this fact, by expanding $\sin(z\cos\theta)$ in (16.20) into a series with respect to increasing powers of z and then making the calculations similar to those given in Section 14 where we considered the Poisson integral.

Using this expansion, we can obtain the following recurrence relations:

$$\mathbf{H}_{\nu-1}(z) + \mathbf{H}_{\nu+1}(z) = \frac{2\nu}{z}\mathbf{H}_\nu(z) + \frac{(z/2)^\nu}{\Gamma(\nu+3/2)\Gamma(1/2)}, \tag{16.22}$$

$$\mathbf{H}_{\nu-1}(z) - \mathbf{H}_{\nu+1}(z) = 2\mathbf{H}'_\nu(z) - \frac{(z/2)^\nu}{\Gamma(\nu+3/2)\Gamma(1/2)}, \tag{16.23}$$

$$\left(\frac{d}{dz} + \frac{\nu}{z}\right)\mathbf{H}_\nu(z) = \mathbf{H}_{\nu-1}(z), \tag{16.24}$$

$$\left(\frac{d}{dz} - \frac{\nu}{z}\right)\mathbf{H}_\nu(z) = -\frac{(z/2)^\nu}{2\Gamma(\nu+3/2)\Gamma(1/2)} - \mathbf{H}_{\nu+1}(z), \tag{16.25}$$

which are, obviously, special cases of the corresponding recurrence relations for the Lommel function.

These formulae imply that the Struve function satisfies the inhomogeneous Bessel equation

$$\nabla_\nu \mathbf{H}_\nu(z) = \frac{4(z/2)^{\nu+1}}{\Gamma(1/2)\Gamma(\nu+1/2)}, \tag{16.26}$$

where the right-hand side is a power function whose exponent exceeds the index of the Bessel operator by one. Thus, the Struve function can be considered as a special case of the Lommel function. If all terms in the expansions of $\mathbf{H}_\nu(z)$ are taken with the sign "plus", then this series represents a function $\mathbf{L}_\nu(z)$ which is studied (in case $\nu = 0$) by Nikolson.

4. As a definition of the *Anguer function* we can take the Bessel integral

$$\mathbf{J}_\nu(z) = \frac{1}{\pi} \int_0^\pi \cos(\nu\theta - z\sin\theta)\, d\theta. \tag{16.27}$$

If the value of ν is an integer n, then the Anguer function coincides with the Bessel function $J_n(z)$, and if ν is non-integer, then $\mathbf{J}_\nu(z)$ is different from the Bessel function of the index ν.

The function

$$\mathbf{E}_\nu(z) = \frac{1}{\pi} \int_0^\pi \sin(\nu\theta - z\sin\theta)\, d\theta \tag{16.28}$$

is called the *Weber function.*

As a result of transformations of the integrals associated with the expansion of the integrand into a series in increasing powers of z we obtain the following equalities

$$J_\nu(z) = \cos \frac{1}{2}\nu\pi \sum_{m=0}^{\infty} \frac{(-1)^m (z/2)^{2m}}{\Gamma\left(m - \frac{1}{2}\nu + 1\right)\Gamma\left(m + \frac{1}{2}\nu + 1\right)}$$
$$+ \sin \frac{1}{2}\nu\pi \sum_{m=0}^{\infty} \frac{(-1)^m (z/2)^{2m+1}}{\Gamma\left(m - \frac{1}{2}\nu + \frac{3}{2}\right)\Gamma\left(m + \frac{1}{2}\nu + \frac{3}{2}\right)}, \tag{16.29}$$

$$E_\nu(z) = \sin \frac{1}{2}\nu\pi \sum_{m=0}^{\infty} \frac{(-1)^m (z/2)^{2m}}{\Gamma\left(m - \frac{1}{2}\nu + 1\right)\Gamma\left(m + \frac{1}{2}\nu + 1\right)}$$
$$- \cos \frac{1}{2}\nu\pi \sum_{m=0}^{\infty} \frac{(-1)^m (z/2)^{2m+1}}{\Gamma\left(m - \frac{1}{2}\nu + \frac{3}{2}\right)\Gamma\left(m + \frac{1}{2}\nu + \frac{3}{2}\right)}. \tag{16.30}$$

The functions $J_\nu(z)$ and $E_\nu(z)$ satisfy the equations

$$\nabla_\nu J_\nu(z) = \frac{(z - \nu)\sin\nu\pi}{\pi}, \tag{16.31}$$

$$\nabla_\nu E_\nu(z) = -\frac{(z + \nu)}{\pi} - \frac{(z - \nu)\cos\nu\pi}{\pi}. \tag{16.32}$$

Thus, it is evident that both these functions are connected with the Lommel functions because they represent the sum of solutions of the inhomogeneous equation for the cases when the exponent of the power function is equal to zero or one. This fact implies the formulae

$$J_\nu(z) = \frac{\sin\nu\pi}{\pi} s_{0,\nu}(z) - \frac{\nu\sin\nu\pi}{\pi} s_{-1,\nu}(z), \tag{16.33}$$

$$E_\nu(z) = -\frac{1 + \cos\nu\pi}{\pi} s_{0,\nu}(z) - \frac{\nu(1 - \cos\nu\pi)}{\pi} s_{-1,\nu}(z). \tag{16.34}$$

17. On integration of the inhomogeneous Bessel equation. Cauchy functions

In solving inhomogeneous linear differential equations an important role is played along with the Green functions, by the fundamental Cauchy functions whose use facilitates the obtaining of the solutions.

In Subsection 1 the Cauchy functions for the Bessel equations and the system of two Bessel equations are considered. In Subsection 2 we consider the solution of the Bessel equation with a special right-hand side which is based on using the fundamental Cauchy functions.

1. Consider the solution of the inhomogeneous equation

$$\Delta_\nu u + u = \frac{1}{z^2}\nabla_\nu u = f(z) \tag{17.1}$$

with the initial conditions

$$u(z_0) = u_0, \qquad \frac{du}{dz}\bigg|_{z=z_0} = v_0. \tag{17.2}$$

The particular solution which was sought in Section 16 (p. 46) has the form

$$u^*(z) = \frac{\pi}{2\sin\nu\pi} \int_{z_0}^{z} x f(x)[J_\nu(x)J_{-\nu}(z) - J_{-\nu}(x)J_\nu(z)]\, dx,$$

where ν is non-integer. The general solution of equation (17.1) can be expressed in the form

$$u(z) = A_1 J_\nu(z) + A_2 J_{-\nu}(z) + u^*(z),$$

where A_1 and A_2 should be determined from initial conditions (17.2).

Let us introduce the Cauchy function

$$Y_1(z,x) = \begin{cases} 0, & z < x; \\ \frac{\pi x}{2\sin\nu\pi} \left[J'_{-\nu}(x)J_\nu(z) - J'_\nu(x)J_{-\nu}(z) \right], & z \geq x; \end{cases} \tag{17.3}$$

$$Y_2(z,x) = \begin{cases} 0, & z < x; \\ \frac{\pi x}{2\sin\nu\pi} \left[J_\nu(x)J_{-\nu}(z) - J_{-\nu}(x)J_\nu(z) \right], & z \geq x. \end{cases} \tag{17.4}$$

Obviously, each of these functions satisfies the Bessel equation with index ν; moreover, one can easily check that at $z = x$ these functions and their derivatives have the following values:

	$Y_1(x,x)$	$Y_2(x,x)$
Y	1	0
Y'	0	1

Let us give other expressions for the Cauchy functions, which can also be used for integer values of the index:

$$Y_1(z,x) = \begin{cases} 0, & z < x; \\ \frac{\pi x}{2} \left[Y'_n(x)J_n(z) - J'_n(x)Y_n(z) \right], & z \geq x; \end{cases} \tag{17.5}$$

$$Y_2(z,x) = \begin{cases} 0, & z < x; \\ \frac{\pi x}{2} \left[J_n(x)Y_n(z) - Y_n(x)J_n(z) \right], & z \geq x. \end{cases} \tag{17.6}$$

A solution of equation (17.1) which satisfies conditions (17.2) can now be written in the form

$$u(z) = u_0 Y_1(z, z_0) + v_0 Y_2(z, z_0) + \int_{z_0}^z Y_2(z,x)f(x)\, dx.$$

Consider a particular solution of the equation

$$\frac{d^2 u}{dz^2} + \frac{1}{z}\frac{du}{dz} + \left(1 - \frac{\nu^2}{z^2}\right) u = f(z),$$

whose right-hand side is a step function; thus, $f(z) = 0$ for $z < z_1$, $f(z) = q_1$ for $z_1 < z < z_2, \ldots,$ $f(z) = \sum_{i=1}^k q_i$ for $z_k < z < z_{k+1}$, where q_i are constant. One can easily see that for $z_k < z < z_{k+1}$ the particular solution has the following form:

$$u^*(z) = \sum_{i=1}^k q_i \Big\{ s_{1,\nu}(z) - s_{1,\nu}(z_i)Y_1(z, z_i)$$

$$+ \nu \left[z_i^{-1} s_{1,\nu}(z_i) - s_{0,\nu-1}(z_i) \right] Y_2(z, z_i) \Big\}.$$

In [20] a similar problem for the Bessel equation of zero index whose right-hand side is a step function is considered; in this case the particular solution obtained by integration of the Green function contain only the Bessel functions.

If the right-hand side of the equation considered represents a continuous function

$$f(z) = \sum_{i=1}^{k} \beta_i(z - z_i) \quad \text{for} \quad z_k \leq z \leq z_{k+1}, \qquad k = 1, 2, 3, \ldots,$$

then the particular solution can be written in the following form:

$$u^*(z) = \sum_{i=1}^{k} \beta_i \Bigg\{ s_{2,\nu}(z) - s_{2,\nu}(z_i)Y_1(z, z_i)$$

$$+ \left[\frac{2}{z} s_{2,\nu}(z_i) - (1 + \nu)s_{1,\nu-1}(z_i) \right] Y_2(z, z_i) \Bigg\},$$

where β_i are some constants.

Let us pass to the equation of the fourth order which can be reduced to a system of two Bessel equations. First, we consider the inhomogeneous equation

$$L^2(u) - u = f(z), \tag{17.7}$$

where L is a linear differential operator of the second order. One can easily show that the general solution of equation (17.7) has the form

$$u = u_1 + u_2,$$

where the functions u_1 and u_2 satisfy the equations

$$L(u_{1,2}) \pm u_{1,2} = \mp \frac{1}{2} f(z),$$

respectively.

If

$$L = \Delta_\nu = \frac{d^2}{dz^2} + \frac{1}{z}\frac{d}{dz} - \frac{\nu^2}{z^2},$$

then we can obtain a particular solution to each of these two equations as has been shown on page 35. Suppose that the index of the operator is an integer, then, combining the particular solutions u_1^* and u_2^*, we obtain a formula of the form

$$u^* = u_1^* + u_2^* = \int_{z_0}^{z} Y_4(z, x)f(x)\, dx, \tag{17.8}$$

where

$$Y_4(z, x) = \frac{\pi x}{4} \Big\{ J_n(z)Y_n(x) - Y_n(z)J_n(x) + \frac{2}{\pi}\left[I_n(z)K_n(x) - K_n(z)I_n(x)\right] \Big\};$$

if the index is non-integer, then n should be replaced by ν.

One can easily see that

$$Y_4(z, z) = 0, \quad Y_4'(z, z) = 0, \quad \Delta_n Y_4(z, z) = 0, \quad \Delta_n' Y_4(z, z) = 1.$$

These equalities immediately imply that we can take Y_4 as one of the fundamental Cauchy functions and the derivative of this function with respect to x, the result of applying the operator Δ_n and its derivative with respect to x can be taken as the remaining three fundamental Cauchy functions. Thus, we shall express the solution of the formulated problem as a sum of the particular solution u^* and some linear combination of the fundamental Cauchy functions

$$Y_1(z, x), \quad Y_2(z, x), \quad Y_3(z, x), \quad Y_4(z, x),$$

which vanish identically for $z < x$ and satisfy the equation $L^2(u) - u = 0$ for $z \geq x$; at $z = x$ the following conditions hold:

m	Y_m	Y'_m	$\Delta_n Y_m$	$\Delta'_n Y_m$
1	1	0	0	0
2	0	1	0	0
3	0	0	1	0
4	0	0	0	1

It can easily be shown that in the case considered these functions have the form for $z \geq x$

$$
\left.
\begin{aligned}
Y_1(z, x) &= \frac{\pi x}{4}\left\{ - J'_n(x)Y_n(z) + Y'_n(x)J_n(z) \right.\\
&\qquad \left. + \frac{2}{\pi}\left[I'_n(x)K_n(z) - K'_n(x)I_n(z)\right]\right\},\\
Y_2(z, x) &= \frac{\pi x}{4}\left\{ J_n(x)Y_n(z) - Y_n(x)J_n(z) \right.\\
&\qquad \left. - \frac{2}{\pi}\left[I_n(x)K_n(z) - K_n(x)I_n(z)\right]\right\},\\
Y_3(z, x) &= \frac{\pi x}{4}\left\{ J'_n(x)Y_n(z) - Y'_n(x)J_n(z) \right.\\
&\qquad \left. + \frac{2}{\pi}\left[I'_n(x)K_n(z) - K'_n(x)I_n(z)\right]\right\},\\
Y_4(z, x) &= \frac{\pi x}{4}\left\{ - J_n(x)Y_n(z) + Y_n(x)J_n(z) \right.\\
&\qquad \left. - \frac{2}{\pi}\left[I_n(x)K_n(z) - K_n(x)I_n(z)\right]\right\}.
\end{aligned}
\right\}
\tag{17.9}
$$

Let us give the formulae for differentiation of the fundamental functions

m	Y_m	Y'_m	$\Delta_n Y_m$	$\Delta'_n Y_m$
1	Y_1	Y'_1	Y_3	Y'_3
2	Y_2	Y'_2	Y_4	Y'_4
3	Y_3	Y'_3	$-Y_1$	$-Y'_1$
4	Y_4	Y'_4	$-Y_2$	$-Y'_2$

The solution of equation (17.7) with the initial conditions

$$u(z_0) = u_0, \quad u'(z_0) = v_0, \quad \Delta_n u_0 = C_0^*, \quad \Delta'_n u_0 = Q_0$$

can be written in the form

$$u(z) = u_0 Y_1(z, z_0) + v_0 Y_2(z, z_0) + C_0^* Y_3(z, z_0) + Q_0 Y_4(z, z_0) + u^*(z).$$

In the monographs of the author [18], [20], the fundamental Cauchy functions are used in problems on oscillations of rods with variable section, oscillations of a circular plate, equilibrium of a conic shell, the theory of heat waves, as well as in problems on the bend of a circular plate on an elastic foundation and others.

The Cauchy functions for a half-integer index of the Bessel equation can be expressed via elementary functions (see [24]).

When solving some physical problems, we have initial conditions defining the form of the Cauchy functions which contain linear differential operators of the second and third order different from the operators Δ_ν and Δ'_ν considered before; however, this fact does not cause any essential difficulties (see, for example, [20]).

At the beginning of the section we mentioned the Green functions. There exists a direct relation between the fundamental Cauchy functions and the Green functions, which follows immediately from the definition of the Green function of a differential operator.

Consider the solution of the equation

$$\Delta_0 \Delta_0 u + u = 0, \tag{17.10}$$

which has no singularities at $z = 0$ and is continuous with the first derivative and the operator Δ for $z = x$ and such that

$$\Delta'_0 u_{x+0} - \Delta'_0 u_{x-0} = 1$$

and $u \to 0$ as $z \to \infty$. This solution is called the Green function of equation (17.10). From formulae (11.9) we can immediately obtain (the notation u_0, v_0, f_0, and g_0 see on pages 13)

$$K(z,x) = \begin{cases} \dfrac{\pi x}{2}[f_0(x)u_0(z) - g_0(x)v_0(z)], & x \geq z; \\[4mm] \dfrac{\pi x}{2}[u_0(x)f_0(z) - v_0(x)g_0(z)], & z \geq x. \end{cases} \tag{17.11}$$

Let us show that we can obtain the function $Y_4(z,x)$ from $K(z,x)$ in a simple way. We form the difference of the function $K(z,x)$ and the expression which represents the upper row of (17.11) without the restriction $x \geq z$. Obviously, this difference is the function $Y_4(z,x)$ and it has the form

$$Y_4(z,x) = \begin{cases} 0, & x > z; \\[3mm] \dfrac{\pi x}{2}[u_0(x)f_0(z) - v_0(x)g_0(z) - f_0(x)u_0(z) + g_0(x)v_0(z)], & z \geq x. \end{cases}$$

The formulae for the functions $Y_1(z,x)$, $Y_2(z,x)$, and $Y_3(z,x)$, are given in [20].

Consider a particular solution of the inhomogeneous equation

$$\Delta_\nu^2 u - u = f(z).$$

Assume that $f(z) = 0$ for $z \neq z_k$, $k = 1, 2, 3, \ldots$. Moreover, let

$$u\Big|_{z_k+0} = u\Big|_{z_k-0} + u_{0,k};$$

$$u'\Big|_{z_k+0} = u'\Big|_{z_k-0} + v_{0,k};$$

$$\Delta_\nu u\Big|_{z_k+0} = \Delta_\nu u\Big|_{z_k-0} + G^*_{0,k};$$

$$\Delta'_\nu u\Big|_{z_k+0} = \Delta'_\nu u\Big|_{z_k-0} + Q_{0,k}.$$

Then the particular solution for $z_k < z < z_{k+1}$ has the following form

$$u_k^*(z) = \sum_{i=1}^{k} \left\{ u_{0,i} Y_1(z, z_i) + v_{0,i} Y_2(z, z_i) + G_{0,i}^* Y_3(z, z_i) + Q_{0,i} Y_4(z, z_i) \right\}.$$

2. Consider the particular solutions of the inhomogeneous Bessel equation in the cases when the right-hand side represents the product of a cylindrical function and the squared argument. We begin with another simple problem.

Let the right-hand side of the Bessel equation represent a cylindrical function whose index is different from the index of the operator:

$$\frac{d^2 u}{dz^2} + \frac{1}{z}\frac{du}{dz} + \left(1 - \frac{\nu^2}{z^2}\right) u = z^{-2} Z_\mu(z), \qquad \mu \neq \nu.$$

A particular solution of this equation has the form

$$u^* = \frac{1}{\mu^2 - \nu^2} Z_\mu(z).$$

A particular solution of the inhomogeneous equation

$$\frac{d^2 u}{dz^2} + \frac{1}{z}\frac{du}{dz} + \left(1 - \frac{\nu^2}{z^2}\right) u = Z_\mu(\lambda z) \qquad (17.12)$$

has the form

$$u = \frac{Z_\nu(\lambda z)}{1 - \lambda^2}, \quad \lambda \neq 1.$$

This result cannot be used when $\lambda = 1$. The solution of equation (17.12) in this case will be called the *resonance solution*; we use the limit passage to find this solution.

The solution of equation (17.12) with the initial conditions $u(z_0) = 0$, $u'(z_0) = 0$ can be written in the form

$$u = \frac{1}{1 - \lambda^2} \left\{ -Z_\nu(\lambda z_0) Y_1(z, z_0) - \lambda Z_\nu'(\lambda z_0) Y_2(z, z_0) + Z_\nu(\lambda z) \right\},$$

where prime denotes differentiation with respect to the argument λz. For $\lambda = 1$ the denominator and the expression in the braces are equal to zero; using the L'Hospital rule, we obtain

$$u(z) = -\frac{1}{2}\left[-z_0 Z_\nu'(z_0) Y_1(z, z_0) + \left(z_0 - \frac{\nu^2}{z_0}\right) Z_\nu(z_0) Y_2(z, z_0) + z Z_\nu'(z) \right].$$

Consider a simpler special case. The solution of the equation

$$\frac{d^2 u}{dz^2} + \frac{1}{z}\frac{du}{dz} + u = A J_0(\lambda z)$$

with the initial conditions $u(0) = 0$, $u'(0) = 0$, for $\lambda \neq 1$ has the form

$$u = \frac{A}{1 - \lambda^2}[J_0(\lambda z) - J_0(z)].$$

Applying the limit passage $(\lambda \to 1)$, we obtain the resonance solution

$$u = \frac{Az}{2} J_1(z).$$

18. Products of Bessel functions

In this section we consider some properties of products of the Bessel functions which we met before in Sections 11 and 17. We deduce the differential equations which are satisfied by these products here and give integral representations of the products. The problem concerning the expansion of the Bessel functions products into power series is considered in [7].

At the beginning, following Watson [7], Orr and Appel, we consider two differential equations

$$
\begin{aligned}
\frac{d^2 v}{dz^2} + R_1(z)v &= 0, \\
\frac{d^2 w}{dz^2} + R_2(z)v &= 0.
\end{aligned}
\tag{18.1}
$$

Denoting the product vw by y, we obtain

$$ y'' = v''w + 2v'w' + vw'' = -(R_1 + R_2)y + 2v'w'. $$

Hence,

$$ y'' + (R_1 + R_2)y = 2v'w'. $$

Differentiating this equation, we get

$$ y''' + 2(R_1 + R_2)y' + (R_1' + R_2')y = (R_1 - R_2)(v'w - vw'). \tag{18.2} $$

The special case $R_1 = R_2$ results in the differential equation of the third order

$$ y''' + 4R_1 y' + 2R_1' y = 0. \tag{18.3} $$

If $R_1 \neq R_2$, then

$$ \frac{d}{dz}\left\{ \frac{y''' + 2(R_1 + R_2)y' + (R_1' + R_2')y}{R_1 - R_2} \right\} = -(R_1 - R_2)y. \tag{18.4} $$

Consider in detail the case $R_1 = -R_2$ which results in the product containing two functions whose arguments differ by the factor equal to the imaginary unit. We have

$$ \frac{d}{dz}\left(\frac{y'''}{2R_1} \right) = -2R_1 y \quad \text{or} \quad y^{IV} - \frac{R_1'}{R_1}y''' + 4R_1^2 y = 0. \tag{18.5} $$

Suppose $R_1 = c^2 z^{2q-2}$. Then (18.5) takes the form

$$ y^{IV} - 2(q-1)\frac{y'''}{z} + 4c^4 z^{4(q-1)}y = 0. \tag{18.6} $$

For these representations of R_1 and R_2, equations (18.1) can be reduced, as is shown in Section 13, to the Bessel equation. Hence, (see page 33), the solution of (18.6) is

$$ y = z Z^*_{1/(2q)}\left(\frac{cz^q}{q} \right) Z_{1/(2q)}\left(\frac{ciz^q}{q} \right). $$

Here Z^* and Z can be two different cylindrical functions, for instance, Z^* is the Bessel function and Z is the Hankel function. For $q = 1$ we have the trivial case

$y^{IV} + 4c^4 y = 0$. If $q = 3/2$, then

$$y^{IV} - \frac{y'''}{z} + 4c^4 z^2 y = 0,$$

$$y = z Z_{1/3}^* \left(\frac{2cz^{1.5}}{3} \right) Z_{1/3} \left(\frac{2icz^{1.5}}{3} \right). \tag{18.7}$$

For $z = i\xi$ and $q = 3/2$ we have

$$\frac{d^4 y}{d\xi^4} - \frac{1}{\xi} \frac{d^3 y}{d\xi^3} - 4c^4 \xi^2 y = 0, \tag{18.8}$$

$$y = \xi Z_{1/3}^* \left(\frac{2ci\sqrt{i}\xi^{1.5}}{3} \right) Z_{1/3} \left(\frac{2c\sqrt{i}\xi^{1.5}}{3} \right).$$

Setting $y = zt(z)$, we obtain for $q = 3/2$

$$t^{IV} + 3t'''/z - 3t''/z^2 + 4c^4 z^2 t = 0, \qquad t = y/z, \tag{18.9}$$

where y is represented by formula (18.7).

If we assume $R_1 = R_2$, then for the case $R_1 = e^{2z} - \nu^2$ we have (see page 34)

$$v = Z_{\pm\nu}(e^z),$$

hence,

$$y = Z_{\pm\nu}(e^z) Z_{\pm\nu}^*(e^z).$$

The differential equation with this solution can be obtained immediately from (18.3).

$$\frac{d}{dz} \left(\frac{d^2}{dz^2} - 4\nu^2 \right) y + 4e^{2z} \left(\frac{d}{dz} + 1 \right) y = 0. \tag{18.10}$$

If the solution is represented by the product

$$y = z^2 Z_{\pm\nu}(e^{1/z}) Z_{\pm\nu}^*(e^{1/z}),$$

then the corresponding equation has the form

$$\frac{d^3 y}{dz^3} + 4\frac{e^{2/z} - \nu^2}{z^4} y' - 4\frac{e^{2/z} + 2z(e^{2/z} - \nu^2)}{z^6} y = 0. \tag{18.11}$$

If $R_2 = kR_1$, $k \neq 1$, then

$$\frac{d}{dz} \left\{ \frac{\frac{d^3 y}{dz^3} + 2(1+k)R_1 y' + (1+k)R_1' y}{(1-k)R_1} \right\} = -(1-k)R_1 y$$

or

$$\frac{d}{dz} \left\{ \frac{y'''}{R_1} + 2(1+k)y' + (1+k)\frac{R_1'}{R_1} y \right\} + (1-k)^2 R_1 y = 0. \tag{18.12}$$

Consider the problem of construction of the fundamental Cauchy functions for equation (18.5).

As before, we have

$$y = vw, \qquad v'' + R_1 v = 0, \qquad w'' - R_1 w = 0. \tag{18.13}$$

We denote the fundamental Cauchy functions[5] of equations (18.1) by v_1, v_2, w_1, and w_2. In this case the functions v_1, v_2, w_1, and w_2 vanish identically for $z < a$; for $z = a$ we have

$$v_1(a) = 1; \qquad v_1'(a) = 0; \qquad v_2(a) = 0; \qquad v_2'(a) = 1;$$
$$w_1(a) = 1; \qquad w_1'(a) = 0; \qquad w_2(a) = 0; \qquad w_2'(a) = 1.$$

First of all, we construct the function

$$y_1 = v_1 w_1, \qquad z \geq a.$$

Obviously,

$$y_1' = v_1' w_1 + v_1 w_1',$$
$$y_1'' = v_1'' w_1 + v_1 w_1'' + 2v_1' w_1' = 2v_1' w_1',$$
$$y_1''' = 2R_1[-v_1 w_1' + w_1 v_1'],$$
$$y_1^{IV} = -4R_1^2 v_1 w_1 + 2R_1'[-v_1 w_1' + w_1 v_1'].$$

For $z = a$ we have

$$y_1(a) = 1, \quad y_1'(a) = 0, \quad y_1''(a) = 0, \quad y_1'''(a) = 0.$$

Let us introduce the following functions

$$y_2 = \frac{v_2 w_1 + v_1 w_2}{2},$$

$$y_3 = \frac{1}{2} v_2 w_2,$$

$$y_4 = \frac{v_2 w_1 - v_1 w_2}{4R_1(a)} \qquad \text{for} \quad z \geq a,$$

$$y_2 = y_3 = y_4 = 0 \qquad \text{for} \quad z < a.$$

After differentiating these functions and determining the values of y_2, y_3, y_4 and their derivatives at $z = a$, we immediately obtain that y_1, y_2, y_3, and y_4 form a system of fundamental Cauchy functions, i.e. for $z = a$ we have

	y_1	y_2	y_3	y_4
y	1	0	0	0
y'	0	1	0	0
y''	0	0	1	0
y'''	0	0	0	1

Let us go to the integral representations of the Bessel functions products. For the reasoning below, it is essential to use the addition theorems for the Bessel functions.

Let us formulate the Neumann addition theorem (see Section 11)

$$Y_0\left(\sqrt{\alpha^2 + \xi^2 - 2\alpha\xi \cos\theta}\right) = 2 \sum_{m=0}^{\infty}{}' J_m(\alpha) Y_m(\xi) \cos m\theta, \quad \alpha \leq \xi. \qquad (18.14)$$

---

[5]These functions can be easily constructed with the help of the results of item 1. in Section 17.

Let us set $\alpha = \xi$. Then the left-hand side of (18.14) takes the form $Y_0(2\alpha \sin \frac{\theta}{2})$. Denoting $\theta/2$ by φ, after multiplying the equality by $\cos 2n\varphi$ and integrating, we obtain

$$\frac{1}{\pi} \int_0^\pi Y_0(2\alpha \sin \varphi) \cos 2n\varphi \, d\varphi = Y_n(\alpha) J_n(\alpha). \tag{18.15}$$

Similar formulae can also be immediately obtained for functions of a pure imaginary and complex argument from the appropriate addition theorems given in Section 11.

Let us now use the generalization of the Neumann addition theorem (see Section 11)

$$Z_\nu \left(\sqrt{\alpha^2 + \xi^2 - 2\alpha\xi \cos \theta} \right) \cos \nu\psi = 2 \sum_{m=0}^\infty {}' Z_{\nu+m}(\xi) J_m(\alpha) \cos m\theta, \quad \alpha \leq \xi.$$

Setting $\alpha = \xi$, after multiplying both sides of the equality by $\cos n\theta$ and integrating, we obtain

$$\frac{1}{2\pi} \int_0^{2\pi} Z_\nu \left(2\alpha \sin \frac{\theta}{2} \right) \cos \left[\frac{\nu(\pi - \theta)}{2} \right] \cos n\theta \, d\theta = Z_{\nu+n}(\alpha) J_n(\alpha).$$

Similarly, we can obtain for $\alpha \leq \xi$

$$Z_{\nu+n}(\xi) J_n(\alpha) = \frac{1}{2\pi} \int_0^{2\pi} Z_\nu \left(\sqrt{\alpha^2 + \xi^2 - 2\alpha\xi \cos \theta} \right) \cos \psi\nu \cos n\theta \, d\theta.$$

The formula

$$\int_0^{\pi/2} J_{\nu+\mu}(2\alpha \cos \theta) \cos[(\mu - \nu)\theta] \, d\theta = \frac{\pi}{2} J_\nu(\alpha) J_\mu(\alpha), \tag{18.16}$$

$$\Re(\mu + \nu) > -1,$$

can be obtained as a result of the representation of the Bessel functions present in the product in the form of the power series [7].

For products of Bessel functions with different arguments the following formula has been proven [7][6]

$$J_\mu(az) J_\nu(bz) = \frac{(az/2)^\mu (bz/2)^\nu}{\Gamma(\mu + 1)} \sum_{\lambda=0}^\infty \frac{(-1)^\lambda (bz/2)^{2\lambda}}{\lambda! \Gamma(\nu + \lambda + 1)} {}_2F_1(-\lambda, -\nu - \lambda; \mu + 1; a^2/b^2). \tag{18.17}$$

19. Integral representations of Bessel functions

In Sections 11 and 14 above we considered two integral representations of the Bessel function: the Bessel integral and the Poisson integral. In this section we consider several contour integral which are generalizations of the Poisson integral, with the help of the theory of functions of a complex variable. Then, without deduction some results connected with similar generalizations of the Bessel integral will be formulated.

---

[6] see also A. *Kratzer* and V. *Franz* Transcendental functions, IL, 1963.

Recall that the Poisson integral gives the following representation for the Bessel function:

$$J_\nu(z) = \frac{(z/2)^\nu}{\Gamma(\nu+1/2)\Gamma(1/2)} \int_{-1}^{1} (1-t^2)^{\nu-1/2} e^{izt}\, dt.$$

We shall try to generalize this result, by considering contour integrals of the form

$$v(z) = \int_C T(t) e^{izt}\, dt. \tag{19.1}$$

Let us find out what restrictions we should impose on the contour C and the function $T(t)$ in order that the function $z^\nu v$ be a solution of the Bessel equation with index ν.

The function $v(z)$ satisfies the equation

$$zv'' + (2\nu+1)v' + zv = 0. \tag{19.2}$$

After differentiating (19.1), we substitute the expressions obtained into (19.2) and integrate the products zv'' and zv by parts. As a result, we obtain

$$[it^2 T(t) - iT(t)]e^{izt}\big|_C - i \int_C [t^2 T'(t) + 2tT(t) - (2\nu+1)tT(t) - T'(t)]e^{izt}\, dt = 0. \tag{19.3}$$

Here the symbol $\big|_C$ denotes the increase of the quantity before the vertical rule after going around the contour C in the positive direction.

In order that equality (19.3) holds, it is, obviously, sufficient that each of the two terms in the left-hand side is equal to zero. By setting the first term equal to zero, we obtain

$$[T(t)(t^2-1)]e^{izt}\big|_C = 0,$$

and taking into account the fact that the second term vanishes if the expression in the bracket is equal to zero, we have

$$\frac{d}{dt}[T(t)(t^2-1)] = (2\nu+1)T(t)t. \tag{19.4}$$

Integrating (19.4), we obtain that up to a constant factor

$$T(t) = (t^2-1)^{\nu-1/2}. \tag{19.5}$$

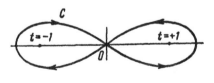

C

$t=-1$ $t=+1$

Figure 2

We assume that the index ν does not accept the values $\nu = n + 1/2$ (where $n = 0,1,2,\ldots$), because in this case the integrand is an analytic function at $t = \pm 1$ and the integrals of type (19.1) vanish.

Under the conditions formulated, (19.5) is a multifunction.

If the contour is a closed curve inside which one of the points $t = \pm 1$ lies, then after going along the contour in the positive direction, the argument increases by $(\nu - 1/2)\pi i$.

Taking the contour C in the form presented in Fig. 2, we bypass both marked points in different directions; therefore, the function does not change its value after going around the whole contour. Hence, applying the integration along this closed contour, we can obtain the representation of the integral of the Bessel equation in the form

$$u = z^\nu \int_C (t^2 - 1)^{\nu - 1/2} e^{zti}\, dt.$$ (19.6)

Giving one or another form to the contour, we can obtain different integral representations of the Bessel functions.

The second group of the integral representations can be obtained, if the contour is a non-closed curve and at the infinite points the product $(t^2 - 1)^{\nu - 1/2} e^{zti}$ tends to zero.

If the integration is performed along the contour presented in Fig. 3, then for the point $b + iy$ of the contour, the exponential function present in the integrand has the form

$$\exp[iz(b + iy)], \qquad \Re z > 0.$$

The integrand tends to zero as $y \to \infty$, and the integral (19.6) converges absolutely.

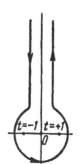

Figure 3

We turn our attention to the case, where the contour has the form shown in Fig. 2.

We agree to assign to the multifunction $(t^2 - 1)^{\nu - 1/2}$ for a contour of this form the sign plus for $t > 1$.

Integral (19.6) converges for $\Re z > 0$. Differentiation under the integral sign is legitimate, and the integral is an analytic function of ν for all values of ν.

Let us show that for an appropriate choice of the constant factor, the integral (19.6) represents the Bessel function $J_\nu(z)$. We shall prove below that the integral along the figure eight curve, which is Hankel's generalization of the Poisson integral, can be reduced to the ordinary Poisson integral (considered in Section 14); hence, it represents the Bessel function $J_\nu(z)$.

We integrate along the symmetrical contour[7] shown in Fig. 4. It is sufficient to consider the integral along the half-contour which lies in the right half-plane and consists of two sloping segments OA and OE, two horizontal segments AB and ED and the arc BCD with the center at the point t $(1, 0)$. Let ρ denote the radius of the arc, $\rho \leq 1$. Obviously, on the arc the following inequality holds:

$$\left| (t^2 - 1)^{\nu - 1/2} e^{zti} \right| < C\rho^{\nu - 1/2},$$

where C is a constant.

This implies that the integral along the arc cannot exceed $C\rho^{\nu - 1/2} 2\pi\rho = 2\pi C\rho^{\nu + 1/2}$. Thus, for $\nu + 1/2 > 0$, the integral along the arc tends to zero as $\rho \to 0$. Obviously, as ρ tends to zero, the lengths of the segments OE and OA simultaneously tend to zero, and on the segment AB the function T has the modulus $(t^2 - 1)^{\nu - 1/2}$ and the argument $-\pi i(\nu - 1/2)$.

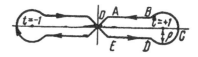

Figure 4

The modulus of the function $T(t)$ on the segment ED remains equal to $(t^2 - 1)^{\nu - 1/2}$, and its argument on this segment is equal to $-\pi i(\nu - 1/2)$. Obviously, the modulus of the integrand on the sloping segments is less than one. Hence, the integral along the right half-loop is equal to the sum of the integrals along the

---

[7]See [28].

segments AB and CD, whose lengths tends to one as $\rho \to 0$. Therefore,

$$u = 2z^\nu \left[e^{-\pi(\nu-1/2)i} \int_0^1 (1-t^2)^{\nu-1/2} e^{itz}\, dt - e^{\pi i(\nu-1/2)} \int_0^1 (1-t^2)^{\nu-1/2} e^{itz}\, dt \right]$$

$$= 2z^\nu i \sin(\nu - 1/2)\pi \int_{-1}^1 (1-t^2)^{\nu-1/2} e^{itz}\, dt.$$

Comparing this result with the Poisson integral, we finally obtain

$$J_\nu(z) = \frac{u}{2^{\nu+1} i \sqrt{\pi} \cos \nu \pi \Gamma(\nu + 1/2)}$$

$$= \frac{(z/2)^\nu}{2i\sqrt{\pi} \cos \nu \pi \Gamma(\nu + 1/2)} \int_C (t^2 - 1)^{\nu-1/2} e^{izt}\, dt, \qquad (19.7)$$

which is Hankel's generalization of the Poisson integral.

The restriction $\nu + 1/2 > 0$ is not necessary, because both parts of the equality are analytic functions of ν and according to the principle of analytic continuation, the result is valid for any ν.

We now consider a contour in the form of a loop (see Fig. 3) and we shall show that, by integrating along this loop, we obtain a representation of the Bessel function $J_{-\nu}(z)$. Assume that the circle $|t| = 1$ lies inside the contour, then $(t^2 - 1)^{\nu-1/2}$ can be expanded in a series in decreasing powers of t which uniformly converges on the contour:

$$(t^2 - 1)^{\nu-1/2} = \sum_{m=0}^\infty \frac{\Gamma\left(\frac{1}{2} - \nu + m\right)}{m!\, \Gamma\left(\frac{1}{2} - \nu\right)} t^{2\nu-1-2m}.$$

Here t is a complex variable with the argument φ $(-3\pi/2 < \varphi < \pi/2)$, and its modulus is greater than one. Then the termwise integration is admissible; therefore,

$$z^\nu \int_{\infty i}^{(-1+,1+)} e^{izt}(t^2 - 1)^{\nu-1/2} dt = \sum_{m=0}^\infty \frac{z^\nu \Gamma\left(\frac{1}{2} - \nu + m\right)}{m!\, \Gamma\left(\frac{1}{2} - \nu\right)} \int_{\infty i}^{(-1+,1+)} t^{2\nu-1-2m} e^{izt}\, dt.$$

It is known [7], that

$$\int_{C_1} t^{2\nu-2m-1} e^{zti}\, dt = \frac{(-1)^m 2\pi e^{-\nu\pi i} z^{2m-2\nu} i}{\Gamma(2m - 2\nu + 1)},$$

where C_1 is the contour shown in Fig. 3; hence,

$$z^\nu \int_{\infty i}^{(-1+,1+)} e^{izt}(t^2 - 1)^{\nu-1/2} dt = \sum_{m=0}^\infty \frac{(-1)^m 2\pi i z^{-\nu+2m} \Gamma\left(\frac{1}{2} - \nu + m\right)}{m!\, \Gamma\left(\frac{1}{2} - \nu\right) \Gamma(2m - 2\nu + 1)}$$

$$= \frac{2^{\nu+1} \pi i e^{-\nu\pi i} \Gamma(1/2)}{\Gamma\left(\frac{1}{2} - \nu\right)} J_{-\nu}(z). \qquad (19.8)$$

This generalization of the Poisson integral is obtained under the following restrictions: $\Re z > 0$ and $\nu + 1/2$ is not a positive integer.

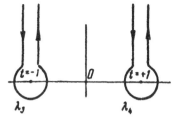

λ_3 λ_4

Figure 5

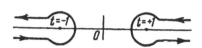

Figure 6

We can express the functions $H_\nu^{(1)}(z)$ and $H_\nu^{(2)}(z)$ via the functions $J_\nu(z)$ and $J_{-\nu}(z)$. We write only the formulae without their deduction:

$$H_\nu^{(1)}(z) = \frac{(z/2)^\nu}{i\sqrt{\pi}\cos\nu\pi\Gamma(\nu + 1/2)} \int_{\lambda_4} (t^2 - 1)^{\nu - 1/2} e^{zti}\, dt, \qquad (19.9)$$

$$H_\nu^{(2)}(z) = \frac{(z/2)^\nu}{i\sqrt{\pi}\cos\nu\pi\Gamma(\nu + 1/2)} \int_{\lambda_3} (t^2 - 1)^{\nu - 1/2} e^{zti}\, dt; \qquad (19.10)$$

here ν is non-integer, and the contours λ_3 and λ_4 are shown in Fig. 5 (the formulae are also valid for non-integer indices).

From these formulae we can obtain the generalized Meler–Sonin integrals. Changing the contour of integration in integrals (19.9) and (19.10), we obtain the integrals along the contours shown in Fig. 6.

Figure 7

Deforming these contours into the infinite half-segments $(-\infty, -1]$ and $[+1, \infty)$, respectively, of the real axes, each of which is taken twice, we can obtain the formulae

$$H_\nu^{(1)}(x) = \frac{\Gamma(1/2 - \nu)(x/2)^\nu}{\pi i \Gamma(1/2)}[1 - e^{-2(\nu - 1/2)\pi i}] \int_1^\infty e^{ixt}(t^2 - 1)^{\nu - 1/2}\, dt; \qquad (19.11)$$

$$H_\nu^{(2)}(x) = \frac{\Gamma(1/2 - \nu)(x/2)^\nu}{\pi i \Gamma(1/2)}[1 - e^{2(\nu - 1/2)\pi i}] \int_1^\infty e^{-ixt}(t^2 - 1)^{\nu - 1/2}\, dt. \qquad (19.12)$$

Generalizations of the Bessel integral are obtained in papers of Schlafli and Sonin. Without calculations, we write the formula

$$J_\nu(z) = \frac{(z/2)^\nu}{2\pi i} \int_{-\infty}^{(0+)} t^{-\nu - 1} \exp\left(t - \frac{z^2}{4t}\right) dt, \qquad |\arg z| < \pi/2. \qquad (19.13)$$

Setting $t = \frac{1}{2}ze^w$ in this formula, we obtain the expression

$$J_\nu(z) = \frac{1}{2\pi i} \int_{\infty - \pi i}^{\infty + \pi i} e^{z\sinh w - \nu w}\, dw. \qquad (19.14)$$

In this case the contour of integration represents three sides of the rectangle shown in Fig. 7 with the corners at the points $(\infty; -\pi i)$, $(0; -\pi i)$, $(0; \pi i)$, and $(\infty; +\pi i)$.

CHAPTER 2

Definite and improper integrals. Series in Bessel functions

20. Definite integrals

In this section we consider the Sonin integrals and some of their generalizations. Some very brief discussion concerning trigonometric Captain's integrals is also given here.

The role of the definite Sonin integrals in the following part of the course is essential. They are used in the theory of improper integrals (Section 21) as well as in the study of dual integral equations (Section 22). The definite Sonin integrals of the Struve and Lommel functions are also used in Section 21 in order to find improper integrals of these functions. The reader can pay attention to the relation of some computations given in this section with the contents of Section 14. This connection is explained by the fact that the Poisson integral is a special case of the first Sonin integral.

One can note that the heuristic approaches which were discussed when considering the Bessel integrals, can also be used, to a considerable extent, when explaining the theory of the Sonin integrals. In particular, it is easy to show that in the simplest case, the integral (20.1) considered below gives the image of a spherical wave symmetrical with respect to the center as a superposition of axi-symmetrical cylindrical waves.

First, consider the *first Sonin integral*

$$\int_0^{\pi/2} J_\mu(z \sin \theta) \sin^{\mu+1} \theta \cos^{2\nu+1} \theta \, d\theta = \frac{2^\nu \Gamma(1 + \nu)}{z^{\nu+1}} J_{\mu+\nu+1}(z), \qquad (20.1)$$

$$\Re \mu > -1, \qquad \Re \nu > -1.$$

The deduction of this formula can be performed by the same method that was used in Section 14 when the Poisson integral was considered. Expanding the integrand (20.1) into a series in powers of z and using formula (8), we obtain

$$\int_0^{\pi/2} J_\mu(z\sin\theta)\sin^{\mu+1}\theta\cos^{2\nu+1}\theta\,d\theta$$

$$= \sum_{m=0}^\infty \frac{(-1)^m z^{\mu+2m}}{m!\Gamma(\mu+m+1)2^{2m+\mu}} \int_0^{\pi/2} \sin^{2\mu+2m+1}\theta\cos^{2\nu+1}\theta\,d\theta$$

$$= \frac{2^\nu\Gamma(1+\nu)}{z^{\nu+1}} \sum_{m=0}^\infty \frac{(-1)^m(z/2)^{\mu+\nu+2m+1}}{m!\Gamma(\mu+\nu+m+2)}$$

$$= \frac{2^\nu\Gamma(1+\nu)}{z^{\nu+1}} J_{\mu+\nu+1}(z).$$

Obviously, if $\mu=-1/2$, then the first Sonin integral becomes the Poisson integral.

After a calculation similar to the one given above, we use the expansion of the Lommel function $s_{\mu,\nu}(z)$ into a power series and obtain

$$\int_0^{\pi/2} J_\mu(z\sin\theta)\sin^{1-\mu}\theta\cos^{2\nu+1}\theta\,d\theta = \frac{1}{2^{\mu-1}z^{\nu+1}\Gamma(\mu)}s_{\mu+\nu,\nu-\mu+1}(z). \qquad (20.2)$$

For $\nu=-1/2$ the equality (20.2) takes the form

$$\int_0^{\pi/2} J_\mu(z\sin\theta)\sin^{1-\mu}\theta\,d\theta = (\pi/(2z))^{1/2}\mathbf{H}_{\mu-1/2}(z). \qquad (20.3)$$

Consider the first Sonin integral of the Struve functions. For this purpose, we expand the following integral, using (16.21), into a series in powers of z:

$$\int_0^{\pi/2} \mathbf{H}_\mu(z\sin\theta)\sin^{k+1}\theta\cos^{2\nu+1}\theta\,d\theta$$

$$= \sum_{m=0}^\infty \frac{(-1)^m(z/2)^{\mu+2m+1}}{\Gamma(m+3/2)\Gamma(\mu+m+3/2)} \int_0^{\pi/2} \sin^{\mu+k+2m+2}\theta\cos^{2\nu+1}\theta\,d\theta$$

$$= \sum_{m=0}^\infty \frac{(-1)^m(z/2)^{\mu+2m+1}}{\Gamma(m+3/2)\Gamma(\mu+m+3/2)} \frac{\Gamma(\frac{\mu}{2}+m+\frac{k}{2}+\frac{3}{2})\Gamma(\nu+1)}{2\Gamma(\frac{\mu}{2}+\nu+m+\frac{k}{2}+\frac{5}{2})};$$

then, setting $k=\mu$, we obtain

$$\int_0^{\pi/2} \mathbf{H}_\mu(z\sin\theta)\sin^{\mu+1}\theta\cos^{2\nu+1}\theta\,d\theta = \frac{\Gamma(\nu+1)}{2(z/2)^{\nu+1}}\mathbf{H}_{\mu+\nu+1}(z). \qquad (20.4)$$

If we set $k = -\mu$, then after some simple transformations, denoting $\mu_1 = \mu + \nu + 1$ and $\nu_1 = \nu - \mu + 1$, we have

$$\int_0^{\pi/2} H_\mu(z \sin \theta) \sin^{1-\mu} \theta \cos^{2\nu+1} \theta \, d\theta$$

$$= \frac{1}{2} \sum_{m=0}^\infty \frac{(-1)^m (z/2)^{\mu+2m+1} \Gamma(\nu + 1)}{\Gamma((\mu_1 - \nu_1 + 2m + 3)/2) \Gamma((\mu_1 + \nu_1 + 2m + 3)/2)}$$

$$= \frac{\Gamma(\nu + 1)}{2^\mu \Gamma(\mu + 1/2) \Gamma(\nu + 3/2)}. \quad (20.5)$$

Let us deduce the Sonin integral for the Lommel functions, by expanding the following integral into a series in powers of z

$$\int_0^{\pi/2} s_{\mu,\nu}(z \sin \theta) \sin^{\mu+1+k_1} \theta \cos^{2\nu+k} \theta \, d\theta$$

$$= \sum_{m=0}^\infty \frac{(-1)^m 2^{\mu-1} (z/2)^{2m+1+\mu} \Gamma((\mu - \nu + 1)/2) \Gamma((\mu + \nu + 1)/2)}{\Gamma((\mu - \nu + 2m + 3)/2) \Gamma((\mu + \nu + 2m + 3)/2)}$$

$$\times \int_0^{\pi/2} \sin^{2m+2\mu+2+k_1} \theta \cos^{2\nu+k} \theta \, d\theta$$

$$= \sum_{m=0}^\infty \frac{(-1)^m \left(\frac{z}{2}\right)^{2m+1+\mu} 2^{\mu-1} \Gamma\left(\frac{\mu-\nu+1}{2}\right) \Gamma\left(\frac{\mu+\nu+1}{2}\right)}{2\Gamma\left(\frac{\mu-\nu+2m+3}{2}\right) \Gamma\left(\frac{\mu+\nu+2m+3}{2}\right)}$$

$$\times \frac{\Gamma\left(m + \mu + \frac{3}{2} + \frac{k_1}{2}\right) \Gamma\left(\nu + \frac{k}{2} + \frac{1}{2}\right)}{\Gamma\left(m + \mu + \nu + \frac{k}{2} + \frac{k_1}{2} + 2\right)}.$$

Setting $k_1 = \nu - \mu$ here, we obtain after some transformations

$$\int_0^{\pi/2} s_{\mu,\nu}(z \sin \theta) \sin^{\nu+1} \theta \cos^{2\nu+k} \theta \, d\theta$$

$$= \frac{\Gamma((\mu + \nu + 1)/2) \Gamma((2\nu + k + 1)/2)}{2\Gamma((3\nu + \mu + k + 2)/2)} z^{-\nu-k/2-1/2}$$

$$\times s_{\mu+\nu+k/2+1/2, 2\nu+k/2+1/2}(z). \quad (20.6)$$

The integrals, in which the argument of the Bessel function is proportional to the squared sines, are of interest. One can easily show [73] that

$$\int_0^{\pi/2} J_\nu(z \sin^2 \theta) \sin^{2\nu+3} \theta \cos \theta \, d\theta = \frac{1}{2z} J_{\nu+1}(z), \quad (20.7)$$

$$\int_0^{\pi/2} H_\nu(z \sin^2 \theta) \sin^{2\nu+3} \theta \cos \theta \, d\theta = \frac{1}{2z} H_{\nu+1}(z), \quad (20.8)$$

$$\Re \nu > -1.$$

Now we give an important formula proven by Rutgers and afterwards by Watson:

$$\int_0^{\pi/2} J_\nu(z \sin^2 \theta) J_\nu(z \cos^2 \theta) \sin^{2\nu+1} \theta \cos^{2\nu+1} \theta \, d\theta = \frac{\Gamma(\nu+1/2) J_{2\nu+1/2}(z)}{2^{2\nu+3/2} \Gamma(\nu+1) z^{1/2}}, \quad (20.9)$$

$$\Re\nu > -1/2.$$

For references to publications, which contain generalizations of this formula, see [4].

Consider the *second Sonin integral*

$$\int_0^{\pi/2} J_\mu(aq \cos \varphi) J_\nu(az \sin \varphi) \cos^{\mu+1} \varphi \sin^{\nu+1} \varphi \, d\varphi = \frac{q^\mu z^\nu}{a} \frac{J_{\mu+\nu+1}\left(a\sqrt{q^2+z^2}\right)}{\left(\sqrt{z^2+q^2}\right)^{\nu+\mu+1}},$$

$$(20.10)$$

$$\Re\mu > -1, \quad \Re\nu > -1.$$

In order to prove this formula, we expand the integrand into a series in powers of aq, change the order of integration and summation and use the first Sonin integral. The series obtained can be summed with the help of the Lommel expansion

$$\int_0^{\pi/2} J_\mu(aq \cos \varphi) J_\nu(az \sin \varphi) \cos^{\mu+1} \varphi \sin^{\nu+1} \varphi \, d\varphi$$

$$= \sum_{m=0}^\infty \left(\frac{aq}{2}\right)^{\mu+2m} \frac{(-1)^m}{m! \Gamma(\mu+m+1)} \int_0^{\pi/2} J_\nu(az \sin \varphi) \sin^{\nu+1} \varphi \cos^{2\mu+2m+1} \varphi \, d\varphi$$

$$= \sum_{m=0}^\infty J_{\nu+\mu+m+1}(az) \frac{2^{\nu+m} \Gamma(\mu+m+1)}{(az)^{\mu+m+1}} \frac{(-1)^m}{m! \Gamma(\mu+m+1)} \left(\frac{aq}{2}\right)^{\mu+2m}$$

$$= \frac{q^\mu z^\nu}{a} \cdot \frac{J_{\nu+\mu+1}\left(a\sqrt{q^2+z^2}\right)}{\left(\sqrt{q^2+z^2}\right)^{\nu+\mu+1}}.$$

Let us briefly consider some *trigonometric Captain integrals*.
The solution of the differential equation

$$\frac{d^2 u}{dz^2} + u = \varphi(z), \quad (20.11)$$

with zero initial conditions has the form

$$u = \int_0^z \varphi(t) \sin(z-t) \, dt.$$

The computation of this integral in the case when the right-hand side of (20.11) is expressed via Bessel functions, leads to the Captain integrals. Setting $\varphi(z) = J_0(z)$, one can easily show that

$$\int_0^z J_0(t) \sin(z-t) \, dt = z J_1(z). \quad (20.12)$$

Setting $\varphi(z) = J_1(z)$, when computing u we have the integral

$$\int_0^z J_1(t) \sin(z - t) \, dt = \sin z - z J_0(z). \tag{20.13}$$

The validity of these equalities can easily be established by direct verification, after introducing them into equation (20.11).

Let us give one more Captain integral

$$\int_0^z J_0(t) \cos(z - t) \, dt = z J_0(z). \tag{20.14}$$

Now we shall write two Captain integrals containing the Struve functions

$$\int_0^z \mathbf{H}_0(t) \sin(z - t) \, dt = z \mathbf{H}_1(z) - \frac{2z}{\pi}, \tag{20.15}$$

$$\int_0^z \mathbf{H}_1(t) \sin(z - t) \, dt = \frac{4}{\pi} - z \mathbf{H}_0(z)., \tag{20.16}$$

whose verification does not lead to difficulties.

A very important integral obtained by Bateman

$$\frac{J_{\nu+\rho+1}(z)}{\rho + 1} = \int_0^z J_\nu(z - a) J_{\rho+1}(a) \frac{da}{a}, \tag{20.17}$$

is given in [7]. Using this integral, Rutgers has obtained a great number (more than four hundreds) of different definite integrals containing Bessel functions [81].

21. Improper integrals

In this section we consider improper integrals, whose integrands contain the Bessel functions and contiguous functions. Many of these integrals can be encountered, when obtaining the integral Bessel transforms [5, 6, 13].

In Subsection 1 of this section we consider improper integrals of the Bessel functions; in Subsection 2 we calculate integrals of the Struve and Lommel functions. The question concerning the application of integral Bessel transforms to the calculation of some integrals is briefly discussed in Subsection 3. In Subsection 4 we consider the integrals with respect to the index. Subsection 5 is devoted to the problem of refinement of the convergence of some improper integrals.

1. First, consider the *Lipschitz integral*

$$\int_0^\infty e^{-at} J_0(bt) \, dt = \frac{1}{\sqrt{a^2 + b^2}}, \tag{21.1}$$

where $\Re a > 0$ and both the numbers $\Re(a \pm ib)$ are positive.

The value of the root is chosen such that $|a + \sqrt{a^2 + b^2}| > |b|$.

In order to prove equality (21.1) we replace the Bessel function under the integral sign by the Bessel integral, which in the case considered becomes the Parseval integral, and change the order of integration. Thus, we obtain

$$\int_0^\infty e^{-at} J_0(bt)\, dt = \frac{1}{\pi} \int_0^\infty e^{-at}\, dt \int_0^\pi e^{ibt\cos\theta}\, d\theta$$

$$= \frac{1}{\pi} \int_0^\pi \frac{d\theta}{a - ib\cos\theta} = \frac{1}{\sqrt{a^2 + b^2}}.$$

Similarly, we can prove a more general formula

$$\int_0^\infty e^{-at} J_n(bt)\, dt = \frac{1}{\sqrt{a^2 + b^2}} \left\{ \frac{\sqrt{a^2 + b^2} - a}{b} \right\}^n, \tag{21.2}$$

where n is an integer (see [11]).

In order to prove this equality, we use the Bessel integral in the form

$$J_n(x) = \frac{(-i)^n}{\pi} \int_0^\pi e^{ix\cos\varphi} \cos n\varphi\, d\varphi,$$

and then change the order of integration. Thus, we have

$$\int_0^\infty e^{-at} J_n(bt)\, dt = \frac{(-i)^n}{\pi} \int_0^\infty e^{-at}\, dt \int_0^\pi e^{ibt\cos\varphi} \cos n\varphi\, d\varphi$$

$$= \frac{(-i)^n}{\pi} \int_0^\pi \cos n\varphi\, d\varphi \int_0^\infty e^{-(a-ib\cos\varphi)t}\, dt$$

$$= \frac{(-i)^n}{\pi} \int_0^\pi \frac{\cos n\varphi\, d\varphi}{a - ib\cos\varphi}$$

$$= \frac{1}{\sqrt{a^2 + b^2}} \left[\frac{\sqrt{a^2 + b^2} - a}{b} \right]^n. \tag{21.3}$$

When deducing this formula, we assume that n is an integer; however, this formula also holds for non-integer indices ν, if $\Re\nu > -1$.

Consider a *generalization of the Lipschitz integral* due to Hankel.

If we expand the Bessel function in a power series and change the order of integration and summation, then, as a result of term-by-term integration, we obtain

$$\int_0^\infty e^{-at} J_\nu(bt) t^{\mu-1}\, dt = \sum_{m=0}^\infty \frac{(-1)^m (b/2)^{\nu+2m}}{m!\,\Gamma(\nu + m + 1)} \cdot \frac{\Gamma(\mu + \nu + 2m)}{a^{\mu+\nu+2m}}$$

$$= \frac{(b/(2a))^\nu \Gamma(\mu + \nu)}{a^\mu \Gamma(\nu + 1)} {}_2F_1\left(\frac{\mu + \nu}{2}, \frac{\mu + \nu + 1}{2}; \nu + 1, -\frac{b^2}{a^2} \right), \tag{21.4}$$

$$\Re(\nu + \mu) > 0.$$

The equality (21.4) holds under the condition that $\dfrac{|b|}{|a|} < 1$, because the term-by-term integration is only admissible in this case. Moreover, for convergence of the integral in (21.4) it is necessary that the inequality $\Re(a) > 0$ holds. With the help of analytical continuation one can prove that (21.4) holds for a larger domain of values of b, if the conditions $\Re(a \pm ib)$ hold simultaneously. Nielsen [76] has called this integral the *integral representation of the hypergeometric function*.

Formula (21.4) can be rewritten in the form

$$\int_0^\infty e^{-at} J_\nu(bt) t^{\mu-1}\, dt$$

$$= \frac{(b/2)^\nu \Gamma(\mu+\nu)}{\Gamma(\nu+1)(a^2+b^2)^{\frac{1}{2}(\mu+\nu)}}\, {}_2F_1\left(\frac{\mu+\nu}{2}, \frac{1-\mu+\nu}{2}; \nu+1, \frac{b^2}{a^2+b^2}\right). \quad (21.5)$$

Similar results can also be obtained in the case when the cylindrical function under the sign of integral is the Neumann function; for this purpose one should use formula (2.1).

This general result contains many important special cases; some of them were considered above.

Interesting results can be obtained for the cases, when the hypergeometric function can be reduced to elementary ones; in particular, for μ equal to $\nu+1$ and $\nu+2$ we have the formulae which were obtained for the first time by Gegenbauer:

$$\int_0^\infty e^{-at} J_\nu(bt) t^\nu\, dt = \frac{(2b)^\nu \Gamma(\nu+1/2)}{(a^2+b^2)^{\nu+1/2}\sqrt{\pi}}, \quad (21.6)$$

$$\int_0^\infty e^{-at} J_\nu(bt) t^{\nu+1}\, dt = \frac{2a(2b)^\nu \Gamma(\nu+3/2)}{(a^2+b^2)^{\nu+3/2}\sqrt{\pi}}. \quad (21.7)$$

Here $\Re\nu > -1/2$ and $\Re\nu > -1$, respectively. Let us write without deduction (see [7]) the formula

$$\int_0^\infty e^{-at} K_0(t)\, dt = \frac{\arccos a}{\sqrt{1-a^2}}, \quad (21.8)$$

where $\Re a > -1$. If we set $a = ib$, $|\Im b| < 1$ in (21.8), then

$$\int_0^\infty \cos bt\, K_0(t)\, dt = \frac{1}{2}\frac{\pi}{\sqrt{1+b^2}}, \quad (21.9)$$

$$\int_0^\infty \sin bt\, K_0(t)\, dt = \frac{\operatorname{arcsinh} b}{\sqrt{1+b^2}}. \quad (21.10)$$

Let us go to consideration of the *Weber integral*

$$\int_0^\infty J_\nu(bt) t^{-\nu+\mu+1}\, dt = \frac{\Gamma(\mu/2)}{b^{\mu-\nu} 2^{\nu-\mu+1}\Gamma(\nu-\mu/2+1)}, \quad (21.11)$$

where $b > 0$, $0 < \Re\mu < \Re(\nu)+3/2$. Weber proved this result only for integer values of n; for arbitrary values of ν this result was generalized by N.Ya. Sonin. Formula (21.11) can be obtained from (21.5) as a result of the passage to the limit, setting $a \to 0$.

Let us give the deduction of this formula, using the Poisson integral and change of the order of integration and assuming that $0 < \Re\mu < \Re\nu + 1/2$.

$$\int\limits_0^\infty J_\nu(bt)t^{-\nu+\mu+1}\,dt$$

$$= \frac{2}{\sqrt{\pi}\,\Gamma\left(\nu+\frac{1}{2}\right)}\left(\frac{b}{2}\right)^\nu \int\limits_0^\infty t^{\mu-1}\,dt \int\limits_0^{\pi/2} \cos(bt\cos\varphi)(\sin\varphi)^{2\nu}\,d\varphi$$

$$= \frac{2}{\sqrt{\pi}\,\Gamma\left(\nu+\frac{1}{2}\right)}\left(\frac{b}{2}\right)^\nu \int\limits_0^{\pi/2} (\sin\varphi)^{2\nu}\,d\varphi \int\limits_0^\infty t^{\mu-1}\cos(bt\cos\varphi)\,dt$$

$$= \frac{2}{\sqrt{\pi}\,\Gamma\left(\nu+\frac{1}{2}\right)}\left(\frac{1}{2}\right)^\nu \frac{\Gamma(\mu)\cos(\pi\mu/2)}{b^{\mu-\nu}} \int\limits_0^{\pi/2} (\sin\varphi)^{2\nu}(\cos\varphi)^{-\mu}\,d\varphi.$$

Using formulae (7) and (8) from the Appendix, we obtain (21.11).
Consider now the *first exponential Weber integral*

$$\int\limits_0^\infty J_0(at)\exp(-p^2t^2)t\,dt = \frac{1}{2p^2}\exp(a^2/(4p^2)). \qquad (21.12)$$

This integral converges only if $|\arg p| < \pi/4$.

In order to prove this formula, we should expand the Bessel function into a power series and change the order of summation and integration.

A *generalization of the Weber integral* is the formula

$$\int\limits_0^\infty J_\nu(at)\exp(-p^2t^2)t^{\nu+1}\,dt = \frac{a^\nu}{(2p^2)^{\nu+1}}\exp\left(-\frac{a^2}{4p^2}\right), \qquad (21.13)$$

$$\Re\nu > -1.$$

This can be proved just in the same manner as (21.12), namely:

$$\int\limits_0^\infty J_\nu(at)\exp(-p^2t^2)t^{\nu+1}\,dt = \sum_{k=0}^\infty \frac{(-1)^k(a/2)^{\nu+2k}}{k!\Gamma(\nu+k+1)} \int\limits_0^\infty e^{-p^2t^2}t^{2\nu+2k+1}\,dt$$

$$= \sum_{k=0}^\infty \frac{(-1)^k(a/2)^{\nu+2k}}{k!\Gamma(\nu+k+1)} \cdot \frac{\Gamma(\nu+k+1)}{2p^{2(\nu+k+1)}}$$

$$= \frac{a^\nu}{(2p^2)^{\nu+1}} \sum_{k=0}^\infty \frac{(-1)^k(a/2p)^{2k}}{k!}$$

$$= \frac{a^\nu}{(2p^2)^{\nu+1}}\exp\left(-\frac{a^2}{4p^2}\right).$$

A more general formula, which can be proved similarly, has the form

$$\int\limits_0^\infty J_\nu(at)\exp(-p^2t^2)t^{\mu-1}\,dt$$

$$= \frac{a^\nu\Gamma((\mu+\nu)/2)}{2^{\nu+1}p^{\mu+\nu}\Gamma(\nu+1)}{}_1F_1\left(\frac{1}{2}\nu+\frac{1}{2}\mu;\nu+1;\frac{-a^2}{4p^2}\right). \quad (21.14)$$

Let us give a special case of formula (21.14)

$$\int\limits_0^\infty J_{2\nu}(at)\exp(-p^2t^2)\,dt = \frac{\sqrt{\pi}}{2p}\exp\left(-\frac{a^2}{8p^2}\right)I_\nu\left(\frac{a^2}{8p^2}\right). \quad (21.15)$$

Replacing ν by $-\nu$ and combining these two solutions, we obtain

$$\int\limits_0^\infty Y_{2\nu}(at)\exp(-p^2t^2)\,dt$$

$$= -\frac{\sqrt{\pi}}{2p}\exp\left(-\frac{a^2}{8p^2}\right)\left[I_\nu\left(\frac{a^2}{8p^2}\right)\tan\nu\pi+\frac{1}{\pi}K_\nu\left(\frac{a^2}{8p^2}\right)\sec\nu\pi\right], \quad (21.16)$$

for $|\Re\nu| < 1/2$.

If we use the addition theorems (see Section 11), replacing a by

$$\tilde\omega = \sqrt{a^2+b^2-2ab\cos\varphi}$$

in (21.14), and set $\mu = 2$, then we have, as it is shown in [7], that

$$\int\limits_0^\infty \exp(-p^2t^2)J_\nu(at)J_\nu(bt)t\,dt = \frac{1}{2p^2}\exp\left(-\frac{a^2+b^2}{4p^2}\right)J_\nu\left(\frac{ab}{2p^2}\right). \quad (21.17)$$

This formula holds for $|\arg p| < \pi/4$, $\Re\nu > -1$.

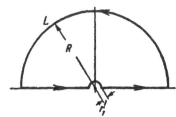

Figure 8

Consider now the *Hankel integrals*

$$\frac{1}{2\pi i}\int\limits_L \frac{z^{\rho-1}H_\nu^{(1)}(az)}{(z^2-r^2)^{m+1}}\,dz. \quad (21.18)$$

The closed contour L is chosen so that is as shown in Fig. 8. For $z = x$ we have

$$z^{\rho-1}H_\nu^{(1)}(az) = x^{\rho-1}H_\nu^{(1)}(ax).$$

For $z = xe^{\pi i}$, i.e., on the left-hand horizontal semiaxis, we have

$$z^{\rho-1}H_\nu^{(1)}(az) = -x^{\rho-1}e^{\rho\pi i}H_\nu^{(1)}(axe^{\pi i}).$$

If $R\to\infty$, $r_1\to 0$, then

$$\int\limits_0^\infty \frac{x^{\rho-1}[H_\nu^{(1)}(ax)-e^{\rho\pi i}H_\nu^{(1)}(axe^{\pi i})]dx}{(x^2-r^2)^{m+1}} = 2\pi i\frac{1}{2m!}\left(\frac{d}{dr^2}\right)^m\{r^{\rho-2}H_\nu^{(1)}(ar)\},$$

$$(21.19)$$

where the expression on the right-hand side is the residue at $z = r$ multiplied by $2\pi i$.

From (21.19), using the equality $H_\nu^{(1)}(z) = J_\nu(z) + iY_\nu(z)$ and setting $r = ik$, we obtain for $\Re k > 0$

$$\int\limits_0^\infty \left[\cos\left[\frac{1}{2}(\rho - \nu)\pi\right] J_\nu(ax) + \sin\left[\frac{1}{2}(\rho - \nu)\pi\right] Y_\nu(ax)\right] \frac{x^{\rho-1}dx}{(x^2 + k^2)^{m+1}}$$

$$= \frac{(-1)^{m+1}}{2^m m!} \left(\frac{d}{k\,dk}\right)^m [k^{\rho-2} K_\nu(ak)]. \quad (21.20)$$

Meler has indicated a particular case of this formula

$$\int\limits_0^\infty \frac{xJ_0(ax)dx}{x^2 + k^2} = K_0(ak). \quad (21.21)$$

A great number of integrals, which are actually generalizations of (21.21), can be obtained if we assume that the numerator is a product of two Bessel functions and the denominator is an even polynomial of the fourth degree[1]. Here we can use the following approach: first, without changing the numerator of (21.21) we assume that the denominator contains an even polynomial of the fourth degree. Here we can use the integration along the same contour. Then with the help of the addition formulae we can obtain the solution for the case when the numerator of the integrand is a product of two Bessel functions. Furthermore, we can introduce power functions under the integral sign or replace one of the Bessel functions by the Lommel, Struve or Anguer function. Notice that the author has actually calculated some of these integrals by another method in [20]: all these integrals are solutions of the problem of the integration of the equation

$$(\Delta_\nu \Delta_\nu + b\Delta_\nu + 1)w = q(\xi), \quad (21.22)$$

where $q(\xi)$ is either a delta-function or a finite function, for instance, a power function on a segment and identically zero everywhere outside this segment.

Let us give only one of the integrals of this type, when the argument x and the numbers a and b are real (see 156, 214):

$$\int\limits_0^\infty \frac{xJ_0(ax)\,dx}{1 + bx^2 + x^4} = \frac{\pi}{2\sin 2\varphi} f_0(\rho); \quad (21.23)$$

several integrals of this type are given in [20], as well as in the second part of this book.

Now let us give the *discontinuous integrals*[2] of *Weber–Sonin–Schafheitlin* which are important in applications

$$\int\limits_0^\infty \frac{J_\mu(at)J_\nu(bt)}{t^\lambda}dt. \quad (21.24)$$

For some special cases these integrals were calculated by Weber; for all values of the parameters λ, μ, ν such that this integral converges it has been calculated by Sonin; Schafheitlin in his investigation has noted that at $a = b$ this integral has

---

[1]The question concerning the integrals (which are another generalization of (21.21)), when the condition of the evenness of the polynomial which is the denominator of the integrand, does not hold, is considered in Subsection 5 of this section.

[2]Watson calls them the integrals of Weber and Schafheitlin.

a discontinuity. In the general case, the Weber–Sonin–Schafheitlin integral can be expressed via the hypergeometric functions

$$\int\limits_0^\infty \frac{J_\mu(at)J_\nu(bt)}{t^\lambda}dt = \frac{b^\nu\Gamma\left(\frac{1}{2}\mu+\frac{1}{2}\nu-\frac{1}{2}\lambda+\frac{1}{2}\right)}{2^\lambda a^{\nu-\lambda+1}\Gamma(\nu+1)\Gamma\left(\frac{1}{2}\lambda+\frac{1}{2}\mu-\frac{1}{2}\nu+\frac{1}{2}\right)}$$

$$\times {}_2F_1\left(\frac{\mu+\nu-\lambda+1}{2},\frac{\nu-\lambda-\mu+1}{2};\nu+1;\frac{b^2}{a^2}\right)\quad (21.25)$$

for $0 < b < a$, and

$$\int\limits_0^\infty \frac{J_\mu(at)J_\nu(bt)}{t^\lambda}dt = \frac{a^\mu\Gamma\left(\frac{\mu+\nu-\lambda+1}{2}\right)}{2^\lambda b^{\mu-\lambda+1}\Gamma\left(\frac{\nu+\lambda-\mu+1}{2}\right)\Gamma(\mu+1)}$$

$$\times {}_2F_1\left(\frac{\mu+\nu-\lambda+1}{2},\frac{\mu-\nu-\lambda+1}{2};\mu+1;\frac{a^2}{b^2}\right)\quad (21.26)$$

for $0 < a < b$. It is assumed that the integrals (21.25) and (21.26) converge (convergence conditions see in [7]).

The proof of formulae (21.25) and (21.26) can be obtained, by using the integral representation of the product of Bessel functions with different indices and arguments, which has the form

$$J_\mu(az)J_\nu(bz) = \frac{1}{\pi}(2az)^\mu(2bz)^\nu\int\limits_{-\pi/2}^{\pi/2} e^{i\theta(\mu-\nu)}(\cos\theta)^{\nu+\mu}(\lambda z)^{-\mu-\nu}J_{\mu+\nu}(\lambda z)\,d\theta,$$

$$(21.27)$$

$$\Re(\nu+\mu) > -1,\quad \lambda = \sqrt{2\cos\theta(a^2e^{i\theta}+b^2e^{-i\theta})}.$$

If we set $z = t$, put (21.27) into the integral (21.24), change the order of integration and afterwards use the Weber integral (21.11), then, as a result, we obtain a definite integral which can be expressed via the hypergeometric function[3] (see [4]).

Consider special cases of formulae (21.25) and (21.26), setting $\nu = 1/2$ and $\lambda = -1/2$ in them:

$$\int\limits_0^\infty J_\mu(at)\sin bt\,dt = \begin{cases} \dfrac{\sin\{\mu\arcsin(b/a)\}}{\sqrt{a^2-b^2}}, & b < a, \\ \infty \quad\text{or}\quad 0, & b = a, \\ \dfrac{a^\mu\cos(\mu\pi/2)}{\sqrt{b^2-a^2}\{b+\sqrt{b^2-a^2}\}^\mu}, & b > a, \end{cases}$$

$$\Re(\mu) > -2;\quad (21.28)$$

$$\int\limits_0^\infty J_\mu(at)\cos bt\,dt = \begin{cases} \dfrac{\cos\{\mu\arcsin(b/a)\}}{\sqrt{a^2-b^2}}, & b < a, \\ \infty \quad\text{or}\quad 0, & b = a, \\ \dfrac{-a^\mu\sin(\mu\pi/2)}{\sqrt{b^2-a^2}\{b+\sqrt{b^2-a^2}\}^\mu}, & b > a, \end{cases}$$

$$\Re(\mu) > -1.\quad (21.29)$$

---

[3]Compare with page 70.

Setting $\mu = 0$, we obtain

$$\int_0^\infty J_0(at) \sin bt \, dt = \begin{cases} 0, & b < a, \\ \infty, & b = a, \\ 1/\sqrt{b^2 - a^2}, & b > a, \end{cases} \qquad (21.30)$$

$$\int_0^\infty J_0(at) \cos bt \, dt = \begin{cases} 1/\sqrt{b^2 - a^2}, & b < a, \\ \infty, & b = a, \\ 0, & b > a. \end{cases} \qquad (21.31)$$

The last two integrals are called the *discontinuous factors of Weber*. At the end of this section we will give a simple deduction for another special case of the Weber–Sonin–Schafheitlin integral (see formula (21.43)).

Let us proceed to consider the *improper Sonin integrals*. First, we consider the Sonin integrals which contain one Bessel function under the integral sign, then we consider the discontinuous Sonin integral whose integrand contains a product of two Bessel functions of different arguments.

First consider the formula due to Sonin

$$\int_0^\infty \frac{J_\nu(a\sqrt{t^2 + z^2})}{(t^2 + z^2)^{\nu/2}} t^{2\mu+1} \, dt = \frac{2^\mu \Gamma(\mu + 1)}{a^{\mu+1} z^{\nu-\mu-1}} J_{\nu-\mu-1}(az), \qquad (21.32)$$

$$a \geq 0, \quad \Re\left(\frac{\nu}{2} - \frac{1}{4}\right) > \Re\mu > -1.$$

This formula can be proven, if we use the Lommel expansion (Section 12), with the help of which the ratio under the integral sign

$$\frac{J_\nu(a\sqrt{t^2 + z^2})}{(t^2 + z^2)^{\nu/2}}$$

can be represented in the form of a series, and then we use the Weber integral (21.11). In this case we obtain

$$\int_0^\infty \frac{J_\nu(a\sqrt{t^2 + z^2})}{(t^2 + z^2)^{\nu/2}} t^{2\mu+1} \, dt = \sum_{m=0}^\infty \frac{(-1)^m (z^2 a^2/2)^m a^{-m}}{m!} \int_0^\infty J_{\nu+m}(at) t^{2\mu-\nu-m+1} \, dt$$

$$= \sum_{m=0}^\infty \frac{(-1)^m (z^2 a^2/2)^m a^{-m}}{m! a^{2\mu+2-\nu-m}} \cdot \frac{\Gamma(1 + \mu)}{2^{-1+m+\nu-2\mu} \Gamma(\nu - \mu + m)}$$

$$= \frac{2^\mu \Gamma(\mu + 1)}{a^{\mu+1} z^{\nu-\mu-1}} \sum_{m=0}^\infty \frac{(-1)^m (az/2)^{\nu-\mu-1+2m}}{m! \Gamma(\nu - \mu + m - 1 + 1)}$$

$$= \frac{2^\mu \Gamma(\mu + 1)}{a^{\mu+1} z^{\nu-\mu-1}} J_{\nu-\mu-1}(az).$$

Consider another integral which is a generalization of the preceding one and can be obtained in a similar manner:

$$\int_0^\infty \frac{J_\nu(a\sqrt{t^2 + z^2})}{(t^2 + z^2)^{\nu/2}} t^{2\mu+k} \, dt = 2^{\mu+(k-1)/2} z^{\mu-\nu+(k+1)/2} a^{-\mu-(k+1)/2}$$

$$\times \Gamma(\mu + (k+1)/2) J_{\nu-\mu-(k+1)/2}(az),$$

where $\mu > -(k+1)/2$, $\nu > 2\mu + k - 1/2$.

Remembering that the Lommel expansion can easily be generalized to arbitrary cylindrical function, including, obviously, the Hankel function, passing from a real argument to imaginary one and repeating the calculations given in page 76, we obtain the important formula

$$\int_0^\infty \frac{K_\nu(a\sqrt{t^2+z^2})}{(t^2+z^2)^{\nu/2}} t^{2\mu+1}\, dt = \frac{2^\mu \Gamma(\mu+1)}{a^{\mu+1} z^{\nu-\mu-1}} K_{\nu-\mu-1}(az), \qquad (21.33)$$

where $a > 0$, $\Re\mu > -1$.

Consider the *discontinuous Sonin integral*

$$\int_0^\infty J_\mu(bt)\frac{J_\nu\left[a\sqrt{t^2+z^2}\right]}{(t^2+z^2)^{\nu/2}} t^{\mu+1}\, dt$$

$$= \begin{cases} 0, & a < b \\[2mm] \dfrac{b^\mu}{a^\nu}\left\{\dfrac{\sqrt{a^2-b^2}}{z}\right\}^{\nu-\mu-1} J_{\nu-\mu-1}\{z\sqrt{a^2-b^2}\}, & a > b \end{cases} \qquad (21.34)$$

for $\Re\nu > \Re\mu > -1$, $a > 0$, $b > 0$.

In order to prove this equality, we use the Lommel expansion (page 31), change the order of integration and summation and use the discontinuous Weber–Sonin–Schafheitlin integral (21.24) in the special case[4] $\lambda = \nu + m - \mu - 1$. As a result, we obtain

$$\int_0^\infty J_\mu(bt)\frac{J_\nu\left[a\sqrt{t^2+z^2}\right]}{(t^2+z^2)^{\nu/2}} t^{\mu+1}\, dt$$

$$= \sum_{m=0}^\infty a^{-m}\frac{(-1)^m(az)^{2m}}{2^m m!}\int_0^\infty J_\mu(bt) J_{\nu+m}(at) t^{\mu-\nu-m+1}\, dt$$

$$= \begin{cases} \displaystyle\sum_{m=0}^\infty \frac{(-1)^m a^m z^{2m} 2^{\mu-\nu-m+1} b^\mu a^{-\nu-m}(a^2-b^2)^{\nu+m-\mu-1}}{m! 2^m \Gamma(\nu-\mu+m)}, & a > b, \\[2mm] 0, & a < b, \end{cases}$$

$$= \begin{cases} \dfrac{b^\mu}{a^\nu} J_{\nu-\mu-1}(z\sqrt{a^2-b^2})\left\{\dfrac{\sqrt{a^2-b^2}}{z}\right\}^{\nu-\mu-1}, & a > b \\[2mm] 0, & a < b. \end{cases}$$

2. Consider improper integrals of the Struve functions. In order to obtain these integrals, we can use (see [7]) the same methods which were applied before in this section, namely, the expansion of the integrand into a power series, or using an integral representation of the Poisson type (see Section 16) which is followed by the change of the order of integration.

Here the result will be obtained with the help of the representation of the Struve function given by formula (20.3), which is followed by the change of the order of integration. First, we will, obviously, use the integrals of the Bessel functions obtained before in this section, and then use the definite integrals considered in Section 20.

---

[4]See formula (21.43).

Let us set

$$\int_0^\infty J_\nu(az)\Phi(bz)z^\mu\, dz = f(a,b,\mu,\nu),$$

where the expressions Φ and f should satisfy the conditions which guarantee the convergence of the integrals and legitimacy of the change of the order of integration, which follows the application of formula (20.3). Then, obviously,

$$\int_0^\infty \mathbf{H}_\nu(az)\Phi(bz)z^\mu\, dz$$

$$= \int_0^\infty \int_0^{\pi/2} \left(\frac{2az}{\pi}\right)^{1/2} J_{\nu+1/2}(az\sin\theta)\Phi(bz)z^\mu(\sin\theta)^{1-\nu-1/2}\, dz\, d\theta$$

$$= \sqrt{\frac{2a}{\pi}} \int_0^{\pi/2} f\left(a\sin\theta, b, \mu+\frac{1}{2}, \nu+\frac{1}{2}\right)\sin^{1/2-\nu}\theta\, d\theta. \quad (21.35)$$

As the first example, we consider the *Weber integral of the Struve function*

$$\int_0^\infty \frac{\mathbf{H}_\nu(z)\, dz}{z^{\nu-\mu+1}} = \sqrt{\frac{2}{\pi}} \int_0^\infty \int_0^{\pi/2} J_{\nu+1/2}(z\sin\theta)\sin^{1/2-\nu}\theta\, z^{-\nu+\mu-1/2}\, dz\, d\theta$$

$$= \frac{\Gamma((\mu+1)/2)}{\Gamma(\nu+1-\mu/2)2^{\nu-\mu}} \cdot \frac{1}{\sqrt{\pi}} \int_0^{\pi/2} \sin^{-\mu}\theta\, d\theta$$

$$= \frac{\Gamma(\mu/2)\tan(\pi\mu/2)}{\Gamma(\nu-\mu/2+1)2^{\nu-\mu+1}}, \quad \Re\mu < \Re\left(\nu+\frac{3}{2}\right). \quad (21.36)$$

In these calculations we used formulae (20.3), (21.11) and relation (6).
    Consider now the integral

$$\int_0^\infty J_\nu(at)\mathbf{H}_{\nu-3/2}(bt)t^{-1/2}\, dt = \int_0^\infty \int_0^{\pi/2} \left(\frac{2bt}{\pi}\right)^{1/2} t^{-1/2}J_{\nu-1}(bt\sin\theta)\sin^{2-\nu}\theta\, d\theta\, dt$$

$$= \sqrt{\frac{2}{\pi}}\frac{b^{\nu-1/2}}{a^\nu}, \quad \Re\nu > 0,\ b/a < 1. \quad (21.37)$$

In the calculations we used the formula

$$\int_0^\infty J_\nu(at)J_{\nu-1}(bt)\, dt = b^{\nu-1}/a^\nu, \quad \text{if}\ \Re\nu > 0,\ b/a < 1,$$

which is a special case of (21.25). This formula will be deduced below (page 80).

Consider the *Sonin integral for the Struve function*

$$\int\limits_0^\infty \frac{\mathbf{H}_\nu\left(a\sqrt{t^2+z^2}\right) dt}{\left(\sqrt{t^2+z^2}\right)^{\nu+1} t}$$

$$= \sqrt{\frac{2a}{\pi}} \int\limits_0^\infty \int\limits_0^{\pi/2} \frac{J_{\nu+1/2}\left(a\sin\theta\sqrt{t^2+z^2}\right)}{\left(\sqrt{t^2+z^2}\right)^{\nu+1/2} t} \sin^{1/2-\nu}\theta \, d\theta \, dt$$

$$= \frac{1}{2} z^{-\nu-1} \mathbf{H}_\nu(az). \quad (21.38)$$

Similarly, we can calculate improper integrals whose integrands contain the Lommel function. So, for instance, using formula (20.2), one can obtain after simple transformations

$$\int\limits_0^\infty s_{\mu+\nu,\ \nu-\mu+1}(az) J_k(az) z^{\mu-\nu-k} dz = \frac{2^{2\mu-k-1}\Gamma(\mu)a^{\nu+k-\mu-1}}{(\nu+k-\mu)\Gamma(k-\mu)}, \quad (21.39)$$

$$\Re\nu > -1, \ \Re\mu > \Re k > 0.$$

In a special case

$$\int\limits_0^\infty s_{\mu+\nu,\ \nu-\mu+1}(az) J_{\mu-\nu-1}(az) z \, dz = -\frac{2^{\mu+\nu}\Gamma(\mu)a^{-2}}{\Gamma(-\nu-1)}, \quad (21.40)$$

$$\Re\nu > -1, \ \Re(\mu-\nu-1) > 0.$$

This approach can also be applied in order to obtain the integral of the product of two Bessel functions of the first kind, which have the same arguments but different indices. For this purpose we use the Neumann formula (18.16).

Using different generalizations of this formula, we can obtain integrals of the product of two different cylindrical functions of the same argument. If the arguments of the cylindrical functions are different, then we can use the Neumann formula (18.17).

The improper integral of the Lommel function of two variables (see page 116) will be computed below in a similar manner, in Section 27.

3. Consider some examples of applications of the integral Bessel transforms to problems of the computation of improper integrals.

If in an integral transform

$$f^*(\lambda) = \int\limits_0^\infty f(t) K(\lambda t) \, dt$$

the kernel $K(\lambda t)$ is a cylindrical function, then the transform is called the *Bessel transform*. The most important among the Bessel transforms is the *Hankel transform*

$$f^*(u) = \int\limits_0^\infty t f(t) J_\mu(ut) \, dt \quad 0 < u < +\infty, \ \mu > -1/2, \quad (21.41)$$

which has the following inversion formula:

$$f(t) = \int\limits_0^\infty u f^*(u) J_\mu(ut)\, du \quad 0 < t < +\infty. \tag{21.41'}$$

The properties of the integral Hankel transform are briefly formulated in [13] and in more detail in [47]. Tables of the integral Hankel transforms are given in [47, 5, 6]. Let us obtain, using the Hankel transforms and the definite Sonin integrals considered in Section 20, some discontinuous improper integrals.

Consider the function

$$f(t) = \begin{cases} (a^2 - t^2)^\nu t^\mu, & 0 < t < a, \\ 0, & a < t < \infty. \end{cases} \tag{21.42}$$

The Hankel transform of the function $f(t)$ has the form

$$f^*(u) = \int\limits_0^a (a^2 - t^2)^\nu t^{\mu+1} J_\mu(ut)\, dt.$$

Setting $t = a \sin\theta$, we reduce the integral to the form

$$f^*(u) = \int\limits_0^{\pi/2} a^{2\nu+\mu+2} \cos^{2\nu+1}\theta \sin^{\mu+1}\theta J_\mu(au \sin\theta)\, d\theta.$$

Thus, the problem is reduced to computation of the Sonin integral (20.1), hence,

$$f^*(u) = a^{2\nu+\mu+2} \frac{2^\nu \Gamma(\nu+1)}{(au)^{\nu+1}} J_{\mu+\nu+1}(au)$$

$$= \frac{2^\nu \Gamma(\nu+1) a^{\nu+\mu+1}}{u^{\nu+1}} J_{\mu+\nu+1}(au).$$

The inversion formula immediately gives the integral

$$\int\limits_0^\infty u^{-\nu} J_{\mu+\nu+1}(au) J_\mu(ut)\, du = f(t) \frac{1}{2^\nu \Gamma(1+\nu) a^{\nu+\mu+1}}, \tag{21.43}$$

which is a special case of the Weber–Sonin–Schafheitlin integral.

As the second example, we consider the H-transform (see [6])

$$f_\nu^*(u) = \int\limits_0^\infty \mathbf{H}_\nu(tu)(tu)^{1/2} f(t)\, dt \tag{21.44}$$

and the corresponding inversion formula

$$f(t) = \int\limits_0^\infty Y_\nu(tu)(tu)^{1/2} f_\nu^*(u)\, du. \tag{21.45}$$

If we assume that $f(t)$ is expressed by the formula

$$f(t) = \begin{cases} (a^2 - t^2)^\mu t^{\nu+1/2}, & 0 < t < a, \\ 0, & a < t < \infty, \end{cases}$$

then, using the first Sonin integral for the Struve function, we obtain

$$f_\nu^*(u) = \mathbf{H}_{\nu+\mu+1}(au)\frac{\Gamma(\mu+1)}{u^{\mu+1/2}}2^\mu a^{\nu+\mu+1}$$

and from the inversion formula (21.45) we immediately obtain

$$\int\limits_0^\infty \mathbf{H}_{\nu+\mu+1}(au)Y_\nu(tu)u^{-\mu}\,du = \begin{cases} \dfrac{t^\nu(a^2-t^2)^\mu}{\Gamma(\mu+1)a^{\nu+\mu+1}}, & 0 < t \le a, \\ 0, & a < t < \infty. \end{cases} \qquad (21.46)$$

This discontinuous integral is analogous to the Weber–Sonin–Schafheitlin integral.

4. Let us consider *improper integrals with respect to the index*.

First, recall that the problem of free vibrations of the membrane of the form of a circular sector is, usually, reduced to the determination of a function which is represented by a series whose general term is the product of a Bessel function by a trigonometric function of the angle θ; making θ fixed, we obtain the Fourier–Bessel series; passage to the limit allows one to go from the series to an integral with infinite limits. If we make r fixed, then the function of the angle θ will be represented by a Fourier series whose coefficients are the Bessel function depending on the index. Here one can also make the passage to the limit, which allows one to pass to the integrals with respect to the indices. Obviously, when obtaining the integrals with respect to the indices, the Fourier transform can be used in many cases.

At present, the question concerning the integrals with respect to the index became very important in connection with different applications of the Kontorovich–Lebedev transformation.

Consider several integrals with respect to the index. As the basis for obtaining these integrals, we first, take the inversion formula of Fourier applied to different definite integrals whose integrands contain as parameters the argument and the index of the Bessel function. The idea of the calculations consists in the fact that the integral is multiplied by an appropriate Fourier kernel, where the variable is the index, then we integrate and use the inversion formula.

Let us use formula (18.16)

$$\int\limits_0^{\pi/2} J_{\nu_1+\mu_1}(2z\cos\theta)\cos[(\mu_1-\nu_1)\theta]\,d\theta = \frac{\pi}{2}J_{\nu_1}(z)J_{\mu_1}(z),$$

$$\Re(\nu_1+\mu_1) > -1.$$

We denote $\mu_1 - \nu_1 = \xi$, $\mu_1 + \nu_1 = \mu$ and introduce the function

$$f_\mu(2z\cos\theta) = \begin{cases} J_{2\mu}(2z\cos\theta), & 0 < |\theta| < \pi/2, \\ 0, & \pi/2 \le |\theta| < \infty \end{cases}$$

and rewrite the above integral in the form

$$\int\limits_{-\infty}^{\infty} f_\mu(2z\cos\theta)\cos(2\xi\theta)\,d\theta = \pi J_{\mu+\xi}(z)J_{\mu-\xi}(z).$$

Thus, up to a constant factor, the right-hand side is the cosine-transform of the function $f_\mu(2z\cos\theta)$.

Using the inversion formula, we obtain

$$\int_0^\infty J_{\mu+\xi}(z) J_{\mu-\xi}(z) \cos(2\xi\theta) \, d\theta = \frac{1}{2} J_{2\mu}(2z \cos\theta), \qquad (21.47)$$

$$0 < |\theta| < \frac{\pi}{2}.$$

Quite a number of integrals of functions contiguous to the Bessel functions can be obtained by using the generalizations of the Bessel integral (see formulae of Section 16 as well as [4]).

As an example, let us consider the integral (see [3])

$$\int_6^\pi \sin(z \sin\theta) \cos(\nu\theta) \, d\theta = (1 + \cos\nu\pi) s_{0,\nu}(z).$$

Using the inversion formula, we immediately obtain

$$\int_0^\infty (1 + \cos\nu\pi) s_{0,\nu}(z) \cos(\nu\theta) \, d\theta = \frac{\pi}{2} \sin(z \sin\theta), \qquad (21.48)$$

$$0 < \theta < \pi.$$

Another method of obtaining the integrals with respect to the index is connected with using the integral representations obtained as a result of the transformation of the Bessel integral (see Section 19).

For instance, if we take the integral

$$K_\nu(z) = \int_0^\infty e^{-z \cosh t} \cosh(\nu t) \, dt, \quad \Re z > 0,$$

then, setting $\nu = ix$ and using the same inversion formula, we obtain

$$\int_0^\infty K_{ix}(z) \cos(xt) \, dx = \frac{\pi}{2} e^{-z \cosh t}, \quad 0 < t < \infty. \qquad (21.49)$$

Some more complicated integrals can be comparatively easy obtained as a result of application of the inversion formula to improper integrals, whose integrands contain the functions $\cosh(\nu x)$ or $\sinh(\nu x)$ as well as $\cosh(\nu \pm \mu)x$ and $\sinh(\nu \pm \mu)x$.

As an example, we can, for instance, take the following integral given in [4]:

$$[K_\nu(x)]^2 \sin(\nu\pi) = \pi \int_0^\infty J_0(2x \sinh t) \sinh(2\nu t) \, dt,$$

$$|\Re\nu| < 3/4, \quad x > 0.$$

Using the change of the variable $\nu = ix$ and the inversion formula, we obtain

$$\int_0^\infty [K_{i\nu}(x)]^2 \sinh(\nu\pi) \sin(2\nu t) \, dt = \frac{\pi^2}{4} J_0(2x \sinh t), \qquad (21.50)$$

$$0 < t < \infty.$$

The question concerning the integrals with respect to the index is considered in detail in the book by N.N. Lebedev [34].

5. Consider the question of the refinement of the convergence of some improper integrals. For this purpose we use the method of contour integration here. As a result of its application, the integral is represented as a sum of a simple expression containing the Bessel functions and another improper integral which converges better than the original one [22].

Consider the improper integral

$$I_1(\xi) = \int\limits_0^\infty \frac{J_n(\lambda\xi)\, d\lambda}{F(\lambda)},$$ (21.51)

where $F(\lambda)$ is a function which is analytic on the whole plane; the integrand represents an one-valued function of the argument λ. With the help of the formula

$$J_n(x) = \frac{1}{2}\left[H_n^{(1)}(x) + H_n^{(2)}(x)\right]$$

we represent the integral as the sum

$$I_1(\xi) = \frac{1}{2}[I_2(\xi) + I_3(\xi)],$$

where

$$I_2(\xi) = \int\limits_0^\infty \frac{H_n^{(1)}(\lambda\xi)\, d\lambda}{F(\lambda)}, \quad I_3(\xi) = \int\limits_0^\infty \frac{H_n^{(1)}(\lambda\xi)\, d\lambda}{F(\lambda)}.$$

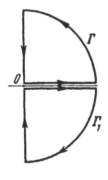

Figure 9

We transform $I_2(\xi)$ and $I_3(\xi)$, by integrating in the plane of the complex variable $z = \lambda + it$ along the contours shown in Fig. 9, and assuming the radius of the arc tending to infinity. Furthermore, taking into account that the integral along the arc vanishes and using the relation, given in Section 4, between the Hankel function of an imaginary argument and the Macdonald function, we obtain

$$I_2(\xi) = \frac{2}{\pi}\int\limits_0^\infty \frac{K_n(t\xi)\, dt}{F(it)} + 2\pi i \sum \mathrm{res}\,\psi_1(z),$$

$$\psi_1(z) = \frac{H_n^{(1)}(z)}{F(z)};$$ (21.52)

the integral $I_3(\xi)$ should be transformed in a similar way.

As a result, we obtain

$$I_1(\xi) = \frac{1}{\pi}\int\limits_0^\infty \frac{K_n(t\xi)[F(it) + F(-it)]\, dt}{F(t)F(it)}$$

$$+ 2\pi i \sum \mathrm{res}\,\psi_1(z) + 2\pi i \sum \mathrm{res}\,\psi_2(z),$$ (21.53)

$$\psi_2(z) = \frac{H_n^{(2)}(z)}{F(z)}.$$

Consider in more detail the calculation of this integral in the case, when $F(z)$ is a polynomial.

This problem is of interest in connection with the calculation of an unbounded plate which lies on an elastic uniform isotropic half-space and loaded by a point force P. In this case the deflection $w(r)$ and the reactive pressure $p(r)$ are expressed in the following form:

$$w(\xi) = \frac{Pl^2}{2\pi D} \int\limits_0^\infty \frac{J_0(\lambda\xi)\, d\lambda}{1 + \lambda^3},$$

$$p(\xi) = \frac{P}{2\pi l^2} \int\limits_0^\infty \frac{\lambda J_0(\lambda\xi)\, d\lambda}{1 + \lambda^3}, \qquad (21.54)$$

where $\xi = r/l$; E and σ are the modulus of elasticity and Poisson's ratio of the material of the plate; E_0 and σ_0 are, respectively, the modulus of elasticity and Poisson's ratio of the foundation; h is the thickness of the plate; D is the flexical rigidity;

$$l = \sqrt[3]{D/k}, \quad k = E_0/[2(1 - \sigma_0^2)].$$

We denote

$$I_{(1)}(\xi) = \int\limits_0^\infty \frac{J_0(\lambda\xi)\, d\lambda}{1 + \lambda^3}. \qquad (21.55)$$

Using the formula

$$J_0(x) = \frac{1}{2}[H_0^{(1)}(x) + H_0^{(2)}(x)],$$

we obtain

$$I_{(1)}(\xi) = \frac{1}{2}[I_{(2)}(\xi) + I_{(3)}(\xi)], \qquad (21.56)$$

where

$$I_{(2)}(\xi) = \int\limits_0^\infty \frac{H_0^{(1)}(\lambda\xi)\, d\lambda}{1 + \lambda^3},$$

$$I_{(3)}(\xi) = \int\limits_0^\infty \frac{H_0^{(2)}(\lambda\xi)\, d\lambda}{1 + \lambda^3}. \qquad (21.57)$$

In order to compute $I_2(\xi)$, we will integrate the function $\psi(z) = H_0^{(1)}(\xi z)/(1 + z^3)$ in the plane of the complex variable $z = \lambda + it$ along the closed contour Γ (see Fig. 9), going around it in the positive direction.

It can easily be seen that the integral along the arc vanishes as $R \to \infty$; therefore

$$\int\limits_0^\infty \frac{H_0^{(1)}(\lambda\xi)\, d\lambda}{1 + \lambda^3} - i \int\limits_0^\infty \frac{H_0^{(1)}(it\xi)\, dt}{1 + (it)^3} = 2\pi i \sum \text{res } \psi. \qquad (21.58)$$

The function ψ has only one pole at the point $z_1 = 1/2 + i\sqrt{3}/2$ in the domain bounded by the contour considered. Using simple calculations, one can show that

$$\text{res } \psi(z_1) = \frac{H_0^{(1)}[\xi(1/2 + i\sqrt{3}/2)]}{-3/2 + i3\sqrt{3}/2}.$$

Thus, from (21.58) we obtain

$$\mathbf{I}_{(2)}(\xi) = \frac{2}{\pi} \int\limits_0^\infty \frac{K_0(t\xi)\,dt}{1+(it)^3} + 4\pi i \frac{H_0^{(1)}[(1/2 + i\sqrt{3}/2)\xi]}{3(-1+i\sqrt{3})}. \tag{21.59}$$

Similarly, we transform the second integral (21.57), taking the contour Γ_1 shown in Fig. 9 as the path of integration and assuming the radius of the arc to be tending to infinity. In contrast to the first case, we shall go around the contour in the negative direction. As a result, using the relation $H_0^{(2)}(-x) = H_0^{(1)}(x)$, we obtain

$$\mathbf{I}_{(3)}(\xi) = \frac{2}{\pi} \int\limits_0^\infty \frac{K_0(t\xi)\,dt}{1-(it)^3} - 4\pi i \frac{H_0^{(2)}[(1/2 - i\sqrt{3}/2)\xi]}{3(-1-i\sqrt{3})}. \tag{21.60}$$

Substituting (21.59) and (21.60) into (21.56) and (21.55), we find that

$$w(\xi) = \frac{Pl^2}{2\pi D} \left\{ \frac{2}{\pi} \int\limits_0^\infty \frac{K_0(t\xi)\,dt}{1+t^6} + \frac{\pi}{3} \left[\tilde{f}_0(\xi)\sqrt{3} + \tilde{g}_0(\xi) \right] \right\}, \tag{21.61}$$

where

$$\tilde{f}_0(\xi) = \Re H_0^{(1)} \left[\xi(1/2 + i\sqrt{3}/2) \right] = \Re H_0^{(1)}(\xi e^{i\varphi_0}),$$

$$\tilde{g}_0(\xi) = \Im H_0^{(1)} \left[\xi(1/2 + i\sqrt{3}/2) \right] = \Im H_0^{(1)}(\xi e^{i\varphi_0}),$$

$$\varphi_0 = \pi/3.$$

From the computational viewpoint the solution obtained is more convenient than the initial (21.54). Actually, (21.61) consists of the sum of two terms. One of them represents a linear combination of tabulated Bessel functions. The second one is an integral which converges much faster than the initial one. For the convenience of the computations, the integral present in (21.61) can be represented as a sum of two integrals with the limits from 0 to α and from α to ∞, where α is small compared to one. Then the first of these integrals can be expressed via the tabulated function.

If, for instance, we take $\alpha = 0.2$, then

$$\int\limits_0^{0.2} \frac{K_0(t\xi)\,dt}{1+t^6} \approx \int\limits_0^{0.2} K_0(t\xi)\,dt.$$

In this case the maximal error in the value of the integrand is equal to 1/15625; introducing the new variable $x = t\xi$, we reduce the integral to the form

$$\int\limits_0^\alpha K_0(t\xi)\,dt = \frac{1}{\xi} \int\limits_0^{x=\alpha\xi} K_0(x)\,dx.$$

The integral obtained is tabulated in [7, Vol. 2]. Thus, it remains only to calculate the integral with the limits α, ∞, using the formulae of mechanical quadratures. The upper limit of this integral can be taken approximately equal to $3 \div 4$ because of the very fast growth of the denominator and the even faster decrease of the Macdonald function in the numerator.

Let us note that, when differentiating (21.61) (with the purpose of obtaining the slopes and internal forces in the plate), the integral in the right-hand side remains sufficiently well convergent.

The method of the transformation of the integral which was described can easily be generalized to the case when the numerator of the integrand contains a more complicated function; the calculations in the case when the denominator of the integrand is a polynomial of a more complicated form are more tedious.

As an example, we consider the problem of the calculation of the integral present in the calculation of an unbounded plate which lies on elastic half-space and suppose that the plate in the middle plane is stretched by a uniform effort p_0. In this case, the deflection and the reactive pressure have the form

$$\left.\begin{array}{c} w(\xi) = \dfrac{Pl^2}{2\pi D} \displaystyle\int_0^\infty \dfrac{J_0(\lambda\xi)\,d\lambda}{1 + a\lambda + \lambda^3}, \\[4mm] p(\xi) = \dfrac{P}{2\pi l^2} \displaystyle\int_0^\infty \dfrac{\lambda J_0(\lambda\xi)\,d\lambda}{1 + a\lambda + \lambda^3}, \\[4mm] a = p_0/kl. \end{array}\right\} \tag{21.62}$$

The determination of the first integral (21.62) is reduced to the calculation of two integrals along the contours shown in Fig. 9.

We write only the final result

$$\int_0^\infty \frac{J_0(\lambda\xi)\,d\lambda}{1 + a\lambda + \lambda^3} = \frac{2}{\pi} \int_0^\infty \frac{K_0(t\xi)\,dt}{1 + (at - t^3)^2} + G(\xi), \tag{21.63}$$

where

$$G(\xi) = 4\pi \frac{\delta \tilde{f}_0(\xi) - \gamma \tilde{g}_0(\xi)}{\gamma^2 + \delta^2},$$

$$\tilde{f}_0(\xi) + i\tilde{g}_0(\xi) = H_0^{(1)}[(\alpha_1 + i\beta_1)\xi] = H_0^{(1)}(\rho e^{i\varphi_0}); \quad \rho = \xi/b,$$

$$\tan\varphi_0 = \beta_1/\alpha_1, \quad \alpha_1 = b/2, \quad \beta_1 = \sqrt{1/b - b^2/4},$$

$$\gamma = b^2/2 - 2/b, \quad \delta = 3b\beta_1,$$

where $b = -z_0$, z_0 is the real root of the equation $z^3 + az + 1 = 0$.

Let us compute the integrals in the case when the loading is distributed on a circular domain and can be represented in the form

$$q(\xi, \theta) = \sum_0^\infty F_n(\xi)\cos n\theta.$$

For the sake of brevity, we consider only integrals of the type (21.56). Let the loading $q_n \cos n\theta$ be distributed along the circumference of the radius R.

Using the superposition principle, we find the deflection at the point with the coordinates $\xi_1 = R/l$, φ. The distance between the points in the accepted dimensionless coordinate system (ξ, φ) and $\xi_1, \theta)$ will be denoted by

$$\zeta = \sqrt{\xi^2 + \xi_1^2 - 2\xi\xi_1\cos(\theta - \varphi)}.$$

Using the addition formulae for the Bessel functions, we can, integrating with respect to θ, immediately obtain for $\xi_1 < \xi$

$$w(\xi, \varphi) = \frac{q_n l^3 \xi_1 \cos n\varphi}{D} \left\{ \frac{2}{\pi} \int_0^\infty \frac{K_n(\xi t) I_n(\xi_1 t)}{1 + t^6} dt \right.$$

$$\left. + \frac{\pi}{3} [\sqrt{3}(\tilde{f}_n(\xi)\tilde{u}_n(\xi_1) - \tilde{g}_n(\xi)\tilde{v}_n(\xi_1)) + \tilde{g}_n(\xi)\tilde{u}_n(\xi_1) + \tilde{f}_n(\xi)\tilde{v}_n(\xi_1)] \right\}, \quad \xi > \xi_1,$$

$$(21.64)$$

where the notation $\tilde{u}_n(\xi)$, $\tilde{v}_n(\xi)$, $\tilde{f}_n(\xi)$, and $\tilde{g}_n(\xi)$, is given on page 14; for $\xi < \xi_1$ the expression for w is obtained from (21.64) by interchanging ξ and ξ_1 in the brackets.

It should be noted that although the function $I_n(x)$ as $x \to \infty$ increases as $e^x/\sqrt{x}$, the asymptotic representation of the product $I_n(t\xi)K_n(t\xi_1)$ contains the exponential factor $e^{-t(\xi_1-\xi)}$, which provides the required decreasing of the numerator for $\xi_1 > \xi$. The same is valid for $\xi > \xi_1$. It is clear that at $\xi = \xi_1$ the advantages caused by the presence of the exponential factor in the numerator, disappear.

For the loading $q(\xi_1)\cos n\theta$ distributed on a circular ring, it is necessary to integrate with respect to the argument ξ_1. If $q(\xi_1) = A\xi_1^n$ $(n = 1, 2, 3, \ldots)$, then the result will be expressed via the cylindrical functions. If $q(\xi_1) = A\xi_1^\mu$, $\mu \neq n$, then the result contains the Lommel functions of the complex argument.

22. Dual integral equations

It is known that many problems of mathematical physics can be reduced to solving dual integral equations. Different contact problems of the theory of elasticity, problems on strained states near cracks, some problems of electrostatics and other topics are among them. The theory of dual equations recently became very developed and is the subject of numerous investigations.

It is interesting to note that today, when solving dual integral equations with Bessel kernels, different Sonin integrals and a series of other results of the Bessel function theory are widely used,

First, consider the dual equations

$$\int_0^\infty t^\alpha A(t) J_\nu(\rho t)\, dt = f(\rho), \qquad\qquad 0 \le \rho < d, \qquad (22.1a)$$

$$\int_0^\infty A(t) J_\nu(\rho t)\, dt = g(\rho), \qquad\qquad \rho > d, \qquad (22.1b)$$

where $A(t)$ is the unknown function, $f(\rho)$ and $g(\rho)$ are known functions and α is a given constant.

Equations (22.1a) and (22.1b) have been considered by many authors. We should note the first works due to Titchmarsh and Basbridge. Titchmarsh's results were published in Russian in his monograph [51], and Basbridge's results are presented in the book by Sneddon [47]. Both these authors obtained the solutions by performing rather complicated calculations. Recently, some papers were published, in which the solution was obtained in a simpler way.

Below, the method due to Noble [77] will be presented, which is based on using the Sonin integrals.

First, we write the first definite Sonin integral

$$J_{\nu+\xi+1}(z) = \frac{z^{\xi+1}}{2^\xi \Gamma(\xi+1)} \int\limits_0^{\pi/2} J_\nu(z \sin\theta) \sin^{\nu+1}\theta \cos^{2\xi+1}\theta \, d\theta,$$

where $\nu > -1$, $\xi > -1$.

Substituting $z = xt$, $\rho = x \sin\theta$ in this integral, we obtain

$$t^{-\xi-1} J_{\nu+\xi+1}(xt) = \frac{x^{-\xi-\nu-1}}{2^\xi \Gamma(\xi+1)} \int\limits_0^x J_\nu(\rho t) \rho^{\nu+1} (x^2 - \rho^2)^\xi \, d\rho. \tag{22.2}$$

After multiplying both sides of the integral equation (22.1a) by

$$\frac{x^{-\xi-\nu-1}}{2^\xi \Gamma(\xi+1)} \rho^{\nu+1} (x^2 - \rho^2)^\xi,$$

integrating with respect to ρ from 0 to x, changing the order of integration in the left-hand side and using the transformed Sonin formula (22.2), we obtain

$$\int\limits_0^\infty t^{\alpha-\xi-1} A(t) J_{\nu+\xi+1}(xt) \, dt = \frac{x^{-\xi-\nu-1}}{2^\xi \Gamma(\xi+1)} \int\limits_0^x f(\rho) \rho^{\nu+1} (x^2 - \rho^2)^\xi \, d\rho, \tag{22.3}$$

$$0 < x < d, \quad \nu > -1, \quad \xi > -1.$$

Now we formulate another Sonin's result related with improper integrals:

$$t^{-\eta-1} J_{\nu-\eta-1}(xt) = \frac{x^{\nu-\eta-1}}{2^\eta \Gamma(\eta+1)} \int\limits_0^\infty J_\nu[t(s^2 + x^2)^{1/2}](s^2 + x^2)^{-\frac{1}{2}\nu} s^{2\eta+1} \, ds,$$

where $\nu/2 - 1/4 > \eta > -1$. Setting $s^2 + x^2 = \rho^2$ here, we obtain

$$t^{-\eta-1} J_{\nu-\eta-1}(xt) = \frac{x^{\nu-\eta-1}}{2^\eta \Gamma(\eta+1)} \int\limits_x^\infty J_\nu(\rho t) \rho^{-\nu+1} (\rho^2 - x^2)^\eta \, d\rho. \tag{22.4}$$

After multiplying the left- and right-hand sides of equation (22.1b) by

$$\frac{x^{\nu-\eta-1}}{2^\eta \Gamma(\eta+1)} \rho^{-\nu+1} (\rho^2 - x^2)^\eta,$$

integrating with respect to ρ from x to ∞, changing the order of integration and using the improper Sonin integral (22.4), we obtain

$$\int\limits_0^\infty t^{-\eta-1} A(t) J_{\nu-\eta-1}(xt) \, dt = \frac{x^{\nu-\eta-1}}{2^\eta \Gamma(\eta+1)} \int\limits_x^\infty g(\rho) \rho^{-\nu+1} (\rho^2 - x^2)^\eta \, d\rho, \tag{22.5}$$

where $d < x < \infty$, $\nu/2 - 1/4 > \eta > -1$.

We shall try to make the index of the Bessel function and the exponent of t in the left-hand sides of the equalities (22.3) and (22.5) equal. Then, combining these two equalities, we can immediately obtain the Hankel transform of the desired function. Thus, we obtain the following equations:

$$\nu + \xi + 1 = \nu - \eta - 1,$$
$$\alpha - \xi - 1 = -\eta - 1,$$

hence, $\xi = -1 + \alpha/2$, $\eta = -1 - \alpha/2$.

Since $\xi + \eta = -2$, it is impossible that ξ and η are simultaneously greater than -1. We shall go to further calculations, considering two special cases separately.

a) $0 < \alpha < 2$. The restrictions imposed on integrals (22.3) and (22.5) imply that $\nu > 1/2 - \alpha$. We set $\xi = -1 + \alpha/2$ and $\eta = -\alpha/2$. Substituting the values of ξ and η into the formulae obtained above, we find

$$\int_0^\infty t^{\alpha/2} A(t) J_{\nu+\alpha/2}(xt)\, dt = \frac{x^{-\nu-\alpha/2}}{2^{-1+\alpha/2}\Gamma(\alpha/2)} \int_0^x f(\rho)\rho^{\nu+1}(x^2 - \rho^2)^{-1+\alpha/2}\, d\rho,$$

$$(22.6)$$

$$0 \le x < d,$$

$$\int_0^\infty t^{\alpha/2-1} A(t) J_{\nu+\alpha/2-1}(xt)\, dt = \frac{x^{\nu+\alpha/2-1}}{2^{-\alpha/2}\Gamma(1-\alpha/2)} \int_x^\infty g(\rho)\rho^{-\nu+1}(\rho^2 - x^2)^{-\alpha/2}\, d\rho,$$

$$(22.7)$$

$$x > d,$$

Multiplying the left- and right-hand sides of equation (22.7) by $x^{-\nu-\alpha/2+1}$ and differentiating with respect to x, we obtain

$$\int_0^\infty t^{\alpha/2} A(t) J_{\nu+\alpha/2}(xt)\, dt = -\frac{2^{\alpha/2} x^{\nu+\alpha/2-1}}{\Gamma(1-\alpha/2)} \frac{d}{dx} \int_x^\infty g(\rho)\rho^{-\nu+1}(\rho^2 - x^2)^{-\alpha/2}\, d\rho,$$

$$(22.8)$$

$$x > d,$$

Using the fact that the left-hand sides of equations (22.6) and (22.8) are identical, we can rewrite these equations in the form of one equation

$$\int_0^\infty t^{\alpha/2} A(t) J_{\nu+\alpha/2}(xt)\, dt = \psi(x), \quad 0 \le x < \infty, \qquad (22.9)$$

where

$$\psi(x) = \begin{cases} \dfrac{x^{-\nu-\alpha/2}}{2^{-1+\alpha/2}\Gamma(\alpha/2)} \displaystyle\int_0^x f(\rho)\rho^{\nu+1}(x^2 - \rho^2)^{-1+\alpha/2}d\rho, & 0 \le x < d, \\[4mm] -\dfrac{2^{\alpha/2} x^{\nu+\alpha/2-1}}{\Gamma(1-\alpha/2)} \dfrac{d}{dx} \displaystyle\int_x^\infty g(\rho)\rho^{-\nu+1}(\rho^2 - x^2)^{-\alpha/2}\, d\rho, & x > d. \end{cases}$$

One can easily see that, up to a factor $\sqrt{x}$, the function $\psi(x)$ represents the Hankel transform of the unknown function $A(t)$. Using the inversion theorem, we immediately obtain the answer for the case $\nu > 1/2 - \alpha$, $\alpha > 0$,

$$A(t) = \frac{(2t)^{1-\alpha/2}}{\Gamma(\alpha/2)} \int_0^d x^{-\nu-\alpha/2+1} J_{\nu+\alpha/2}(xt) F_1(x)\, dx$$

$$+ \frac{2^{\alpha/2} t^{1-\alpha/2}}{\Gamma(1-\alpha/2)} \int_d^\infty x^{\nu+\alpha/2} J_{\nu+\alpha/2}(xt) G_1(x)\, dx, \qquad (22.10)$$

where

$$F_1(x) = \int\limits_0^x f(\rho)\rho^{\nu+1}(x^2-\rho^2)^{-1+\alpha/2}d\rho,$$

$$(22.11)$$

$$G_1(x) = -\frac{d}{dx}\int\limits_x^\infty g(\rho)\rho^{-\nu+1}(\rho^2-x^2)^{-\alpha/2}\,d\rho.$$

Formula (22.10) gives the result by Titchmarsh [51], who considered the case $0 < \alpha < 2$.

b) Let $-2 < \alpha < 0$, $\nu < \max(-3/2-\alpha, -1-\alpha/2)$. Suppose that $\xi = \alpha/2$, $\eta = -1-\alpha/2$. This implies that in (22.3) and (22.5) we should have $-2 < \alpha < 0$, $\nu > -3/2-\alpha$. Replacing ξ and η in (22.3) and (22.5) by their values, we obtain

$$\int\limits_0^\infty t^{-1+\alpha/2}A(t)J_{\nu+\alpha/2+1}(xt)\,dt = \frac{x^{-\nu-\alpha/2-1}}{2^{\alpha/2}\Gamma(1+\alpha/2)}\int\limits_0^x f(\rho)\rho^{\nu+1}(x^2-\rho^2)^{\alpha/2}\,d\rho,$$

$$(22.12)$$

$$0 \le x < d,$$

$$\int\limits_0^\infty t^{\alpha/2}A(t)J_{\nu+\alpha/2}(xt)\,dt = \frac{x^{\nu+\alpha/2}}{2^{-1-\alpha/2}\Gamma(-\alpha/2)}\int\limits_x^\infty g(\rho)\rho^{-\nu+1}(\rho^2-x^2)^{-1-\alpha/2}\,d\rho,$$

$$(22.13)$$

$$x > d,$$

We transform the first of these equations. Multiplying both its sides by $x^{\nu+\alpha/2+1}$ and differentiating with respect to x we obtain as a result

$$\int\limits_0^\infty t^{\alpha/2}A(t)J_{\nu+\alpha/2}(xt)\,dt = \frac{x^{-\nu-\alpha/2-1}}{2^{\alpha/2}\Gamma(1+\alpha/2)}\frac{d}{dx}\int\limits_0^x f(\rho)\rho^{\nu+1}(x^2-\rho^2)^{\alpha/2}\,d\rho,$$

$$(22.14)$$

$$0 \le x < d.$$

The left-hand side of equation (22.14) has the same form as the left-hand side of equation (22.13). This has fulfilled the aim of the transformations.

It remains now, similarly to the preceding case, to make the inverse transformation. As a result, we obtain

$$A(t) = \frac{2^{-\alpha/2}t^{1-\alpha/2}}{\Gamma(1+\alpha/2)}\int\limits_0^d x^{-\nu-\alpha/2}J_{\nu+\alpha/2}(xt)F_2(x)\,dx$$

$$+ \frac{2^{1+\alpha/2}t^{1-\alpha/2}}{\Gamma(1-\alpha/2)}\int\limits_d^\infty x^{\nu+\alpha/2+1}J_{\nu+\alpha/2}(xt)G_2(x)\,dx,$$

$$(22.15)$$

where

$$F_2(x) = \frac{d}{dx} \int_0^x f(\rho)\rho^{\nu+1}(x^2 - \rho^2)^{\alpha/2} d\rho,$$

$$G_2(x) = \int_x^\infty g(\rho)\rho^{-\nu+1}(\rho^2 - x^2)^{-1-\alpha/2} d\rho.$$

Integrating by parts, we can reduce the result to the form, which coincides with the solution obtained by Basbridge [52]. Here we finish consideration of equations (22.1a) and (22.1b) and pass to another, more complicated problem.

Consider the dual equations

$$\int_0^\infty t^\alpha [1 + H(t)]B(t)J_\nu(\rho t)\, dt = f(\rho), \qquad 0 \le \rho \le d, \qquad (22.16)$$

$$\int_0^\infty B(t)J_\nu(\rho t)\, dt = g(\rho), \qquad \rho > d, \qquad (22.17)$$

where $H(t)$ is a function decreasing at infinity.

In contrast to equations (22.1a) and (22.1b), we cannot obtain the solution here in quadratures. The final aim is the reduction of the problem to a solution of an integral Fredholm equation of the second kind.

If $0 < \alpha < 2$, then equations (22.16) and (22.17) can be transformed in the same way, which allowed us to obtain equations (22.6) and (22.8) from (22.1a) and (22.1b):

$$\int_0^\infty t^{\alpha/2} [1 + H(t)]B(t)J_{\nu+\alpha/2}(xt)\, dt = \frac{x^{-\nu-\alpha/2}}{2^{-1+\alpha/2}\Gamma(\alpha/2)} F_1(x), \qquad (22.18)$$

$$0 \le x < d,$$

$$\int_0^\infty t^{\alpha/2} B(t)J_{\nu+\alpha/2}(xt)\, dt = 0, \quad x > d, \qquad (22.19)$$

where $F_1(x)$ is defined by formula (22.11).

Let us denote

$$\int_0^\infty t^{\alpha/2} B(t)J_{\nu+\alpha/2}(xt)\, dt = \frac{x^{-1/2}}{2^{-1+\alpha/2}\Gamma(\alpha/2)}\theta(x), \quad 0 \le x < d. \qquad (22.20)$$

Taking into account (22.19) and using the inverse Hankel transformation, we obtain

$$B(t) = \frac{t^{1-\alpha/2}}{2^{-1+\alpha/2}\Gamma(\alpha/2)} \int_0^d \xi^{1/2}\theta(\xi)J_{\nu+\alpha/2}(\xi t)\, d\xi. \qquad (22.21)$$

Substituting expression (22.21) instead of $B(t)$ into the left-hand side of (22.17) (assuming $0 \le \rho < d$), which we denote by $h(\rho)$, and using the discontinuous improper Weber–Sonin–Schafheitlin integral

$$\int_0^\infty J_\mu(at)J_k(bt)t^{k-\mu+1}\,dt = \begin{cases} \dfrac{(a^2-b^2)^{\mu-k-1}b^k}{2^{\mu-k-1}a^\mu\Gamma(\mu-k)}, & a > b, \\[3mm] 0, & a < b, \end{cases} \tag{22.22}$$

as a result, we obtain

$$h(\rho) = \frac{\rho^\nu 2^{2-\alpha}}{[\Gamma(\alpha/2)]^2}\int_\rho^d \xi^{-\nu-\frac{\alpha}{2}+\frac{1}{2}}\theta(\xi)(\xi^2-\rho^2)^{\alpha/2-1}\,d\xi. \tag{22.23}$$

In order to obtain an equation with respect to $\theta(\xi)$, we substitute (22.20) into (22.18)

$$\theta(x) + x^{1/2}2^{-1+\alpha/2}\Gamma(\alpha/2)\int_0^\infty t^{\alpha/2}H(t)B(t)J_{\nu+\alpha/2}(xt)\,dt = x^{1/2-\nu-\alpha/2}F_1(x), \tag{22.24}$$

$$0 \le x < d.$$

Using (22.21) now, we reduce equation (22.24) to the Fredholm equation of the second kind

$$\theta(x) + \frac{1}{\pi}\int_0^d M(x,\xi)\theta(\xi)\,d\xi = x^{1/2-\nu-\alpha/2}F_1(x), \tag{22.25}$$

$$0 \le x < d,$$

with the kernel

$$M(x,\xi) = \pi(x\xi)^{1/2}\int_0^\infty tH(t)J_{\nu+\alpha/2}(xt)J_{\nu+\alpha/2}(\xi t)\,dt. \tag{22.26}$$

In order for the analysis above to be justified, the integral (22.26) should be convergent.

Let us find what conditions the function $H(t)$ should satisfy in order for $M(x,\xi)$ to be a convergent integral.

Obviously, it is necessary that $H(t)$ tends to zero like t^λ as $t \to 0$, where $\lambda > -(2\nu+\alpha+2)$ and $H(t) \to 0$ as $t \to \infty$. If $H(t)$ tends to zero like t^η, $\eta < 0$, as $t \to \infty$, then in order for the kernel to have no singularity at $x = \xi$, it is necessary that $\eta < -1$; if $0 > \eta \ge -1$, then the kernel has a singularity at $x = \xi$.

If $-2 < \alpha < 0$, then in the same way we obtain

$$\left.\begin{array}{l} \theta(x) + \dfrac{1}{\pi}\displaystyle\int_0^d M(x,\xi)\theta(\xi)\,d\xi = x^{-\nu-\alpha/2-1/2}F_2(x), \\[5mm] \displaystyle\int_0^\infty t^{\alpha/2}B(t)J_{\nu+1/2}(xt)\,dt = \dfrac{x^{-1/2}}{2^{\alpha/2}\Gamma(1+\alpha/2)}\theta(x), \quad 0 \le x < d. \end{array}\right\} \tag{22.27}$$

The deduction of the solution of the dual integral equation with the Bessel kernels described above is due to Noble [77]. Such a type of equations was also considered in a series of other papers[5]. Similar results for dual integral equations with kernels in the form of a linear combination of cylindrical functions of the first and second kinds are obtained by Srivastav [83]; dual equations with kernels in the form of the Struve functions were considered by A.I. Tseitlin, who discovered a method of solution of dual integral equations and equations with dual series of a rather large class, based on the application of so-called operators of transformation [56, 57].

23. Roots of the Bessel functions

The problems of the determination of the roots of the Bessel functions play a very important role in applications. As an example, we consider the displacements of a circular membrane which performs free axi-symmetrical oscillations: $w = AJ_0(\lambda r)\sin(\omega t + \varphi_0)$, where ω is the circular frequency of oscillations; $\lambda = \omega\sqrt{\rho/T}$, ρ is the surface density; T is the tension.

This solution should vanish on the contour of the membrane: $w(R) = 0$. This implies that for the existence of a non-trivial solution it is necessary that

$$J_0(\lambda R) = 0. \tag{23.1}$$

It is well known that equation (23.1) has infinitely many simple real roots and has neither multiple, nor complex roots. From physical reasonings, it is obvious that, for sufficiently large numbers of the roots, the problem reduces to the consideration of the oscillations of a membrane which has a rather large number of nodal curves representing concentric circumferences whose center coincides with the center of the contour. Therefore, we should expect some analogy with the problem of oscillations of a string. Hence, we have reasons to assume that the large roots of the Bessel function will be distant from one another by a distance which tends to a constant equal to π

Simultaneous solution of two and more Bessel equations leads to problems with the roots of more complicated expressions, where Bessel functions of different arguments are present. So, for instance, the problem of oscillations of a circular plate without an aperture with a fixed border, which is one of the simplest problems of this kind, leads to investigation of the roots of the function

$$J_n(x)I_n'(x) - J_n'(x)I_n(x).$$

Investigation of the properties of solutions of appropriate problems of mathematical physics allows one in some cases to facilitate the study of the roots of the Bessel functions. This circumstance has been indicated, in particular, by G.N. Watson, who writes that a very large part of the results related with the study of the roots of Bessel functions, which are given in his monograph, can immediately be obtained from the results of Rayleigh associated with the solution of some particular problems of mathematical physics.

In the book by G.N. Watson [7], the proofs of numerous theorems related with the investigation of the properties of the Bessel functions are considered (pages 525–573 in the Russian edition). A series of interesting results is contained in the monograph by A. Kratzer and V. Franz cited above (pages 374–416).

---

[5]see Cooke J.C., A solution of Tranter's dual integral equations problem, *Quarterly Journ. Mech. and Applied. Math.* **9**, no. 1 (1956); Lebedev N.N., The distribution of electricity on a thin paraboloidal segment, *DAN SSSR* **114**, no. 3 (1957); Lebedev N.N. and Ufland Ya.S., Axis-symmetrical contact problem for the elastic layer, *PMM* **22**, no. 3 (1958).

We give several very simple conclusions relating to the case, when the index ν is real.

THEOREM 1. *If the index $\nu > -1$, then the function $J_\nu(z)$ has only real roots.*

Since

$$J_\nu(z) = \left(\frac{z}{2}\right)^\nu \sum_{m=0}^{\infty} \frac{(-1)^m (z/2)^{2m}}{m!\Gamma(\nu+m+1)},$$

assuming that $z = i\alpha$, we obtain a series which contains only positive terms. Hence, the function $J_\nu(z)$ cannot have pure imaginary roots.

Let us prove that $J_\nu(z)$ has no complex roots. Suppose that α is a complex root of the function considered. Since $J_\nu(z)$ can be expanded into a power series with real coefficients, $\bar{\alpha}$ should also be its root. Now we shall use the integral

$$\int_0^x t J_\nu(\beta t) J_\nu(\bar{\beta} t)\, dt = \frac{x}{\beta^2 - \bar{\beta}^2}\left[J_\nu(\beta x)\frac{dJ_\nu(\bar{\beta}x)}{dx} - J_\nu(\bar{\beta}x)\frac{dJ_\nu(\beta x)}{dx}\right], \qquad (23.2)$$

whose proof is given in Section 15 [see (15.21)].

Setting $x = 1$ in formula (23.2) and taking into account that β and $\bar{\beta}$ are roots of the function $J_\nu(z)$, we obtain

$$\int_0^1 t J_\nu(\beta t) J_\nu(\bar{\beta} t)\, dt = 0.$$

However, this is impossible, because $J_\nu(\bar{\beta}t) = \overline{J_\nu(\beta t)}$; hence,

$$\int_0^1 t J_\nu(\beta t) J_\nu(\bar{\beta} t)\, dt = \int_0^1 t |J_\nu(\beta t)|^2\, dt > 0.$$

This implies that $J_\nu(z)$ has no complex roots. Hence, for $\nu > -1$ the function $J_\nu(z)$ has only real roots.

THEOREM 2. *All roots of the function $Z_\nu(z)$, except $z = 0$, are simple.*

Consider the cylindrical function $Z_\nu(z)$. Suppose that z_0 is a root of order $n > 1$. Then $Z_\nu(z_0)$ and $Z_\nu'(z_0)$ are equal to zero. Using the differential Bessel equation, we immediately obtain that $Z_\nu''(z_0)$ is equal to zero. Now differentiating the Bessel equation and substituting the values $Z_\nu''(z_0)$ and $Z_\nu'(z_0)$, we obtain $Z_\nu'''(z_0) = 0$. In the same way we can show that all derivatives of the function $Z_\nu(z)$ vanish at $z = z_0$. This implies that $Z_\nu(z) \equiv 0$, that is false. Hence, the function $Z_\nu(z)$ can have only simple roots.

THEOREM 3. *The roots of the functions $Z_p(z)$ and $Z_{p+1}(z)$ alternate each other.*

Let us write the formula for the differentiation

$$\frac{d}{dz}\left(z^{-\nu} Z_\nu(z)\right) = -z^{-\nu} Z_{\nu+1}(z).$$

Consider two consecutive roots of the function $Z_\nu(z)$, which are, as has been shown, real. Then, according to the Rolle theorem, at least one root of the derivative should be located between these values; hence, a root of the function $Z_{\nu+1}(z)$ is also located between these values.

THEOREM 4. *The roots of two cylindrical functions, which have one and the same index, alternate each other.*

Consider two different cylindrical functions $Z_\nu^*(x)$ and $Z_\nu(x)$. Since they are linearly independent, their Wronski determinant is non-zero, i.e.,

$$Z_\nu Z_\nu^{*\prime} - Z_\nu' Z_\nu^* = \frac{c}{z},$$

where $c \neq 0$.

Let x_1 and x_2 be two consecutive roots of the function $Z_\nu(x)$. As has been proved, x_1 and x_2 are real and

$$Z_\nu'(x_1) \neq 0, \quad Z_\nu'(x_2) \neq 0.$$

Assume that the function $Z_\nu(x)$ is positive in the interval $x_1 < x < x_2$. Obviously, we have

$$Z_\nu'(x_1) > 0,$$
$$Z_\nu'(x_2) < 0.$$

Using the Wronski determinant and consecutively setting $z = x_1$ and $z = x_2$, we obtain

$$Z_\nu^*(x_1) = -\frac{c}{x_1 Z_\nu'(x_1)},$$
$$Z_\nu^*(x_2) = -\frac{c}{x_2 Z_\nu'(x_2)}.$$

This implies that the signs of $Z_\nu^*(x_1)$ and $Z_\nu^*(x_2)$ are distinct. One can easily deduce that the same conclusion will be obtained, if $Z_\nu(x) < 0$ for $x_1 < x < x_2$. Hence, the function $Z_\nu^*(x)$ has at least one root between x_1 and x_2.

Since at least one root of the function $Z_\nu(x)$ lies between two roots of the function $Z_\nu^*(x)$, this implies that $Z_\nu^*(x)$ has only one root between x_1 and x_2.

Let us give some properties of roots of the cylindrical functions without proofs. More detailed information is given in [7] and [4] (see also the footnote on page 59).

1. *The functions $J_\nu(z)$ and $J_{\nu+m}(z)$, $m = 1, 2, 3, \ldots$, have no common roots except $z = 0$ (the Bourger conjecture).*

In this connection we note one result due to Siegel, who has proved that if ν is a rational number and z in a non-zero algebraic number, then $J_\nu(z)$ is not an algebraic number.

2. *For the smallest roots γ_ν, γ_ν', and γ_ν'' of the functions $J_\nu(x)$, $J_\nu'(x)$, and $J_\nu''(x)$, we have*

$$\sqrt{\nu(\nu+2)} < \gamma_\nu < \sqrt{2(\nu+1)(\nu+3)}, \qquad \nu > 0,$$
$$\sqrt{\nu(\nu+2)} < \gamma_\nu' < \sqrt{2\nu(\nu+1)}, \qquad \nu > 0,$$
$$\sqrt{\nu(\nu-1)} < \gamma_\nu'' < \sqrt{\nu^2-1}, \qquad \nu > 0.$$

3. *The function*

$$J_\nu(ax)Y_\nu(bx) - J_\nu(bx)Y_\nu(ax), \quad a > 0, \ b > 0,$$

has only real simple roots. The set of these roots is infinite.

4. For $\nu \geq 0$ the function $K_\nu(z)$ has no roots such that $|\arg z| \leq \pi/2$. It has real roots for purely imaginary indices.

5. The Hankel functions with real indices have no positive roots.

6. Large roots of the function $Z_\nu(z) = J_\nu(z)\cos\alpha - Y_\nu(z)\sin\alpha$ can be calculated with the help of the following asymptotic expansion

$$\left(n + \frac{1}{2}\nu - \frac{1}{4}\right)\pi - \alpha - \frac{4\nu^2 - 1}{8\{(n + \nu/2 - 1/4)\pi - \alpha\}}$$
$$- \frac{(4\nu^2 - 1)(28\nu^2 - 31)}{384\{(n + \nu/2 - 1/4)\pi - \alpha\}^3} - \cdots$$

7. Any positive root of the function $Z_\nu(z)$ can be considered as a continuous increasing function of the real argument ν.

8. Any positive root of the transcendental equation

$$J_\nu(kz)Y_\nu(z) - J_\nu(z)Y_\nu(kz) = 0$$

is a continuous increasing function of the real argument ν under the condition that ν is positive.

24. Series of Fourier–Bessel and Dini

In this section we consider the Fourier–Bessel series which plays an important role in applications as well as some other similar series which are their natural generalization.

If the functions $u(z)$ and $v(z)$ satisfy the equations

$$\frac{d^2u}{dz^2} + Pu = 0, \qquad \frac{d^2v}{dz^2} + Qv = 0,$$

then

$$\int_a^b (P - Q)uv\,dz = \left(u\frac{dv}{dz} - v\frac{du}{dz}\right)\bigg|_a^b. \qquad (24.1)$$

Setting $P = \alpha^2 - (4\nu^2 - 1)/z^2$, $Q = \beta^2 - (4\nu^2 - 1)/z^2$, we have

$$u = z^{1/2}Z_\nu(\alpha z), \quad v = z^{1/2}Z_\nu(\beta z).$$

These relations can easily be checked by differentiating. From (24.1) we obtain

$$(\alpha^2 - \beta^2)\int_a^b zZ_\nu(\alpha z)Z_\nu(\beta z)\,dz$$

$$= -|\alpha z Z_\nu'(\alpha z)Z_\nu(\beta z) - \beta z Z_\nu'(\beta z)Z_\nu(\alpha z)|_a^b. \qquad (24.2)$$

Let us prove that the Bessel functions satisfy the orthogonality property which can be written in the following form:

$$\int_0^1 x J_\nu(j_m x)J_\nu(j_k x)\,dx = 0, \qquad k \neq m, \ \nu > -1, \qquad (24.3)$$

where x is the real variable; j_m and j_k are real roots of the Bessel function.

Setting $Z_\nu(\alpha z) = J_\nu(\alpha x)$, $Z_\nu(\beta z) = J_\nu(\beta x)$, $a = 0$, $b = 1$ in (24.2), we see that the right-hand side of (24.2) vanishes at $a = 0$ for $\nu > -1$.

This implies [see also (15.21)] that

$$\int_0^1 x J_\nu(\alpha x) J_\nu(\beta x)\, dx = \frac{\alpha J_\nu(\beta) J_\nu'(\alpha) - \beta J_\nu(\alpha) J_\nu'(\beta)}{\beta^2 - \alpha^2}. \qquad (24.4)$$

The right-hand side of (24.4) is ambiguous at $\beta = \alpha$. Expanding it by the L'Hospital rule and using (6.3), we obtain

$$\int_0^1 x J_\nu^2(\alpha x)\, dx = J_{\nu+1}^2(\alpha)/2. \qquad (24.5)$$

If α and β are distinct roots of the Bessel function of index ν, i.e.,

$$\alpha = j_m, \quad \beta = j_k, \quad m \neq k,$$

then the right-hand side of (24.4) vanishes. Hence, the Bessel functions are orthogonal with the weight x.

The functions

$$\frac{\sqrt{2}}{\mathbf{J}_{\nu+1}(j_m)} \mathbf{J}_\nu(j_m x)$$

are normed on the interval $(0, 1)$ as one can see from (24.5).

The functions $J_m(\gamma_m x)$, $m = 1, 2, 3, \ldots$, form a complete system. A proof of this fact can be found in different courses, in particular, see [30], [7] and so on.

The properties of the Bessel functions formulated above allow one to obtain the following expansion of the function

$$f(x) = \sum_{m=1}^\infty a_m J_\nu(j_m x), \qquad (24.6)$$

where j_m, $m = 1, 2, 3, \ldots$, are as before the roots of the Bessel function.

If, assuming that an expansion of this kind exists and admits a term-by-term integration, we multiply the right- and left-hand sides of (24.6) by $x J_\nu(j_k x)$ and integrate, then we obtain the following expression for the coefficient a_k:

$$a_k = \frac{2}{J_{\nu+1}^2(j_k)} \int_0^1 x f(x) J_\nu(j_k x)\, dx. \qquad (24.7)$$

The series (24.6), whose coefficients are defined by the formula (24.7), is called the *Fourier-Bessel series*. Let us formulate the following theorem without proof.

THEOREM. *Let $f(t)$ be an arbitrary function given on the interval $(0, 1)$ such that the integral $\int_0^1 t^{1/2} f(t)\, dt$ exists and converges absolutely. Suppose that a_k are given by the formula (24.7), where $\nu \geq -1/2$. Let x be an internal point of the interval (a, b), $0 < a < b < 1$, and suppose that $f(t)$ is of bounded total variation. Then the series*

$$\sum_{m=1}^\infty a_m J_\nu(j_m x)$$

converges and its sum is equal to $\frac{1}{2}\{f(x+0)+f(x-0)\}$.

This theorem was proven by Gobson and can be found in [7].

If $f(x)$, satisfying the conditions formulated above, is a function continuous in the interval (a, b), then the Fourier–Bessel expansion of the function $f(x)$ converges uniformly to the function $f(x)$ in any interval $(a+\Delta, b-\Delta)$, where Δ is an arbitrary positive number [7].

For the uniform convergence near $x = 1$ it is necessary that $f(1-0) = 0$.

The question on uniform convergence at $x = 0$ requires a change in the expansion, namely, it should be multiplied by $\sqrt{x}$ (see [7]).

If $t^{-\nu}f(t)$ is continuous in the interval $(0, b)$ and $t^{1/2}f(t)$ is of bounded total variation on $(0, b)$, then [7] the series

$$\sum_{m=1}^{\infty} a_m x^{1/2} J_\nu(j_m x) \tag{24.8}$$

converges uniformly on $(0, b - \Delta)$ and its sum is equal to $x^{1/2}f(x)$.

If λ_m are the positive roots of the function $xJ'_\nu(x) + HJ_\nu(x)$, where H does not depend on x and ν, then the series

$$\sum_{m=1}^{\infty} b_m J_\nu(\lambda_m x), \tag{24.9}$$

where

$$b_m = \frac{2\lambda_m^2}{(\lambda_m^2 - \nu^2)J_\nu^2(\lambda_m) + \lambda_m^2 J_\nu'^2(\lambda_m)} \int_0^1 tf(t)J_\nu(\lambda_m t)\,dt, \tag{24.10}$$

is called the *Dini series*.

The functions $J_m(\lambda_m x)$ satisfy the orthogonality property

$$\int_0^1 x J_\nu(\lambda_m x)J_\nu(\lambda_k x)\,dx = 0,$$

which immediately follows from (24.4) and the condition

$$\lambda_m J'_\nu(\lambda_m) + HJ_\nu(\lambda_m) = 0.$$

Setting $\alpha = \beta = \lambda_m$ in (24.4) and expanding the indeterminacy obtained, we obtain the condition of the norming used in (24.10). If the series (24.9) converges to the function $f(x)$ at any point of the interval $(0, 1)$, then it is called the *Dini expansion* of the function $f(x)$. If $H = -\nu$, then we should add to the Dini expansion (24.10) the initial term

$$B_0(x) = 2(1+\nu)x^\nu \int_0^1 t^{\nu+1}f(t)\,dt.$$

If $H < -\nu$, then the initial term is equal to

$$B_0(x) = \frac{2\lambda_0^2 I_\nu(\lambda_0 x)}{(\lambda_0^2 + \nu^2)I_\nu^2(\lambda_0) - \lambda_0^2 I_\nu'^2(\lambda_0)} \int_0^1 tf(t)I_\nu(\lambda_0 t)\,dt,$$

where $\pm i\lambda_0$ are purely imaginary roots of the function

$$z^{-\nu}[zJ_\nu'(z) + HJ_\nu(z)].$$

As an example we give the expansion of the function x^ν into the Fourier–Bessel series

$$x^\nu = \sum_{m=1}^{\infty} \frac{2J_\nu(j_m x)}{j_m J_{\nu+1}(j_m)}, \quad 0 \le x < 1 \tag{24.11}$$

and into the Dini series

$$x^\nu = \sum_{m=1}^{\infty} \frac{2\lambda_m J_\nu(\lambda_m x) J_{\nu+1}(\lambda_m)}{(\lambda_m^2 - \nu^2)J_\nu^2(\lambda_m) + \lambda_m^2 J_\nu'^2(\lambda_m)},$$

$$0 \le x \le 1 \quad H + \nu > 0. \tag{24.12}$$

In order to expand the power function x^μ into the Fourier–Bessel and Dini series, we should to calculate the integrals

$$\int_0^1 x^{\mu+1} J_\nu(j_k x)\, dx \quad \text{and} \quad \int_0^1 x^{\mu+1} J_\nu(\lambda_k x)\, dx$$

which can be expressed via the Lommel functions by the formulae given in Section 16[6] for $\mu \ne \nu$; in some cases, when the right-hand side contains exponential and trigonometric functions, the coefficients of these series can be expressed via the Lommel functions of two variables and partial cylindrical functions (see Sections 27, 28). The series considered are a special case of series in eigenfunctions of the Sturm–Liouville problem for the Bessel equation.

Consider the expansions into the series in cylindrical functions which are suitable for positive finite integrals.

If the function $f(x)$ is defined in the interval $a < x < b$, $a > 0$, then the expansion obtained by Titchmarsh is

$$f(x) = \sum_{m=1}^{\infty} a_m[J_\nu(\gamma_m x)Y_\nu(\gamma_m b) - Y_\nu(\gamma_m x)J_\nu(\gamma_m b)]; \tag{24.13}$$

here γ_m is the mth positive root of the equation

$$J_\nu(az)Y_\nu(bz) - Y_\nu(az)J_\nu(bz) = 0,$$

$$a_m = \{[J_\nu(\gamma_m a)]^2 - [J_\nu(\gamma_m b)]^2\}^{-1} \frac{\pi\gamma_m^2}{2}$$

$$\times [J_\nu(\gamma_m a)]^2 \int_a^b [J_\nu(\gamma_m t)Y_\nu(\gamma_m b) - Y_\nu(\gamma_m t)J_\nu(\gamma_m b)]t f(t)\, dt.$$

The boundary conditions of this problem are analogous to the boundary conditions of the problem which generates the Fourier–Bessel series; from the physical viewpoint they correspond, for instance, to the problem of free oscillations of a membrane fixed along the contour, which occupies a domain in the form of a circular rectangle.

In the membrane under consideration the curvilinear part of the contour is represented by segments of arcs whose radii are equal to a and b, respectively. The

---

[6]In some cases (see page 42) they can be expressed via Bessel functions.

segments of the radii which connects the ends of the arcs form an angle which is equal to $\nu\pi$. Only such forms of oscillations are considered the nodal curves of which are arcs of circumferences which are concentric to the circumferences containing the arcs of the contour. The forms of these oscillations coincide up to the factor $C \sin \nu\theta$, where $C = \text{const}$, with the expressions in the brackets in (24.13).

Let us formally obtain the Dini series for a segment. For this purpose we first write the function which we have considered in Section 17:

$$Y_1(\alpha a, \alpha x) = \frac{\pi a \alpha}{2 \sin \nu\pi}[J_\nu'(\alpha a)J_{-\nu}(\alpha x) - J_{-\nu}'(\alpha a)J_\nu(\alpha x)],$$

$$Y_2(\alpha a, \alpha x) = -\frac{\pi a \alpha}{2 \sin \nu\pi}[J_\nu(\alpha a)J_{-\nu}(\alpha x) - J_{-\nu}(\alpha a)J_\nu(\alpha x)];$$

for $x = a$

$$Y_1 = 1, \quad Y_1' = 0, \quad Y_2 = 0, \quad Y_2' = 1.$$

Assume that the function $w(x)$ which is a solution of the Bessel equation with the index ν and parameter α, satisfies the following condition at $x = a$

$$w + H_1 \frac{dw}{dx} = 0, \tag{24.14}$$

and at $x = b$

$$w + H_2 \frac{dw}{dx} = 0. \tag{24.15}$$

Obviously, the first boundary condition is satisfied by the function

$$\Phi_\nu(\alpha a, \alpha x) = \alpha H_1 Y_1(\alpha a, \alpha x) - Y_2(\alpha a, \alpha x). \tag{24.16}$$

From the second boundary condition we obtain the following transcendental equation with respect to α:

$$\alpha H_2[\alpha H_1 Y_1'(\alpha a, \alpha b) - Y_2'(\alpha a, \alpha b)] + \alpha H_1 Y_1(\alpha a, \alpha b) - Y_2(\alpha a, \alpha b) = 0, \tag{24.17}$$

whose roots are denoted by μ_m. Similarly to the reasoning above, in order to prove the orthogonality of the functions $\Phi_\nu(\mu_m x, \mu_m a)$ and $\Phi_\nu(\mu_k x, \mu_k a)$, $k \neq m$, we consider the integral

$$(\mu_m^2 - \mu_k^2) \int_a^b x \Phi_\nu(\mu_m a, \mu_m x)\Phi_\nu(\mu_k a, \mu_k x)\,dx$$

$$= \left| \mu_m x \Phi_\nu'(\mu_m a, \mu_m x)\Phi_\nu(\mu_k a, \mu_k x) \right.$$

$$\left. - \mu_k x \Phi_\nu'(\mu_k a, \mu_k x)\Phi_\nu(\mu_m a, \mu_m x) \right|_a^b$$

$$= \mu_m b \Phi_\nu'(\mu_m a, \mu_m b)\Phi_\nu(\mu_k a, \mu_k b)$$

$$- \mu_k b \Phi_\nu'(\mu_k a, \mu_k b)\Phi_\nu(\mu_m a, \mu_m b). \tag{24.18}$$

The condition (24.15) implies that the right-hand side of (24.18) is equal to zero.

In order to norm the functions obtained, we consider the integral

$$\int_a^b x[Z_\nu(\alpha x)]^2\,dx = \frac{x^2}{2}\left\{[Z_\nu'(\alpha x)]^2 + \left(1 - \frac{\nu^2}{\alpha^2 x^2}\right)Z_\nu^2(\alpha x)\right\}\Bigg|_a^b. \tag{24.19}$$

Here $Z_\nu(\alpha x)$ is an arbitrary function which satisfies the Bessel equation with the index ν and parameter α.

The expression (24.19) can be obtained either by integration by parts or as a result of the application of the L'Hospital rule [7].

For $Z_\nu(\alpha x = \Phi_\nu(\alpha a, \alpha x)$ we obtain

$$
\int_a^b x[\Phi_\nu(\alpha a, \alpha x)]^2 \, dx
$$

$$
= \frac{b^2}{2} \left\{ [\Phi'_\nu(\alpha a, \alpha b)]^2 + \left(1 - \frac{\nu^2}{\alpha^2 b^2}\right) \Phi_\nu^2(\alpha a, \alpha b)\right\}
$$

$$
- \frac{a^2}{2} \left[\left(1 - \frac{\nu^2}{\alpha^2 a^2}\right) H_1^2 \alpha^2 + 1\right]. \quad (24.20)
$$

Taking into account (24.15), we transform (24.20) to the following form:

$$
\int_a^b x[\Phi_\nu(\alpha a, \alpha x)]^2 \, dx
$$

$$
= \frac{a^2}{2} \left\{ \left(1 + \frac{1}{H_2^2 \alpha^2} - \frac{\nu^2}{\alpha^2 b^2}\right) \frac{[H_1 \alpha J'_\nu(\alpha a) + J_\nu(\alpha a)]^2}{[H_2 \alpha J'_\nu(\alpha b) + J_\nu(\alpha b)]^2} H_2^2 \alpha^2 \right.
$$

$$
\left. -1 - \left(1 - \frac{\nu^2}{\alpha^2 a^2}\right) H_1^2 \alpha^2 \right\} = B(\alpha). \quad (24.21)
$$

Denoting the right-hand side of (24.21) by $B(\alpha)$, we rewrite the expansion into the Dini series in the following form:

$$
f(x) = \sum_{m=1}^\infty \frac{C_m}{B(\mu_m)} \Phi_\nu(\mu_m x, \mu_m a), \quad (24.22)
$$

where

$$
C_m = \int_a^b t f(t) \Phi_\nu(\mu_m t, \mu_m a) \, dt.
$$

Consider the expansion into the series of eigenfunctions of the problem described by the differential equation

$$
\Delta_\nu \Delta_\nu w_\nu - \alpha^4 w_\nu = 0 \quad (24.23)
$$

and the boundary conditions

$$
w_\nu(\alpha a) = 0,
$$
$$
w'_\nu(\alpha a) = 0,
$$
$$
w_\nu(\alpha b) = 0,
$$
$$
w'_\nu(\alpha b) = 0,
$$

where

$$
\Delta_\nu = \frac{d^2}{dx^2} + \frac{1}{x}\frac{d}{dx} - \frac{\nu^2}{x^2}.
$$

The eigenfunctions of this problem are, in many senses, analogues of the beam functions which are considered in the theory of oscillations of elastic systems in connection with the investigation of free oscillations of a rod of constant cross-section and are widely used in different applications.

We need to find the eigenvalues α_k and the eigenfunctions, which will be denoted by $\Phi_\nu(\alpha_k a, \alpha_k x)$.

In order to obtain the eigenfunctions, we will use the Cauchy functions considered above in Section 17, which satisfy the property of the identity matrix. In order to satisfy the boundary conditions at $r = a$, we assume

$$\Phi_\nu(\alpha_k a, \alpha_k x) = Y_3(\alpha_k a, \alpha_k x) + A Y_4(\alpha_k a, \alpha_k x),$$

where A is the value which should be determined.

From the boundary conditions at $x = b$ we obtain a transcendental equation, whose roots are the eigenvalues α_k. Furthermore, from these conditions we find the coefficient of the second term; thus, we obtain

$$\Phi_\nu(\alpha_k a, \alpha_k x) = Y_3(\alpha_k a, \alpha_k x) - \frac{Y_3(\alpha_k a, \alpha_k b)}{Y_4(\alpha_k a, \alpha_k b)} Y_4(\alpha_k a, \alpha_k x). \qquad (24.24)$$

We turn our attention to the proof of the orthogonality of the functions obtained. For this purpose we multiply the equations which are satisfied by the functions $\Phi_\nu(\alpha_m a, \alpha_m x)$ and $\Phi_\nu(\alpha_n a, \alpha_n x)$, by $x\Phi_\nu(\alpha_n a, \alpha_n x)$ and $x\Phi_\nu(\alpha_m a, \alpha_m x)$, respectively, integrate with respect to x from a to b and subtract one equation from the other one:

$$(\alpha_m^4 - \alpha_n^4) \int_a^b x\Phi_\nu(\alpha_m a, \alpha_m x)\Phi(\alpha_n a, \alpha_n x)\, dx$$

$$= \int_a^b x\Phi_\nu(\alpha_m a, \alpha_m x)\Delta_\nu\Delta_\nu\Phi(\alpha_n a, \alpha_n x)\, dx$$

$$- \int_a^b x\Phi_\nu(\alpha_n a, \alpha_n x)\Delta_\nu\Delta_\nu\Phi(\alpha_m a, \alpha_m x)\, dx.$$

Integrating by parts, one can easily obtain

$$\int_a^b x\Phi_\nu(\alpha_m a, \alpha_m x)\Delta_\nu\Delta_\nu\Phi(\alpha_n a, \alpha_n x)\, dx$$

$$= x\Phi_\nu(\alpha_m a, \alpha_m x)\frac{d}{dx}\Delta_\nu\Phi(\alpha_n a, \alpha_n x)\Big|_a^b$$

$$- x\frac{d}{dx}\Phi_\nu(\alpha_m a, \alpha_m x)\Delta_\nu\Phi(\alpha_n a, \alpha_n x)\Big|_a^b$$

$$+ \int_a^b x\Delta_\nu\Phi_\nu(\alpha_m a, \alpha_m x)\Delta_\nu\Phi(\alpha_n a, \alpha_n x)\, dx.$$

For the given boundary conditions, the terms outside the integral vanish.

Taking into account the relation which can be obtained from the previous one by interchanging α_m and α_n, we obtain

$$(\alpha_m^4 - \alpha_n^4) \int_a^b x\Phi_\nu(\alpha_m a, \alpha_m x)\Phi(\alpha_n a, \alpha_n x)\, dx = 0, \quad m \neq n.$$

When calculating the coefficients a_k of the expansion

$$f(x) = \sum_{k=1}^\infty a_k \Phi_\nu(\alpha_k a, \alpha_k x), \tag{24.25}$$

where

$$a_k = \frac{\int_a^b x f(x)\Phi_\nu(\alpha_k a, \alpha_k x)\, dx}{\int_a^b x[\Phi_\nu(\alpha_k a, \alpha_k x)]^2\, dx},$$

the following formula can be useful

$$\int_a^b x[w_\nu(\alpha x)]^2\, dx = \frac{x^2}{4}\Bigg\{[w_\nu(\alpha x)]^2 + [\Delta_\nu w_\nu(\alpha x)]^2$$

$$- 2[\Delta_\nu w_\nu(\alpha x)]' w_\nu'(\alpha x) + 2\frac{\nu^2}{x^2}w_\nu(\alpha x)\Delta_\nu w_\nu(\alpha x)$$

$$+ \frac{2}{\alpha x}[\Delta_\nu' w_\nu(\alpha x) - \Delta_\nu w_\nu(\alpha x)w_\nu'(\alpha x)]\Bigg\}\Bigg|_a^b, \tag{24.26}$$

where the differentiation is fulfilled with respect to the argument αx and, furthermore, we set in (24.26)

$$\Delta_\nu = \frac{\partial^2}{\partial(\alpha x)^2} + \frac{1}{\alpha x}\cdot\frac{\partial}{\partial(\alpha x)} - \frac{\nu^2}{(\alpha x)^2}.$$

This equality is analogous to (24.1) and can be obtained in the same way. Under the given boundary conditions we have

$$\int_a^b x[\Phi_\nu(\alpha_k a, \alpha_k x)]^2\, dx = \frac{x^2}{4}[\Delta_\nu \Phi_\nu(\alpha_k a, \alpha_k x)]^2\, dx\Bigg|_a^b \tag{24.27}$$

Taking into account the relation

$$\Delta_\nu \Phi_\nu(\alpha_k a, \alpha_k x) = -Y_1(\alpha_k a, \alpha_k x) + \frac{Y_3(\alpha_k a, \alpha_k b)}{Y_4(\alpha_k a, \alpha_k b)}Y_2(\alpha_k a, \alpha_k x),$$

we obtain after elementary simplifications

$$\int_a^b x[\Phi_\nu(\alpha_k a, \alpha_k x)]^2\, dx$$

$$= \frac{a^2}{4}\Bigg\{\frac{\pi^2 b^2 \alpha_k^2}{4\sin^2\nu\pi}\left[J_\nu'(\alpha_k b)J_{-\nu}(\alpha_k a) - J_{-\nu}'(\alpha_k a)J_\nu(\alpha_k b)\right]^2$$

$$\times \left[1 - \frac{\pi - 2A_{k\nu}\sin\nu\pi}{\pi - 2B_{k\nu}\sin\nu\pi}\right]^2 - 1\Bigg\},$$

where

$$A_{k,\nu} = \frac{I'_\nu(\alpha_k b)K_\nu(\alpha_k a) - K'_\nu(\alpha_k b)I_\nu(\alpha_k a)}{J'_\nu(\alpha_k b)J_{-\nu}(\alpha_k a) - J'_{-\nu}(\alpha_k b)J_\nu(\alpha_k a)},$$

$$B_{k,\nu} = \frac{I'_\nu(\alpha_k b)K'_\nu(\alpha_k a) - K'_\nu(\alpha_k b)I'_\nu(\alpha_k a)}{J'_\nu(\alpha_k b)J'_{-\nu}(\alpha_k a) - J'_{-\nu}(\alpha_k b)J'_\nu(\alpha_k a)}.$$

The results of this section can be used without essential changes, when considering different particular problems on integrating differential equations which lead to the Bessel equation or to a system of the Bessel equation (see Section 17)[7].

25. Schlömlich series

Usually, a series of the form

$$a_0/2 + \sum_{m=1}^{\infty} a_m J_\nu(mx), \tag{25.1}$$

where the index does not depend on the number and the argument is proportional to the number, is called the *Schlömlich series*.

First, consider the Schlömlich series for zero index of the Bessel function. We prove that a continuous function $f(x)$ for $0 \le x \le \pi$ admits an expansion into the series

$$f(x) = \frac{a_0}{2} + \sum_{m=1}^{\infty} a_m J_0(mx), \tag{25.2}$$

where

$$\left.\begin{array}{l} a_0 = 2f(0) + \dfrac{2}{\pi} \displaystyle\int_0^\pi \int_0^{\pi/2} u f'(u \sin \varphi)\, d\varphi\, du, \\[20pt] a_m = \dfrac{2}{\pi} \displaystyle\int_0^\pi \int_0^{\pi/2} u f'(u \sin \varphi) \cos mu\, d\varphi\, du. \end{array}\right\} \tag{25.3}$$

In this case it is required that $f'(x)$ is continuous and of bounded total variation for $0 \le x \le \pi$.

The proof is based on the fact that a solution of the integral equation

$$f(x) = \frac{2}{\pi} \int_0^{\pi/2} g(x \sin \theta)\, d\theta, \tag{25.4}$$

which is called the *integral Schlömlich equation*, is the function

$$g(x) = f(0) + x \int_0^{\pi/2} f'(x \sin \varphi)\, d\varphi. \tag{25.5}$$

---

[7]See, for instance, *B.G. Korenev, L.M. Reznikov* On oscillations of constructions with dynamic absorbers under stationary random actions, *Stroitel'naya mekhanika i raschet sooruzheniy*, no. 4 (1969).

The equation (25.4) can easily be reduced to the Abel equation which has the form

$$\int\limits_{a}^{y} \frac{u(t)\,dt}{(y-t)^{\alpha}} = f(y), \qquad 0 < \alpha < 1, \quad \alpha = \text{const}, \tag{25.6}$$

and is a special case of the integral Volterra equation of the first kind.. An equation of the following form can easily be reduced to the Abel equation

$$\int\limits_{a}^{y} \frac{u(t)\,dt}{\sqrt{y^2 - t^2}} = f(y). \tag{25.7}$$

This equation has been considered, for instance, in [15] and [43].

The Schlömlich equation (25.4) can be transformed into the Abel equation (25.7) as a result of the following changes:

$$t = x \sin\theta, \quad x = y, \quad u(t) = \frac{2}{\pi} g(t).$$

If we assume that the lower limit of the integration in the Schlömlich equation is equal to α, then in the Abel equation (25.7) we have

$$a = y \sin\alpha.$$

The solution of the Abel equation in this special case is given in [15]. It can be obtained by replacing y by s and multiplying both sides of the equation by $2s(y^2 - s^2)^{-1/2}$. If we now integrate with respect to the variable s from a to x, then we obtain as a result

$$u(y) = \frac{2}{\pi} \frac{d}{dy} \int\limits_{a}^{y} \frac{s f(s)\,ds}{\sqrt{y^2 - s^2}}. \tag{25.8}$$

Setting $s = y \sin\theta$, we obtain for $a > 0$, $a \leq y$

$$u(y) = \frac{2y}{\pi} \left\{ \frac{f(a)}{\sqrt{y^2 - a^2}} + \int\limits_{b}^{\pi/2} f'(y \sin\theta)\,d\theta \right\}, \tag{25.9}$$

$$b = \arcsin(a/y).$$

Replacing y by x, after elementary transformations as $a \to 0$ we come to (25.5).

The function $g(x)$ which plays an important role in obtaining the expansion of $f(x)$, is continuous and of bounded total variation. For $0 \leq x \leq \pi$ it allows the expansion

$$\left. \begin{aligned} g(x) &= \frac{a_0}{2} + \sum_{m=1}^{\infty} a_m \cos mx, \quad 0 \leq x \leq \pi, \\ a_m &= \frac{2}{\pi} \int\limits_{0}^{\pi} g(u) \cos mu\,du \\ &= \frac{2}{\pi} \int\limits_{0}^{\pi} \left[f(0) + u \int\limits_{0}^{\pi/2} f'(u \sin\varphi)\,d\varphi \right] \cos mu\,du. \end{aligned} \right\} \tag{25.10}$$

Taking into account (25.4), we obtain

$$
f(x) = \frac{2}{\pi} \int_0^{\pi/2} g(x \sin \theta) \, d\theta
$$

$$
= \frac{2}{\pi} \int_0^{\pi/2} \left[\frac{a_0}{2} + \sum_{m=1}^{\infty} a_m \cos(mx \sin \theta) \right] d\theta
$$

$$
= \frac{a_0}{2} + \frac{2}{\pi} \sum_{m=1}^{\infty} a_m \int_0^{\pi/2} \cos(mx \sin \theta) \, d\theta
$$

$$
= \frac{a_0}{2} + \sum_{m=1}^{\infty} a_m J_0(mx). \tag{25.11}
$$

Obviously, the Parseval integral was used in this calculation.

A generalization of the Schlömlich series can be obtained if we replace the Bessel function of zero index by the Bessel functions of an arbitrary index ν. Furthermore, we can replace the Bessel functions in the general term of the series by a linear combination of the Bessel function and either the Neumann function or the Struve function. These series can be written in the form

$$
\frac{a_0}{2\Gamma(1+\nu)} + \sum_{m=1}^{\infty} \frac{a_m J_\nu(mx) + b_m Y_\nu(mx)}{(mx/2)^\nu}, \tag{25.12}
$$

$$
\frac{a_0}{2\Gamma(1+\nu)} + \sum_{m=1}^{\infty} \frac{a_m J_\nu(mx) + b_m \mathbf{H}_\nu(mx)}{(mx/2)^\nu}, \quad -\frac{1}{2} < \nu < \frac{1}{2}. \tag{25.13}
$$

The series (25.13), which are studied in more detail, can be obtained by expanding a function $f(x)$ in the interval $(-\pi, \pi)$. These series are called the *generalized Schlömlich series*. The theory of these series, which was considered in detail in the book by Watson, has been developed and simplified by Cooke [63].

In order to expand the function $f(x)$ into the generalized Schlömlich series, one can use the integral equation of the Schlömlich equation type, which is given by Watson and is contained in another form in the book by Nielsen [76]:

$$
f(x) = \frac{2}{\Gamma(\nu + 1/2)\Gamma(1/2)} \int_0^{\pi/2} \cos^{2\nu} \theta \, g(x \sin \theta) \, d\theta, \quad 0 < \nu < 1/2. \tag{25.14}
$$

The continuous solution of the equation (25.14) has the form

$$
g(x) = \Gamma(\nu + 1) f(0)
$$

$$
+ \frac{\Gamma(1/2)}{\Gamma(1/2 - \nu)} \int_0^{\pi/2} \sec^{2\nu+1} \theta \frac{d}{d\theta} [\sin^{2\nu} \theta \{ f(x \sin \theta) - f(0) \}] \, d\theta. \tag{25.15}
$$

The equation (25.14) can be reduced to the Abel equation (25.6) with the help of the changes

$$g(x) = xu(x^2),$$
$$f(x) = [\Gamma(\nu + 1/2)\Gamma(1/2)x^{2\nu}]^{-1}f_1(x^2),$$
$$2\nu = 1 - 2\alpha,$$
$$x^2 = y,$$
$$x^2 \sin^2 \theta = t,$$

and in this case the lower bound in (25.14) can be assumed equal to $\beta = \arcsin \sqrt{a/y}$.

If the function $g(x)$ allows the expansion (25.10), then, using (25.14) and (25.15), we obtain

$$f(x) = \frac{a_0}{2\Gamma(\nu + 1)} + \sum_{m=1}^{\infty} \frac{a_m J_\nu(mx) + b_m \mathbf{H}_\nu(mx)}{(mx/2)^\nu},$$

where

$$\left. \begin{aligned}
a_m &= \int_{-\pi}^{\pi} \int_0^{\pi/2} \frac{\sec^{2\nu+1}\theta}{\Gamma(1/2 - \nu)\Gamma(1/2)} \frac{d}{d\theta}[\sin^{2\nu}\theta\{f(u\sin\theta) - f(0)\}]\cos mu \, d\theta \, du, \\
b_m &= \int_{-\pi}^{\pi} \int_0^{\pi/2} \frac{\sec^{2\nu+1}\theta}{\Gamma(1/2 - \nu)\Gamma(1/2)} \frac{d}{d\theta}[\sin^{2\nu}\theta\{f(u\sin\theta) - f(0)\}]\sin mu \, d\theta \, du.
\end{aligned} \right\}$$

$$(25.16)$$

We have used the Poisson integral here and the similar formula (16.20) which is the integral representation of the Struve function.

Let us note that, following Cooke, one can directly obtain the expansions without using the integral Schlömlich equations, on the basis of the Sonin integrals. This approach will be used later, when considering some generalizations of the Schlömlich series; the latter can easily be obtained in the following way:

- the expansion of the function $g(x)$ in some other series in the functions $g_m(x)$ such that the integrals

$$\int_0^{\pi/2} g_m(x\sin\theta)\varphi(\theta)\, d\theta,$$

where $\varphi(\theta)$ does not depend on the number of the term of the series, can easily enough be expressed via the cylindrical functions;

- the use of the Sonin integrals, with the help of which the expansion in the Bessel functions of the argument $x\sqrt{m^2 + k^2}$ can be obtained;

- the consideration with the help of solutions of the Abel equations with the lower limit $a \neq 0$ of modified Schlömlich equations which results in expansions into series in partial cylindrical functions.

All the calculations formulated above can be fulfilled separately or combined; the latter, in turn, results in another modification of the Schlömlich series.

First, assume[8] that the expansion of the function $g(x)$ into the Fourier–Bessel series has the form

$$g(x) = \sum_{m=1}^{\infty} a_m J_0(j_m x), \tag{25.17}$$

where

$$a_m = \frac{2}{[J_1(j_m)]^2} \int_0^1 u g(u) J_0(j_m u)\, du,$$

$$g(u) = f(0) + u \int_0^{\pi/2} f'(u \sin \theta)\, d\theta.$$

Then from (25.4) we obtain

$$f(x) = \frac{2}{\pi} \sum_{m=1}^{\infty} \int_0^{\pi/2} J_0(j_m x \sin \varphi)\, d\varphi. \tag{25.18}$$

Using the formula

$$\int_0^{\pi/2} J_0(z \sin \varphi)\, d\varphi = \frac{\pi}{2} J_0^2\left(\frac{z}{2}\right),$$

which can be obtained from the formula on page 59 after the substitutions $\nu = 0$, $n = 0$, $Z_\nu(\alpha) = J_\nu(\alpha)$, we finally find

$$f(x) = \sum_{m=1}^{\infty} a_m J_0^2\left(\frac{j_m x}{2}\right). \tag{25.19}$$

However, this expansion in squared Bessel functions which is obtained following the method of Schlömlich, does not satisfy the main property of the Schlömlich series, which consists in the fact that the argument is proportional to the number of the terms of the series.

Let us introduce an auxiliary function

$$F(x) = \frac{a_0}{2} + \sum_{m=1}^{\infty} a_m \cos mx, \qquad 0 \le x \le \pi. \tag{25.20}$$

Then we construct the function $g(x)$:

$$g(x) = \frac{2}{\pi} \int_0^{\pi/2} F(x \sin \theta)\, d\theta, \tag{25.21}$$

$$F(x) = g(0) + x \int_0^{\pi/2} g'(x \sin \theta)\, d\theta. \tag{25.22}$$

---

[8]See *Thielmann H.P.*, Bull. Amer. Math. Soc. 40 (1934).

According to (25.10) and (25.11) the following expansion holds

$$g(x) = \frac{a_0}{2} + \sum_{m=1}^{\infty} a_m J_0(mx).$$

Let the function $f(x)$ be connected with $g(x)$ by the relation

$$g(x) = f(0) + x \int_0^{\pi/2} f'(x \sin \theta) \, d\theta. \tag{25.23}$$

Using (25.4), we obtain the expansion, in which we find a_m from (25.3) after replacing $f(u \sin \varphi)$ by $g(u \sin \varphi)$ according to formula (25.33):

$$f(x) = \frac{2}{\pi} \int_0^{\pi/2} g(x \sin \theta) \, d\theta$$

$$= \frac{2}{\pi} \int_0^{\pi/2} \left[\frac{a_0}{2} + \sum_{m=1}^{\infty} a_m J_0(mx \sin \theta) \right] d\theta$$

$$= \frac{a_0}{2} + \frac{2}{\pi} \sum_{m=1}^{\infty} a_m \int_0^{\pi/2} J_0(mx \sin \theta) \, d\theta$$

$$= \frac{a_0}{2} + \sum_{m=1}^{\infty} a_m J_0^2 \left(\frac{mx}{2} \right). \tag{25.24}$$

Now suppose that the function $g(x)$ can be expanded into the Fourier–Bessel or Dini series on a segment $[a, b]$ (the formulae (24.13) or (24.22)). In this case the function $g(x)$ can be expanded into the series whose general term is a linear combination of cylindrical functions of a common index. Restricting ourselves to the case when this index is equal to zero, we obtain that if $g(x)$ is expanded into a series with the general term

$$C_m J_0(j_m x) + d_m Y_0(j_m x), \tag{25.25}$$

then the expansion of the function $f(x)$ has the form

$$f(x) = \sum_{m=1}^{\infty} \left[C_m J_0^2 \left(\frac{j_m x}{2} \right) + d_m Y_0 \left(\frac{j_m x}{2} \right) J_0 \left(\frac{j_m x}{2} \right) \right]. \tag{25.26}$$

Here we have used the formula

$$\int_0^{\pi/2} Y_0(z \sin \varphi) \, d\varphi = \frac{\pi}{2} J_0 \left(\frac{z}{2} \right) Y_0 \left(\frac{z}{2} \right),$$

which can be obtained from (18.15) at $n = 0$.

Consider the Schlömlich series in which the argument of the cylindrical function is equal to $z = \sqrt{k^2 + m^2 x}$, where m is the number of the term of the series and k is a non-zero constant.

The expansion of the function $f(x)$ into the series

$$f(x) = \frac{1}{2}a_0 J_0(kx) + \sum_{m=1}^{\infty} a_m J_0\left(\sqrt{k^2 + m^2}x\right),$$ (25.27)

$$0 \le x \le \pi,$$

where

$$a_m = \frac{2}{\pi} \int_a^\pi f(x) G_m(x, k)\, dx,$$

$$G_m(x, k) = \frac{(-1)^m x \cosh(k\sqrt{\pi^2 - x^2})}{\sqrt{\pi^2 - x^2}} + \int_x^\pi \frac{mx \cosh(k\sqrt{t^2 - x^2}) \sin mt\, dt}{\sqrt{t^2 - x^2}},$$

has been obtained in [67]. The proof is analogous to the one given before when obtaining the expansion (25.2), the difference consists in the fact that the Schlömlich equation obtained in [67]

$$f(x) = \frac{2}{\pi} \int_0^{\pi/2} g(x \sin \theta) \cos(kx \cos \theta)\, d\theta,$$ (25.28)

which can also be reduced to the Abel equation, is considered instead of (25.4).

Representing the function $g(x)$ in the form of a Fourier series in the cosines and using the Parseval integral, we immediately obtain (25.27).

Let us give following Cooke's method [4], another proof which is based on the using of the Sonin integral.

Consider the Sonin integral (see Section 20):

$$\int_0^{\pi/2} J_m(a \sin \theta) J_\nu(b \cos \theta) \sin^{\mu+1} \theta \cos^{\nu+1} \theta\, d\theta$$

$$= a^\mu b^\nu (a^2 + b^2)^{(-\nu-\mu-1)/2} J_{\nu+\mu+1}\left(\sqrt{a^2 + b^2}\right),$$

$$\Re\mu > -1, \quad \Re\nu > -1.$$

Let us set

$$a = x\sqrt{m^2 + k^2}, \quad b = ikx, \quad \nu = -\mu - 1/2.$$

Then

$$\int_0^{\pi/2} J_\mu\left[\sqrt{m^2 + k^2}\, x \sin \theta\right] J_{-\mu-1/2}(ikx \cos \theta) \sin^{\mu+1} \theta \cos^{-\mu+1/2} \theta\, d\theta$$

$$= \left[x^2(m^2 + k^2)\right]^{\mu/2} (ikx)^{-\mu-1/2}(mx)^{-1/2} J_{1/2}(mx).$$ (25.29)

Using the formulae

$$J_{1/2}(mx) = \sqrt{2/(\pi mx)} \sin mx,$$

$$J_{-\mu-1/2}(ikx \cos \theta) = \exp\left[-\frac{i\pi}{1}\left(\mu + \frac{1}{2}\right)\right] I_{-\mu-1/2}(kx \cos \theta),$$

we rewrite (25.29) in the following form

$$\int_0^{\pi/2} J_\mu \left[\sqrt{m^2 + k^2} x \sin \theta \right] J_{-\mu-1/2}(kx \cos \theta) \sin^{\mu+1} \theta \cos^{-\mu+1/2} \theta \, d\theta$$

$$= [x^2(m^2 + k^2)]^{\mu/2} (kx)^{-\mu-1/2} \sqrt{\frac{2}{\pi}} \frac{\sin mx}{mx}. \quad (25.30)$$

Suppose that we have the following expansion of the function $f(x)$:

$$f(x) = \sum_{m=1}^{\infty} a_m J_\mu \left(\sqrt{m^2 + k^2} x \right) \frac{\sqrt{\pi mx}(kx)^{\mu+1/2}}{\sqrt{2}[x^2(m^2 + k^2)]^{\mu/2}}, \quad (25.31)$$

$$0 \leq x \leq \pi.$$

We replace x by $x \sin \theta$ in the general term of the series and multiply both the sides by

$$(\sin \theta)^{\mu-1/2}(\cos \theta)^{-\mu+1/2} I_{-\mu-1/2}(kx \cos \theta).$$

After integrating from 0 to $\pi/2$ and applying the formula (25.30), we formally obtain

$$\int_0^{\pi/2} f(x \sin \theta)(\sin \theta)^{\mu-1/2}(\cos \theta)^{-\mu+1/2} I_{-\mu-1/2}(kx \cos \theta) \, d\theta$$

$$= \sum_{m=1}^{\infty} a_m \sin mx, \quad (25.32)$$

where

$$a_m = \frac{2}{\pi} \int_0^{\pi} \sin mt \, dt \int_0^{\pi/2} f(t \sin \theta)(\sin \theta)^{-1/2+\mu}(\cos \theta)^{-\mu+1/2} I_{-\mu-1/2}(kt \cos \theta) \, d\theta.$$

Setting $\mu = 0$, we obtain the result which has, as can be shown, a sufficiently simple form.

Nielsen [76] has considered different Schlömlich series whose sum is equal to zero. The simplest of such series has the form

$$\frac{1}{2} + \sum_{m=1}^{\infty}(-1)^m J_0(mx) = 0, \quad 0 < x < \pi. \quad (25.33)$$

This series oscillates at $x = 0$ and diverges at $x = \pi$.

This fact shows that if a function given in the interval (π, π) can be represented by the Schlömlich series convergent in this interval (except, possibly, a finite number of points), then this representation is not unique.

It is shown in [67] that the sum of the following series is also equal to zero

$$\frac{1}{2} J_0(kx) + \sum_{m=1}^{\infty}(-1)^m J_0 \left(\sqrt{m^2 + k^2} x \right), \quad (25.34)$$

$$0 < x < \pi,$$

where k is a real number.

The proof of the formula (25.33) given by Watson can easily be generalized for the sum of the series (25.34).

Let us return to the series (25.2). If we suppose that the lower limit a in the integral, which is present in the Abel equation, is non-zero, hence, the lower limit in the integral Schlömlich equation is equal to $\arcsin \sqrt{a/x}$, then, expanding the function $g(x)$ into a series in cosines and afterward calculating the integrals

$$\int_{\arcsin(a/x)}^{\pi/2} \cos(mx \sin \theta)\, d\theta,$$

we obtain the Bessel function as well as the partial Bessel functions (see Section 28) and the Schlömlich series takes the form

$$f(x) = \frac{a_0}{2} + \sum_{m=1}^{\infty} a_m[J_0(mx) - J_0(w_m, x)], \qquad (25.35)$$

where $w_m = \arcsin \sqrt{a/(mx)}$; see the notation $J_0(w_m, x)$ in Section 28.

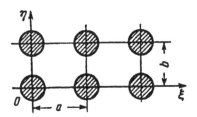

Figure 10

Obviously, we also obtain the appropriate partial cylindrical functions, introducing a non-zero lower limit in other formulae considered above, where the Parseval integrals or the Poisson integrals are present.

In applications we often meet series which are very close to the Schlömlich series considered above, and which are obtained as a result of obvious physical reasoning. They arise when considering the solutions of boundary-value problems for the inhomogeneous Helmholtz equation with a periodic right-hand side which vanishes in a part of the domain and the cells where it is non-zero, are small with respect to the periods of the right-hand side of the equation (see Fig. 10). For definiteness, let us consider an example.

Let a function $w(\xi, \eta)$ satisfy the equation

$$\nabla^2 w - w = q(\xi, \eta),$$

where $q = 0$, if the point lies outside the shaded circles,

$$q = q_{mn} = f_{mn}\left(\sqrt{(\xi - na)^2 + (\eta - mb)^2}\right),$$

where m and n are the numbers which characterize the number of the centre of the circle whose coordinates are $\xi_{mn} = na$, $\eta_{mn} = mb$, if the point belongs to the circle shaded.

Let us find $w((0^0))$, using for this purpose the axially symmetric solutions which correspond to the influence of all of the loadings q_{mn}.

Obviously, the solution corresponding to the influence of the loading located on the circle ($m = 0$, $n = 0$), has a form which is defined by the formulae given in [20] and can be found as a result of the integration of the principal influence function. For all other values of m and n, the contribution given by the loading located on the corresponding circle is approximately equal to

$$w_{m,n} = A_{mn} K_0\left(\sqrt{m^2 a^2 + n^2 b^2}\right).$$

In this case the numbers A_{00} and $A_{m,n}$ should be calculated as is shown in [20], where the influence of a point source is considered and the integration of the corresponding influence function is fulfilled. Thus, the solution has the form

$$w(0,0) = A_0 + \sum_{m=1}^{\infty} \sum_{n=1}^{\infty} A_{mn} K_0 \left(\sqrt{m^2 a^2 + n^2 b^2} \right).$$

This problem can be interpreted as the determination of the deflection of a membrane which lies on an elastic Winkler-type foundation.

The solution of a similar problem for a plate, which lies on an elastic Winkler-type foundation, is also of interest (see [20]).

26. Neumann series

In Sections 24, 25 we considered the series which are in a sense analogous to the Fourier series. Here we shall consider the series in Bessel functions, which are in a sense analogous to the power series.

A series of the form

$$\sum_{n=0}^{\infty} a_n J_{\nu+n}(z) \tag{26.1}$$

is called the *Neumann series*.

In fact, the reader has met the Neumann series, when considering the addition formulae and deducing the Lommel expansion.

Let us give two examples of the expansion of functions into the Neumann series, which are obtained from the addition formulae

$$J_\nu(z + y) = \sum_{m=-\infty}^{\infty} J_{\nu-m}(y) J_m(z), \quad |z| \le |y|$$

$$K_0(z - y) = \begin{cases} 2 \sum_{m=0}^{\infty}{}' I_m(y) K_m(z), & y \le z, \\ 2 \sum_{m=0}^{\infty}{}' I_m(z) K_m(y), & y \ge z. \end{cases}$$

Let us give without deduction (see [7]) the expansion of a function $f(z)$ for which the representation in the form of a Maclaurin series is given

$$f(z) = \sum_{n=0}^{\infty} b_n z^n, \quad |z| \le R,$$

into the Neumann series. The coefficients of the Neumann series (26.2) in this case has the form

$$a_0 = b_0,$$

$$a_n = n \sum_{m=0}^{\le n/2} 2^{n-2m} \frac{(n-m-1)!}{m!} b_{n-2m}.$$

In the case, when the function $f(z)$ can be represented in the form of a Laurent series

$$f(z) = \sum_{n=0}^{\infty} b_n z^n + \sum_{n=1}^{\infty} b_n'/z^n, \quad r \le |z| \le R,$$

the Neumann series has the form

$$f(z) = \sum_{n=0}^{\infty} a_n J_n(z) + \sum_{n=1}^{\infty} a'_n O_n(z), \tag{26.2}$$

where $O_n(z)$ is the Neumann polynomial, which can be expressed by the formula

$$O_n(z) = \frac{1}{4} \sum_{m=0}^{n} \frac{n\Gamma(n/2+m/2)\cos^2[(m\pm n)\pi/2]}{\Gamma(n/2-m/2+1)(z/2)^{m+1}};$$

$a_0, \dots, a_n$ have the same representation as for the Maclaurin series, and the coefficients a'_n are defined by the formula

$$a'_n = \sum_{m=0}^{\infty} \frac{(-1)^m}{2^{n+2m}m!(m+n)!} b'_{n+2m+1}.$$

27. Lommel functions of two variables

The Lommel functions of two variables can be defined with the help of the Neumann series by the following equalities:

$$\left. \begin{aligned} U_n(w,z) &= \sum_{m=0}^{\infty} (-1)^m (w/z)^{n+2m} J_{n+2m}(z), \\ V_n(w,z) &= \sum_{m=0}^{\infty} (-1)^m (w/z)^{-n-2m} J_{-n-2m}(z). \end{aligned} \right\} \tag{27.1}$$

If the index ν is non-integer, then

$$U_\nu(w,z) = \sum_{m=0}^{\infty} (-1)^m (w/z)^{\nu+2m} J_{\nu+2m}(z). \tag{27.2}$$

The following relations hold between the functions U and V when the index is integer

$$\left. \begin{aligned} U_n(w,z) - V_{-n+2}(w,z) &= \cos\left(\frac{w}{2} + \frac{z^2}{2w} - \frac{\pi n}{2}\right), \\ U_{n+1}(w,z) - V_{-n+1}(w,z) &= \sin\left(\frac{w}{2} + \frac{z^2}{2w} - \frac{\pi n}{2}\right). \end{aligned} \right\} \tag{27.3}$$

If the index ν is non-integer, then we have

$$\left. \begin{aligned} V_\nu(w,z) &= U_{-\nu+2}(w,z) + \cos\left(\frac{w}{2} + \frac{z^2}{2w} + \frac{\pi\nu}{2}\right), \\ U_\nu(w,z) + U_{\nu+2}(w,z) &= (w/z)^\nu J_\nu(z), \\ V_\nu(w,z) + V_{\nu+2}(w,z) &= (w/z)^{-\nu} J_{-\nu}(z) \end{aligned} \right\} \tag{27.4}$$

These functions have been introduced by Lommel when investigating problems of diffraction. They can be considered as some particular solutions of the inhomogeneous Bessel equation. One can establish some relations between these functions and the functions contiguous to the Bessel functions which we considered above.

One can easily obtain the differential equation which is satisfied by the Lommel functions of two variables.

Using the formula

$$\frac{d}{dz}[z^{-\nu}Z_\nu(z)] = -z^{-\nu}Z_{\nu+1}(z)$$

and differentiating the series (27.2) with respect to z, we obtain

$$\frac{\partial}{\partial z}U_\nu(w,z) = -\frac{z}{w}U_{\nu+1}(w,z).$$

After differentiating once again, we have

$$\frac{\partial^2}{\partial z^2}U_\nu(w,z) = \frac{z^2}{w^2}U_{\nu+2}(w,z) - \frac{1}{w}U_{\nu+1}(w,z). \tag{27.5}$$

Using the first of the equations (27.1), these formulae allow one to obtain the following result:

$$\left\{\frac{\partial^2}{\partial z^2} - \frac{1}{z}\frac{\partial}{\partial z} + \frac{z^2}{w^2}\right\}U_\nu(w,z) = \frac{z^2}{w^2}\{U_{\nu+2}(w,z) + U_\nu(w,z)\}$$

$$= \left(\frac{w}{z}\right)^{\nu-2} J_\nu(z). \tag{27.6}$$

The Lommel function $U_\nu(w,z)$ is a particular solution of the differential equation obtained, which with the help of the change of variable $z = \sqrt{2wt}$ can be rewritten in the form

$$\frac{d^2y}{dt^2} + y = \left(\frac{w}{2}\right)^{\nu/2} t^{-\nu/2} J_\nu\left(\sqrt{2wt}\right),$$

where $y = U_\nu(w,z)$.

The general solution of the corresponding homogeneous equation has the form

$$y = A\cos\frac{z^2}{2w} + B\sin\frac{z^2}{2w}.$$

One can easily show that $V_\nu(w,z)$ is a particular solution of the equation

$$\frac{\partial^2y}{\partial z^2} - \frac{1}{z}\frac{\partial}{\partial z} + \frac{z^2y}{w^2} = \left(\frac{z}{w}\right)^\nu J_{-\nu+2}(z). \tag{27.7}$$

Let us write the integrals which give the representation of the Lommel functions in the form

$$\left. \begin{aligned} U_\nu(w,z) &= \frac{w^\nu}{z^{\nu-1}}\int_0^1 J_{\nu-1}(zt)\cos\left\{\frac{1}{2}w(1-t^2)\right\}t^\nu\,dt, \\[2mm] U_{\nu+1}(w,z) &= \frac{w^\nu}{z^{\nu-1}}\int_0^1 J_{\nu-1}(zt)\sin\left\{\frac{1}{2}w(1-t^2)\right\}t^\nu\,dt. \end{aligned} \right\} \tag{27.8}$$

These formulae hold for $\Re\nu > 0$; the formulae for other values of ν can be obtained as a result of the integration in the plane of a complex variable. Combining the two formulae given above, we find

$$U_\nu(w,z) \pm iU_{\nu+1}(w,z) = \frac{w^\nu}{z^{\nu-1}}\int_0^1 J_{\nu-1}(zt)\exp\left\{\pm\frac{1}{2}iw(1-t^2)\right\}t^\nu\,dt.$$

As a result of the integration in the plane of the complex variable, we obtain the formulae

$$\frac{w^\nu}{z^{\nu-1}} \int_0^\infty J_{\nu-1}(zt) \frac{\cos}{\sin} \left\{ \frac{1}{2}w(1-t^2) \right\} t^\nu \, dt = \frac{\cos}{\sin} \left(\frac{w}{2} + \frac{z^2}{2w} - \frac{\nu\pi}{2} \right), \qquad (27.9)$$

which are valid for $w > 0$, $z > 0$, $3/2 > \Re(\nu) > 0$.

Let us give some other formulae, for the obtaining of which we should use (27.9), (27.4), and (27.8),

$$\left. \begin{aligned} V_\nu(w,z) &= -\frac{z^{\nu-1}}{w^{\nu-2}} \int_1^\infty J_{1-\nu}(zt) \cos\left\{ \frac{1}{2}w(1-t^2) \right\} t^{2-\nu} \, dt, \\ V_{\nu-1}(w,z) &= -\frac{z^{\nu-1}}{w^{\nu-2}} \int_1^\infty J_{1-\nu}(zt) \sin\left\{ \frac{1}{2}w(1-t^2) \right\} t^{2-\nu} \, dt. \end{aligned} \right\} \qquad (27.10)$$

In the special case, when we set $\nu = 1$ in (27.9), we obtain

$$\int_0^\infty J_0(zt) \frac{\cos}{\sin} \left(\frac{wt^2}{2} \right) t \, dt = \frac{1}{w} \frac{\cos}{\sin} \left(\frac{z^2}{2w} \right). \qquad (27.11)$$

This result is closely related with the first exponential Weber integral.

The formula of differentiation with respect to z given above is a recurrence relation for $U_\nu(w,z)$. In order to obtain other formulae, we should calculate the derivative with respect to the argument w

$$\frac{\partial}{\partial w} U_\nu(w,z) = \frac{1}{2} \left[U_{\nu-1}(w,z) + \left(\frac{z}{w} \right)^2 U_{\nu+1}(w,z) \right].$$

One can obtain similar formulae for the function $V_\nu(w,z)$:

$$\frac{\partial}{\partial z} V_\nu(w,z) = -\frac{z}{w} V_{\nu-1}(w,z),$$

$$2\frac{\partial}{\partial w} V_\nu(w,z) = V_{\nu+1}(w,x) + \frac{z^2}{w^2} V_{\nu-1}(w,z).$$

Consider the improper integral of the Lommel function of two variables, which is analogous to the improper Weber integral of the Bessel function. For this purpose we use the first of the formulae (27.8) and change the order of integration. We obtain

$$\int\limits_0^\infty U_\nu(w,z)z^{-\nu+k}\,dz = \int\limits_0^\infty \frac{w^\nu}{z^{\nu-1}}z^{-\nu+k}\,dz \int\limits_0^1 J_{\nu-1}(zt)\cos\left\{\frac{1}{2}w(1-t^2)\right\}t^\nu\,dt$$

$$= \int\limits_0^1 \frac{\Gamma((k-\nu+1)/2)w^\nu t^{3\nu-k-2}}{2^{2\nu-1-k}\Gamma(\nu-(k-\nu+1)/2)}\cos\left\{\frac{1}{2}w(1-t^2)\right\}\,dt$$

$$= \frac{\Gamma((k-\nu+1)/2)w^\nu}{2^{2\nu-1-k}\Gamma(\nu-(k-\nu+1)/2)}$$

$$\times \frac{2^{(3\nu-k-3)/2}\Gamma((3\nu-k-1)/2)}{w^{(3\nu-k-1)/2}}\cdot U_{(3\nu-k-1)/2}(w,0)$$

$$= \frac{w^{-\nu/2+k/2+1}\Gamma((k-\nu+1)/2)}{2^{\nu/2-k/2+1}\Gamma((3\nu-k-3)/2)}s_{(3\nu-k-4)/2,1/2}\left(\frac{w}{2}\right),$$

$$\Re k+1 > \Re \nu > \frac{\Re k}{2}+\frac{1}{4}.$$

Using the inversion formula for the Mellin transform [5], we can immediately express a transform for the Lommel function of two variables in the form of an integral of the Lommel function $S_{\mu,\nu}$

$$\int\limits_{\sigma-i\infty}^{\sigma+i\infty} s_{(3\nu-k-4)/2,1/2}\left(\frac{w}{2}\right)z^{-k-1}\frac{\Gamma((k-\nu+1)/2)w^{(k+2-\nu)/2}}{\Gamma((3\nu-k-3)/2)2^{(2-k+\nu)/2}}\,dk$$

$$= 2\pi i z^{-\nu}U_\nu(w,z),$$

where $z^{\sigma-1}z^{-\nu}U_\nu(w,z) \in L(0,+\infty)$ $(\sigma+2 > \Re\nu > \sigma/2+3/4)$.

Let us turn our attention to some special cases of the Lommel functions of two variables and their connection with other special functions.

First, consider the Lommel functions for $w = cz$, $c = \text{const}$. In this case, the formulae given above allow one to show that the functions $U_\nu(cz,z)$ and $V_{-\nu+2}(cz,z)$ are particular solutions of the equation

$$4\frac{d^2y}{dz^2}+\left(c+\frac{1}{c}\right)^2 y = c^\nu J_{\nu-2}(z)+c^{\nu-2}J_\nu(z).$$

One can see from the integral representations of the Lommel functions of two variables that at $z = 0$, using the expansion of the Bessel function into a power series, we find (also see (27.17)) that this function differs only by a factor, which has a simple form, from the Lommel function in one variable, which was considered before in Section 16:

$$U_\nu(w,0) = \frac{(w/2)^{1/2}}{\Gamma(\nu-1)}s_{\nu-3/2,1/2}\left(\frac{1}{2}w\right), \tag{27.12}$$

$$V_{-\nu+2}(w,0) = \frac{(w/2)^{1/2}}{\Gamma(\nu-1)}S_{\nu-1/2,1/2}\left(\frac{1}{2}w\right). \tag{27.13}$$

For an integer positive index and $w = z$, the following simple relations hold, for whose obtaining we used the Jacobi expansions (page 23):

$$\left.\begin{aligned}
U_0(z,z) = V_0(z,z) &= \frac{1}{2}(J_0(z) + \cos z),\\
U_{2n}(z,z) = V_{2n}(z,z) &= \frac{1}{2}(-1)^n\left(\cos z - 2\sum_{m=0}^{n-1}(-1)^m J_{2m}(z)\right),
\end{aligned}\right\} \qquad (27.14)$$

$$\left.\begin{aligned}
U_1(z,z) = -V_1(z,z) &= (\sin z)/2,\\
U_{2n+1}(z,z) = -V_{2n+1}(z,z) &= \frac{1}{2}(-1)^n\left(\sin z - 2\sum_{m=0}^{n-1}(-1)^m J_{2m+1}(z)\right),
\end{aligned}\right\} \qquad (27.15)$$

These relations hold under the condition that $n \geq 1$ in (27.14) and $n \geq 0$ in (27.15).

The formulae which connect the Lommel functions of the index $1/2$ and $3/2$ with the functions of the same index, whose second variable vanishes, are of interest

$$U_{1/2}(w,z) \pm iU_{3/2}(w,z) = \frac{1}{2}e^{\mp iz}\{U_{1/2}(2\sigma,0) \pm iU_{3/2}(2\sigma,0)\}$$

$$+ \frac{1}{2}e^{\pm iz}\{U_{1/2}(2\delta,0) \pm iU_{3/2}(2\delta,0)\}, \qquad (27.16)$$

where

$$\sqrt{\sigma} = \frac{w+z}{\sqrt{2w}}, \qquad \sqrt{\delta} = \frac{w-z}{\sqrt{2w}}.$$

Using the integral representation, we have for $z = 0$

$$\left.\begin{aligned}
U_\nu(w,0) &= \frac{w^\nu}{2^{\nu-1}\Gamma(\nu)}\int_0^1 t^{2\nu-1}\cos\left\{\frac{1}{2}w(1-t^2)\right\}\,dt,\\
U_{\nu+1}(w,0) &= \frac{w^\nu}{2^{\nu-1}\Gamma(\nu)}\int_0^1 t^{2\nu-1}\sin\left\{\frac{1}{2}w(1-t^2)\right\}\,dt.
\end{aligned}\right\} \qquad (27.17)$$

For $\nu = 1/2$, denoting $w = \pi u^2$, we obtain

$$\int_0^u \cos\left(\frac{1}{2}\pi t^2\right)\,dt = \frac{1}{2}\int_0^z \left(\frac{2}{\pi t}\right)^{1/2}\cos t\,dt$$

$$= [U_{1/2}(2z,0)\cos z + U_{3/2}(2z,0)\sin z]/\sqrt{2}$$

$$= \frac{1}{2} + [V_{1/2}(2z,0)\sin z + V_{3/2}(2z,0)\cos z]/\sqrt{2};$$

$$\int_0^u \sin\left(\frac{1}{2}\pi t^2\right)\,dt = \frac{1}{2}\int_0^z \left(\frac{2}{\pi t}\right)^{1/2}\sin t\,dt$$

$$= [U_{1/2}(2z,0)\sin z - U_{3/2}(2z,0)\cos z]/\sqrt{2}$$

$$= \frac{1}{2} - [V_{1/2}(2z,0)\cos z - V_{3/2}(2z,0)\sin z]/\sqrt{2}.$$

These formulae establish a connection between the Lommel functions of two variables and the *Fresnel integrals*

$$S(x) = \frac{2}{\sqrt{\pi}} \int\limits_0^x \sin t^2 \, dt,$$

$$C(x) = \frac{2}{\sqrt{\pi}} \int\limits_0^x \cos t^2 \, dt.$$

In applications one often meets the *error function*

$$\Phi(z) = \frac{2}{\sqrt{\pi}} \int\limits_0^z e^{-u^2} \, du,$$

which is connected with Fresnel integrals as well as, obviously, with the Lommel functions of two variables.

It is interesting to note that Lommel functions of two variables have found a wide application in the theory of passing of dynamic systems through resonance with linear variation of the frequency of the disturbing action.

In the problems on passing through resonance, the following integrals can be met, which also arise when considering the Lommel functions of two variables with the indices $3/2$ and $1/2$, if z is an imaginary number:

$$\int\limits_0^t e^{\kappa t} \cos\left(\frac{1}{2}\omega t^2\right) \, dt,$$

$$\int\limits_0^t e^{\kappa t} \sin\left(\frac{1}{2}\omega t^2\right) \, dt.$$

These integrals can be expressed via the functions S and T, where

$$S = 2^{-1}\sqrt{\pi}\exp(x^2 - y^2)(u \sin 2xy + v \cos 3xy),$$
$$T = 2^{-1}\sqrt{\pi}\exp(x^2 - y^2)(1 + v \sin 2xy - u \cos 3xy),$$

and $u(z)$ and $v(z)$ denote the real and imaginary parts of the error function in a complex domain

$$u(z) + iv(z) = W(z).$$

Let us write the following formulae:

$$
\begin{aligned}
A_1(t) &= \int_0^t e^{\kappa t} \cos\left(\frac{1}{2}\omega t^2\right) dt \\
&= -\omega^{-1/2}\{(\sin\beta - \cos\beta)[T(\sqrt{-\beta}) + T(\tau) \\
&\quad - (\sin\beta + \cos\beta)[S(\sqrt{-\beta}) + S(\tau)]\}; \\
A_2(t) &= \int_0^t e^{\kappa t} \sin\left(\frac{1}{2}\omega t^2\right) dt \\
&= \omega^{-1/2}\{(\sin\beta - \cos\beta)[T(\sqrt{-\beta}) + T(\tau)] \\
&\quad + (\sin\beta + \cos\beta)[S(\sqrt{-\beta}) + S(\tau)]\},
\end{aligned}
$$

where $\beta = \kappa^2/(2\omega)$, $\tau = \sqrt{\omega/4}[(t - \kappa/\omega) + i(t + \kappa/\omega)]$.

28. On partial cylindrical functions

In the preceding sections we considered different particular solutions of the inhomogeneous Bessel equation. Some of them, for instance, the Anguer and Weber functions have been obtained in a natural way as a result of a generalization of the Bessel integrals, the other have been obtained as a result of using the method of the variation of arbitrary constants. In this section a large class of solutions of the inhomogeneous Bessel equations will briefly be considered.

These solutions are called the *partial cylindrical functions* and are, usually, defined from the integral representations of Bessel, Poisson and Sonin–Schlafli, in which the limits of the integration are variable. In this case these functions are defined so that they coincide for some values of the limits with ordinary cylindrical functions. The partial cylindrical functions depend on three independent variables: the index, the argument and a parameter of incompleteness, which is denoted by w. This fact allows us to consider a very large class of solutions of ingomogeneous equations in the context of the theory of partial cylindrical functions, which contains, in particular, the particular solutions of the inhomogeneous Bessel equation, which were considered before in Section 16. The theory of these functions has been developed in detail in the monograph by A.A. Agrest and M.Z. Maksimov [2]. In this section we give a very brief review of the theory of partial cylindrical functions, which is based on the monograph cited.

Consider the function

$$
E_\nu^+(w, z) = \frac{2z^\nu}{A_\nu} \int_0^w e^{iz\cos t} \sin^{2\nu} t \, dt, \tag{28.1}
$$

$$
A_\nu = 2^\nu \Gamma(\nu + 1/2)\Gamma(1/2), \qquad \Re(\nu + 1/2) > 0,
$$

which for $w = \pi/2$ coincides with the usual Poisson integral; therefore,

$$
E_\nu^+(\pi/2, z) = J_\nu(z).
$$

Considering z as an argument and w as a parameter, we consecutively find the first and the second derivatives of (28.1). Then we use the expressions obtained when applying the Bessel operator of the index ν to the function $E_\nu^+(w, z)$. As a result of simple calculations, we obtain the following differential equation for the function $E_\nu^+(w, z)$:

$$\nabla_\nu E_\nu^+(w, z) = 2iz^{\nu+1} \sin^{2\nu+1} w A_\nu^{-1} e^{iz \cos w} = \Psi_\nu'(w, z). \qquad (28.2)$$

The formulae, which were multiply used before (see, for instance, Section 16) and are based, in fact, on the application of the Cauchy functions for equations with a right-hand side, imply

$$E_\nu^+(w, z) = J_\nu(w) \left[C_\nu(w) - \frac{\pi}{2} \int_0^z Y_\nu(t) \Psi_\nu(w, t) \frac{dt}{t} \right]$$

$$+ Y_\nu(z) \left[D_\nu(w) + \frac{\pi}{2} \int_0^z J_\nu(t) \Psi_\nu(w, t) \frac{dt}{t} \right], \qquad (28.3)$$

where, as the investigation of the asymptotic of $E_\nu^+(w, z)$ shows, we have

$$C_\nu(w) = 2 \frac{\Gamma(\nu+1)}{\Gamma(\nu+1/2)\Gamma(1/2)} \int_0^w \sin^{2\nu} t \, dt, \qquad (28.4)$$

$$D_\nu(w) = 0.$$

Using these results and replacing $\Psi_\nu(w, t)$ by the expression (28.2), we obtain

$$E_\nu^+(w, z) = 2J_\nu(z) + \frac{\Gamma(\nu+1)}{\Gamma(\nu+1/2)\Gamma(1/2)} \int_0^w \sin^{2\nu} t \, dt$$

$$+ 2\pi i \frac{\sin^{2\nu+1} w}{2^{\nu+1}\Gamma(\nu+1/2)\Gamma(1/2)} \left[Y_\nu(z) \int_0^z J_\nu(t) e^{it \cos w} t^\nu \, dt \right.$$

$$\left. - J_\nu(z) \int_0^z Y_\nu(t) e^{it \cos w} t^\nu \, dt \right]. \qquad (28.5)$$

Let us also introduce the function

$$E_\nu^-(w, z) = \frac{2z^\nu}{A_\nu} \int_0^w e^{-iz \cos u} \sin^{2\nu} u \, du \qquad (28.6)$$

and give the expressions for the partial Bessel function $J_\nu(w, z)$ and the partial Struve function $\mathbf{H}_\nu(w, z)$:

$$J_\nu(w, z) = \frac{E_\nu^+(w, z) + E_\nu^-(w, z)}{2}$$

$$= 2 \frac{z^\nu}{A_\nu} \int_0^w \cos(z \cos \theta) \sin^{2\nu} \theta \, d\theta, \qquad (28.7)$$

$$\mathbf{H}_\nu(w, z) = \frac{E_\nu^+(w, z) - E_\nu^-(w, z)}{2i}$$

$$= 2 \frac{z^\nu}{A_\nu} \int_0^w \sin(z \cos \theta) \sin^{2\nu} \theta \, d\theta.$$

Obviously,

$$J_\nu(\pi/2, z) \equiv J_\nu(z),$$
$$\mathbf{H}_\nu(\pi/2, z) \equiv \mathbf{H}_\nu(z). \qquad (28.8)$$

The change of the argument z by iz results in the following formulae:

$$\left.\begin{aligned} J_\nu(w, iz,) &= e^{i\nu\pi/2} I_\nu(w, z), \\ \mathbf{H}_\nu(w, iz,) &= i e^{i\nu\pi/2} \mathbf{L}_\nu(w, z), \end{aligned}\right\} \qquad (28.9)$$

where

$$\left.\begin{aligned} I_\nu(w, z) &= \frac{2z^\nu}{A_\nu} \int_0^w \cosh(z\cos\theta) \sin^{2\nu}\theta \, d\theta, \\[2mm] \mathbf{L}_\nu(w, z) &= \frac{2z^\nu}{A_\nu} \int_0^w \sinh(z\cos\theta) \sin^{2\nu}\theta \, d\theta, \\[2mm] &\Re(\nu + 1/2) > 0. \end{aligned}\right\} \qquad (28.10)$$

The function $I_\nu(w, z)$ is often called the *partial Bessel function of the purely imaginary argument*, and $\mathbf{L}_\nu(w, z)$ is often called the *partial Struve function of the purely imaginary argument.*, and $\mathbf{L}_\nu(w, z)$

The function $E_\nu^+(-iw, z)$ goes into the Hankel function of the first kind as $w \to -i\infty$; for this reason the function

$$\mathcal{H}_\nu(w, z) = E_\nu^+(-iw, z) \qquad (28.11)$$

is called the *partial Hankel function*. Obviously,

$$\mathcal{H}_\nu(\beta, z) = -\frac{2ie^{-i\pi\nu} z^\nu}{A_\nu} \int_0^\beta e^{iz\cosh u} \sinh^{2\nu} u \, du \qquad (28.12)$$

for real $\cosh\beta > 1$ and $\Re(\nu + 1/2) > 0$.

Then, passing to the complex argument iz, we come to the *partial Macdonald function*

$$\begin{aligned} K_\nu(w, z) &= \frac{i\pi}{2} e^{i\nu\pi/2} E_\nu^+(-iw, iz) \\[2mm] &= \frac{\pi z^\nu}{A_\nu} \int_1^{\cosh w} e^{-zt}(t^2 - 1)^{\nu - 1/2} \, dt, \qquad (28.13) \\[2mm] &\Re(\nu + 1/2) > 0. \end{aligned}$$

Let us now give the recurrence relations between the partial cylindrical functions. For this purpose, we should act on the functions $E_\nu^+(w, z)$ by the following operators:

$$\left.\begin{aligned} L_\nu(f_\nu) &\equiv f_{\nu-1} + f_{\nu+1} - \frac{2\nu}{z} f_\nu, \\[2mm] L'_\nu(f_\nu) &\equiv f_{\nu-1} - f_{\nu+1} - 2\frac{d}{dz} f_\nu, \end{aligned}\right\} \qquad (28.14)$$

which implies the following relations:

$$L_\nu[E_\nu^+(w,z)] = 2z^\nu \sin^{2\nu-1} w \left[\frac{\cos w}{A_\nu z} + i\frac{\sin^2 w}{A_{\nu+1}} \right] e^{iz\cos w},$$
$$L_\nu'[E_\nu^+(w,z)] = 2z^\nu \sin^{2\nu-1} w \left[\frac{\cos w}{A_\nu z} - i\frac{\sin^2 w}{A_{\nu+1}} \right] e^{iz\cos w}. \Bigg\} \quad (28.15)$$

From these formulae we immediately obtain the ordinary recurrence relations for the Hankel functions of the first kind for $\Im z > 0$ and $w \to -i\infty$. We have

$$E_\nu^+(w,z) = J_\nu(w,z) + i\mathbf{H}_\nu(w,z),$$
$$e^{iz\cos w} = \cos(z\cos w) + i\sin(z\cos w).$$

With the help of these formulae, we obtain from (28.15)

$$L_\nu[J_\nu(w,z)] = g_\nu^-(w,z);$$
$$L_\nu'[J_\nu(w,z)] = g_\nu^+(w,z), \quad (28.16)$$

where

$$g^\pm(w,z) = 2z^\nu \sin^{2\nu-1} w \left[\frac{\cos w}{zA_\nu} \cos(z\cos w) \pm \frac{\sin^2 w}{A_{\nu+1}} \sin(z\cos w) \right].$$

Let us give similar formulae for the partial Struve functions

$$L_\nu[\mathbf{H}_\nu(w,z)] = f_\nu^+(w,z),$$
$$L_\nu'[\mathbf{H}_\nu(w,z)] = f_\nu^-(w,z),$$

where

$$f^\pm(w,z) = 2z^\nu \sin^{2\nu-1} w \left[\frac{\cos w}{zA_\nu} \sin(z\cos w) \pm \frac{\sin^2 w}{A_{\nu+1}} \cos(z\cos w) \right].$$

An analogue of the partial cylindrical functions of a half-integer argument are the cylindrical functions of a half-integer argument, for instance:

$$E_{1/2}^+(w,z) = 2\sqrt{2/\pi z}(e^{iz} - e^{iz\cos w})/2i.$$

All the functions considered above are called the *partial cylindrical functions in the form of Poisson*. One can also consider the partial cylindrical functions associated with other integral representations of the Bessel functions.

The *partial cylindrical functions in the form of Bessel* are defined in the following way:

$$\varepsilon_\nu(w,z) = \frac{1}{\pi i} \int_0^w e^{z\sinh t - \nu t} \, dt. \quad (28.17)$$

Similarly to the ordinary cylindrical functions, these functions for $w = 2\pi i$ coincide with the Bessel functions only for integer values of the index: $\varepsilon_n(2\pi i, z) = 2J_n(z)$. This also implies that the partial functions in the form of Bessel and the functions in the form of Poisson belong, as is considered in detail in [2], to different classes of functions. In this case, one can also introduce the partial Bessel functions of a purely imaginary argument in the form of Bessel, the partial Hankel functions in the form of Bessel and so on.

The partial cylindrical functions in the form of Bessel satisfy the inhomogeneous differential equation

$$\nabla_\nu \varepsilon_\nu(w, z) = [(\nu + z \cosh w) \exp(z \sinh w - \nu w) - (\nu + z)]/\pi i. \qquad (28.18)$$

An one-to-one connection between the functions $E_\nu^+(w, z)$ and $\varepsilon_\nu(w, z)$ only exists for integer indices.

Consider the partial cylindrical functions in the Sonin–Schlafli form.

The Sonin–Schlafli integral representation has the form

$$J_\nu(z) = \frac{1}{2\pi i} \left(\frac{z}{2}\right)^\nu \int_{c-i\infty}^{c+i\infty} t^{-\nu-1} \exp\left(t - \frac{z^2}{4t}\right) dt, \qquad (28.19)$$

where $c > 0$, $\Re \nu > -1$.

The contour of integration is open, its ends are two infinite points, located in the right half-plane and whose abscissae are equal to each other[9]. If we fulfil the integration along the contour, whose beginning and end is given by complex numbers p and q, then, obviously, in the cases, when $p = c - i\infty$, $q = c + i\infty$, we obtain, as before, the Bessel function; in all other cases this integral represents a function, which can be called the *partial cylindrical function in the Sonin–Schlafli form*

$$S_\nu(p, q; z) = \frac{1}{2\pi i} \left(\frac{z}{2}\right)^\nu \int_p^q t^{-\nu-1} \exp\left(t - \frac{z^2}{4t}\right) dt. \qquad (28.20)$$

Setting $p = z/2$, $q = (z/2) \exp w$ and making the change of the variable of integration $t = (z/2) \exp u$, we obtain

$$S_\nu\left(\frac{z}{2}, \frac{z}{2} e^w; z\right) = \frac{1}{2\pi i} \int_0^w e^{z \sinh u - \nu u} \, du$$

$$= \frac{1}{2} \varepsilon_\nu(w, z). \qquad (28.21)$$

This shows that the functions $S_\nu(p, q; z)$ are more general than the functions $\varepsilon_\nu(w, z)$, which were considered above.

Acting on $S_\nu(p, q; z)$ by the Bessel operator, we can obtain the inhomogeneous Bessel equation

$$\nabla_\nu S_\nu(p, q; z) = \Psi(z, q) - \Psi(z, p), \qquad (28.22)$$

where

$$\Psi(z, x) = \frac{z}{\pi i} \left(\frac{z}{2x}\right)^{\nu+1} \exp\left(x - \frac{z^2}{4x}\right),$$

which is satisfied by the function $S_\nu(p, q; z)$.

---

[9]See [7], where the contour of integration, similar to (28.19), is a straight line parallel to the imaginary axis.

The integral representations of $S_\nu(p, q; z)$ directly allow one to obtain the recurrence formulae:

$$\left.\begin{aligned} S_{\nu-1} + S_{\nu+1} - \frac{2\nu}{z} S_\nu &= \frac{2}{z^3} [q\Psi(z, q) - p\Psi(z, p)], \\ S_{\nu-1} - S_{\nu+1} - 2\frac{\partial S_\nu}{\partial z} &= \frac{2}{z^3} [q\Psi(z, q) - p\Psi(z, p)]. \end{aligned}\right\} \qquad (28.23)$$

The integral representations of $S_\nu(p, q; z)$, based on obtaining of a solution of the inhomogeneous differential Bessel equation, has the following form:

$$S_\nu(p, q; z) = \frac{1}{2i} Y_\nu(z)[Q_\nu(q, z) - Q_\nu(p, z)]$$
$$+ \frac{1}{2i} J_\nu(z)[P_\nu(q, z) - P_\nu(p, z)], \qquad (28.24)$$

where

$$Q_\nu(x, t) = (2x)^{-\nu-1} e^x \int_0^z t^{\nu+1} J_\nu(t) \exp\left(-\frac{t^2}{4x}\right) dt,$$

$$P_\nu(x, t) = (2x)^{-\nu-1} e^x \int_z^{\infty \exp i\lambda} t^{\nu+1} Y_\nu(t) \exp\left(-\frac{t^2}{4x}\right) dt,$$

$\Re\nu > -1$; for a fixed x the inequality $|2\lambda - \arg x| < \pi/2$ should hold.

The functions $Q_\nu(x, t)$ and $P_\nu(x, t)$ have independent significance and are called the partial Weber integrals from the cylindrical functions of the first and the second kind. A more detailed investigation of the properties of these integrals is given in [2].

The representation of the partial cylindrical functions via the partial Weber integrals intuitively shows a natural connection between the Lommel functions of two variables and the partial cylindrical functions. Comparing the partial Weber integral with the formulae for the Lommel functions given in Section 27, we can obtain the following relations:

$$\left.\begin{aligned} Q_\nu(x, z) &= i^{\nu+1} \left[U_{\nu+1}\left(\frac{z^2}{2ix}, z\right) + iU_{\nu+2}\left(\frac{z^2}{2ix}, z\right)\right] \exp\left(x - \frac{z^2}{4x}\right), \\ \tilde{Q}_\nu(x, z) &= \left[V_{\nu+1}\left(\frac{z^2}{2x}, z\right) + V_{\nu+2}\left(\frac{z^2}{2x}, z\right)\right] \exp\left(x - \frac{z^2}{4x}\right), \end{aligned}\right\} \qquad (28.25)$$

where

$$\tilde{Q}_\nu(x, z) = (2x)^{-\nu-1} e^x \int_0^z t^{\nu+1} I_\nu(t) \exp(-t^2/(4x))\, dt.$$

29. Asymptotic expansion of Bessel functions

As a rule, practical calculations connected with the application of Bessel functions are based on the use of tables of these functions. In some cases, one can perform calculations, based on using the generalized power series, which have been given at the beginning of the book, as well as series containing the factor $\ln z$ in some cases; these series in increasing powers of the argument are convenient for the calculations only for small values of the argument. If the argument is large enough, then a question arises concerning the construction of approximate solutions to the

Bessel equation, which are available for large values of the argument. The problem formulated is connected with the construction of the asymptotic expansions; when solving it, one should operate with divergent series, for which, however, the property holds that they are convenient for the calculations in a certain domain of the values and admit a simple estimate of the error.

From the beginning, we should emphasize the difference in the solution to problems of different classes here, depending on the behavior of the argument and the index ν of the cylindrical function:

1. $z \to \infty$, z is fixed (the Hankel expansion);
2. the argument and the index increase infinitely; moreover, $z - \nu = \gamma$, where γ is a constant;
3. the ration z/ν is fixed, $z \to \infty$, $z/\nu = \cosh b$, $b > 0$.

There is most interest in the first problem, whose solution is the most simple.

In order to obtain the asymptotic expansions for large values of the argument and a fixed index, one can use the generalization of the integral Poisson representation given by Hankel; these expansions are sometimes called the Hankel expansion.

Let us roughly describe the process of obtaining the asymptotic expansion, which is given by Hankel and slightly modified by Watson.

The integral representation of the Hankel function of the first kind

$$H_\nu^{(1)} = \left(\frac{2}{\pi z}\right)^{1/2} \frac{\exp i(z - \pi\nu/2 - \pi/4)}{\Gamma(\nu + 1/2)}$$

$$\times \int_0^{\infty \exp i\beta} e^{-u} u^{\nu - 1/2} \left(1 + \frac{iu}{2z}\right)^{\nu - 1/2} du \quad (29.1)$$

holds for

$$\pi/2 > \beta > -\pi/2,$$

$$-\pi/2 + \beta < \arg z < 3\pi/2 + \beta,$$

$$\Re(\nu + 1/2) > 0.$$

Obtaining the solution is based on the binomial expansion with the remainder of the factor $(1 + iu/(2z))^{\nu - 1/2}$, which is present in the integrand. After introducing this expansion into the integral, integrating term-by-term and estimating the remainder, we obtain

$$H_\nu^{(1)}(z) \sim (2/(\pi z))^{1/2} \exp i(z - \pi\nu/2 - \pi/4) \sum_{m=0}^{\infty} \frac{(-1)^m (\nu, m)}{(2iz)^m}, \quad (29.2)$$

where

$$(\nu, m) = \frac{\Gamma(\nu + m + 1/2)}{m! \Gamma(\nu - m + 1/2)}, \quad -\pi < \arg z < 2\pi.$$

The following formula can be obtained in a similar way

$$H_\nu^{(2)}(z) \sim (2/(\pi z))^{1/2} \exp[-i(z - \nu\pi/2 - \pi/4)] \sum_{m=0}^{\infty} \frac{(\nu, m)}{(2iz)^m}, \quad (29.3)$$

for $-2\pi < \arg z < \pi.$

Keeping only the first term of the asymptotic expansion, which is called the *asymptotic representation*, we obtain

$$H_\nu^{(1)}(z) \sim (2/\pi z)^{1/2} \exp i(z - \nu\pi/2 - \pi/4),$$
$$H_\nu^{(2)}(z) \sim (2/\pi z)^{1/2} \exp[-i(z - \nu\pi/2 - \pi/4)].$$

Having the asymptotic expansions of the Hankel functions, one can by the formulae of Section 3 easily obtain the expansions of the Bessel and Neumann functions which are available for $|\arg z| < \pi$:

$$\left.\begin{aligned}
J_\nu(z) &\sim (2/\pi z)^{1/2} \left[\cos(z - \nu\pi/2 - \pi/4) \sum_{m=0}^{\infty} \frac{(-1)^m (\nu, 2m)}{(2z)^{2m}} \right. \\
&\qquad \left. - \sin(z - \nu\pi/2 - \pi/4) \sum_{m=0}^{\infty} \frac{(-1)^m (\nu, 2m+1)}{(2z)^{2m+1}} \right], \\
Y_\nu(z) &\sim (2/\pi z)^{1/2} \left[\sin(z - \nu\pi/2 - \pi/4) \sum_{m=0}^{\infty} \frac{(-1)^m (\nu, 2m)}{(2z)^{2m}} \right. \\
&\qquad \left. + \cos(z - \nu\pi/2 - \pi/4) \sum_{m=0}^{\infty} \frac{(-1)^m (\nu, 2m+1)}{(2z)^{2m+1}} \right].
\end{aligned}\right\} \quad (29.4)$$

Using the relation between the Hankel function and the Macdonald function, we find

$$K_\nu(z) \sim \left(\frac{\pi}{2z}\right)^{1/2} e^{-z} \sum_{m=0}^{\infty} \frac{(\nu, m)}{(2z)^m}, \quad (29.5)$$

$$|\arg z| < \frac{3\pi}{2}.$$

The dependence between the Bessel functions and the modified Bessel functions implies the following formulae:

$$I_\nu(z) \sim \frac{e^z}{(2\pi z)^{1/2}} \sum_{m=0}^{\infty} \frac{(-1)^m (\nu, m)}{(2z)^m} + \frac{\exp[-z + (\nu + 1/2)\pi i]}{(2\pi z)^{1/2}} \sum_{m=0}^{\infty} \frac{(\nu, m)}{(2z)^m} \quad (29.6)$$

for $3\pi/2 > \arg z > -\pi/2$;

$$I_\nu(z) \sim \frac{e^z}{(2\pi z)^{1/2}} \sum_{m=0}^{\infty} \frac{(-1)^m (\nu, m)}{(2z)^m} + \frac{\exp[-z - (\nu + 1/2)\pi i]}{(2\pi z)^{1/2}} \sum_{m=0}^{\infty} \frac{(\nu, m)}{(2z)^m} \quad (29.7)$$

for $-3\pi/2 < \arg z < \pi/2$[10].

Just in the same way, one can obtain the asymptotic expansions for the real and imaginary parts of functions of the complex argument.

The order of the remainder term of the expansion for the Hankel functions is the same as that of the first terms neglected. For the cylindrical functions of the first and the second kind some sharper estimates hold: for positive z and real ν, the remainder is numerically less that the first term neglected and has the same sign; if for real ν and z the value $2z - m + 1/2$ is small with respect to z (m is the number of the first term neglected), then the remainder is approximately equal to half of the first term neglected.

---

[10] It seems that the formulae (29.6) and (29.7) for $-\pi/2 < \arg z < \pi/2$ are contradictory; this is connected with the Stokes Phenomenon considered in·[7].

The asymptotic expansions can immediately be obtained from consideration of the differential Bessel equation.

If for large values of the argument z we introduce the asymptotic expansion of the Bessel functions into the integral representations of the functions contiguous to the Bessel functions, then the integration immediately gives the asymptotic expansions for the functions indicated.

30. On Bessel functions with a large index

Before passing to a brief exposition of the properties of the Bessel functions for large indices, we make several remarks connected with physical representations, which in a sense clarify the statement of the problem.

First, we restrict ourselves by consideration of Bessel functions $J_n(x)$ assuming that the index is a positive integer and x is a real number. As was shown in the introduction, the problem of free oscillations of a circular uniform membrane with a fixed contour leads to consideration of the transcendental equation $J_n(\lambda_k R) = 0$. Here n is the index of the Bessel function which is equal to the number of the nodal diameters of the form of the eigen-oscillations; k is the ordinal number of the root, which is equal to the number of nodal circumferences, including the boundary one; R is the radius of this circumference, which, without loss of generality, can be taken equal to one.

First, suppose that n is small enough, and k is a large number, and consider the domain bounded by two adjacent nodal circumferences with the radii ρ_l and ρ_{l+1} ($n \ll l+1 < k$) and two adjacent nodal radii of the membrane.

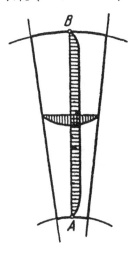

Figure 11

Obviously, the form of the oscillations of the membrane for such a domain is, except for the regions adjacent to the radii, close to the form of the oscillations of a membrane, whose nodal lines are parallel straight lines, the distance between which is equal to $\rho_{l+1} - \rho_l$. Thus, the z-coordinates of the form of the oscillations vary along the radius according to a nearly sinusoidal law. If we take the opposite case, assuming that k is small with respect to n, then the part of the membrane under consideration will be extended in the direction of the radii and the form of the oscillations in this case is such that along AB (Fig. 11) its z-coordinates only vary fast near the points A and B; in the remaining part of the segment they vary sufficiently slowly. Obviously, in the case which is considered in this section (along with the cases when the difference between the index and the argument is small), we should start from the asymptotic representations which are different from those obtained in Section 29. For non-integer values of the index, we should consider a membrane which has the form of a circular sector. The problem of a membrane with a hole leads to similar problems for a function of the second kind.

Here we shall only give the isolated, bare facts related with this rather complicated question, which is considered in detail in the monograph by Watson [7].

The simplest form has the *solution of Meissel*.

We transform the Bessel equation with the help of the change

$$J_\nu(\nu z) = \frac{\nu^\nu}{\Gamma(\nu+1)} \exp\left\{\int^z u(z)\, dz\right\}$$

to the form

$$z^2[u'(z) + \{u(z)\}^2] + zu(z) - \nu^2(1 - z^2) = 0.$$

We represent the solutions of this equation in the form of a series in decreasing powers of ν

$$u = \nu u_0 + u_1 + \frac{u_2}{\nu} + \frac{u_3}{\nu^2} + \dots;$$

here $u_0, u_1, u_2, \dots$ are functions of z. After substituting the series into the equation, comparing the coefficients of the same powers and integrating, we obtain the following result:

$$J_\nu(\nu z) = \frac{(\nu z)^\nu \exp\{\nu\sqrt{1 - z^2}\} \exp(-V_\nu)}{e^\nu \Gamma(\nu+1)(1 - z^2)^{1/4}\{1 + \sqrt{1 - z^2}\}^\nu}, \qquad (30.1)$$

where

$$V_\nu = \frac{1}{24\nu}\left\{\frac{2 + 3z^2}{(1 - z^2)^{3/2}} - 2\right\} - \frac{4z^2 + z^4}{16\nu^2(1 - z^2)^3}$$

$$- \frac{1}{5760\nu^3}\left\{\frac{16 - 1512z^2 - 3654z^4 - 375z^6}{(1 - z^2)^{9/2}} - 16\right\} + \dots$$

The first approximation, obtained from (30.1),

$$J_\nu(\nu z) \sim \frac{z^\nu \exp\left[\nu\sqrt{1 - z^2}\right]}{(2\pi\nu)^{1/2}(1 - z^2)^{1/4}\left[1 + \sqrt{1 - z^2}\right]^\nu}$$

is the *Carlini formula*.

If z is real and greater than one, then the expansion (30.1) is invalid. In this case, the second Meissel expansion (see [7]) is used; keeping only the principal terms in it, we obtain the following approximate formulae for $J_\nu(x)$ and $Y_\nu(x)$:

$$\left.\begin{aligned}
J_\nu(x) &\sim \left(\frac{2}{\pi\sqrt{x^2 - \nu^2}}\right)^{1/2}\cos\theta_\nu, \\
Y_\nu(x) &\sim \left(\frac{2}{\pi\sqrt{x^2 - \nu^2}}\right)^{1/2}\sin\theta_\nu, \\
\theta_\nu &\sim \sqrt{x^2 - \nu^2} - \nu\pi/2 + \nu\arcsin(\nu/x) - \pi/4.
\end{aligned}\right\} \qquad (30.2)$$

Let us go to the asymptotic expansions for the cylindrical functions such that the difference between the argument and the index is a constant number.

Let us give a result related to the case $z \to \infty$, $z - \nu = \gamma$, where γ is a constant.

$$H_{z-\gamma}^{(2)} \sim \frac{1}{\pi\sqrt{3}}\sum_{l=0}^{\infty}\left(\frac{l+1}{3}\right)\left(\frac{6}{z}e^{2\pi i}\right)^{(l+1)/3}\Gamma\left(\frac{l+1}{3}\right)B_l(\gamma), \qquad (30.3)$$

$$\binom{l+1}{3} = \frac{2}{\sqrt{3}} \sin \frac{2\pi(l+1)}{3}, \ B_l(\gamma) \text{ is a polynomial of degree } l; \ B_0(x) = 1,$$

$B_1(x) = x, \ B_2(x) = x^2/2 - 1/20, \ B_3(x) = \frac{1}{6}x^3 - \frac{1}{15}x, \ \dots \ ^{11}.$

If the difference $z - \nu$ is a quantity of order $z^{1/3}$ or $\nu^{1/3}$ as $z \to \infty$, then we can use the approximate Nickolson formulae

$$\left.\begin{aligned}
e^{-\pi i/6} H^{(1)}_{z-\gamma}(z) &\sim \left(\frac{2\gamma}{3z}\right)^{1/2} H^{(1)}_{1/3}\left(\frac{(2\gamma)^{3/2}}{3z^{1/2}}\right), \\
e^{\pi i/6} H^{(2)}_{z-\gamma}(z) &\sim \left(\frac{2\gamma}{3z}\right)^{1/2} H^{(2)}_{1/3}\left(\frac{(2\gamma)^{3/2}}{3z^{1/2}}\right).
\end{aligned}\right\}$$

(30.4)

The problem on asymptotic representations of the cylindrical functions has been considered in detail by Schöbe [82].

In conclusion of this section we give the following formula:

$$2J_\nu(\nu \operatorname{sech} b) \sim \frac{2}{\sqrt{\pi}} e^{\nu(\tanh b - b)} \sum_{l=0}^{\infty} \frac{P_l(\coth^2 b)}{(2\nu \tanh b)^{l+1/2}}$$

(30.5)

as $\nu \to \infty$, $b > 0$, and

$$P_0(\xi) = 1,$$

$$P_1(\xi) = \frac{1}{15}\xi,$$

$$P_2(\xi) = \frac{\xi^5}{7200} + \frac{13}{1260}\xi^2,$$

$$P_3(\xi) = \frac{283}{9\,072\,000}\xi^6 + \frac{463}{226\,800}\xi^3 + \frac{1}{900}, \ \dots \ ^{12}.$$

The series (30.3) and (30.5) are called the Debye series.

The theory of asymptotic expansions of Bessel functions for large values of the index is expounded in detail in [7]. A summary of different asymptotic expansions is given in [4]; a rather detailed information concerning the asymptotic expansion is also contained in the reference book by Janke, Emde, and Lesch [61].

---

^{11}in [82] the question concerning the construction of the polynomials $B_l(\gamma)$ has been considered in detail and tables of these polynomials have been given.

^{12}The construction of the polynomials $P_l(\xi)$ is considered in [82].

Part 2

Applications of Bessel functions

CHAPTER 3

Problems of the theory of plates and shells

31. Oscillations and stability of a circular plate

1. Consider the problem of oscillations of a circular plate. We assume that the deflection of the plate w is small with respect to its thickness h and the thickness, in turn, is small with respect to the radius of the plate. We locate the polar coordinate system r, φ in the median plane and take the center of the circumference along which the side surface of the plate intersect the median plane as the origin.

Let D denote the flexural rigidity of the plate, h denote its thickness and γ denote its density.

First, we consider free oscillations of the plate. Suppose that the plate is fixed along the contour whose radius will be denoted by b.

For $r = b$ we have $W = 0$ and $\frac{\partial W}{\partial r} = 0$, where $W(r, \varphi, t)$ is the dynamic deflection of the plate.

The differential equation of the deflection surface of the plate has the following form

$$D\nabla^2\nabla^2 W + \gamma h \frac{\partial^2 W}{\partial t^2} = 0. \tag{31.1}$$

In order to integrate the equation (31.1) we use the method of separation of variables, taking a particular solution in the form

$$W(r, \varphi, t) = w(r, \varphi)F(t). \tag{31.2}$$

Substituting (31.2) into the equation (31.1), we obtain

$$\frac{D}{\gamma h} \frac{\nabla^2\nabla^2 w}{w} = -\frac{d^2 F}{dt^2} \bigg/ F. \tag{31.3}$$

Since the left-hand side of the equation (31.3) represents a function only of the variables r, φ, and the right-hand side depends only on the time t, it is obvious that each of the quantities $D\nabla^2\nabla^2 w/\gamma h w$ and $-\frac{d^2 F}{dt^2}/F$ should be constant. Denoting this constant by ω^2, we have

$$\frac{d^2 F}{dt^2} = -\omega^2 F,$$

hence,

$$F = A\sin(\omega t + \varphi_0).$$

Thus, ω is the angular frequency of the oscillations. The function w is called the *form of the oscillations*. It satisfies the following differential equation

$$\nabla^2\nabla^2 w - \lambda^4 w = 0, \tag{31.4}$$

where

$$\lambda^4 = w^2 \gamma h / D.$$

Let us pass from the variable r to the dimensionless quantity $\xi = \lambda r$ (below we will call ξ by the *reduced distance*), then the equation (31.4) takes the form

$$\left(\frac{\partial^2}{\partial \xi^2} + \frac{1}{\xi} \frac{\partial}{\partial \xi} + \frac{1}{\xi^2} \frac{\partial^2}{\partial \varphi^2} \right)^2 w - w = 0. \tag{31.5}$$

Let us denote

$$\nabla_\xi^2 = \frac{\partial^2}{\partial \xi^2} + \frac{1}{\xi} \frac{\partial}{\partial \xi} + \frac{1}{\xi^2} \frac{\partial^2}{\partial \varphi^2}.$$

We seek a solution of the equation (31.5) in the form

$$w(r, \varphi) = w_1(r, \varphi) + w_2(r, \varphi),$$

where the functions $w_1(r, \varphi)$ and $w_2(r, \varphi)$ satisfy the equations

$$\nabla_\xi^2 w_1 + w_1 = 0, \tag{31.6}$$

$$\nabla_\xi^2 w_2 - w_2 = 0. \tag{31.7}$$

We again use the method of separation of variables, assuming that the particular solutions of the equation (31.6) and (31.7) have the form

$$w_1 = R_1(r) \Phi_1(\varphi), \tag{31.8}$$

$$w_2 = R_2(r) \Phi_2(\varphi). \tag{31.9}$$

Introducing (31.8) and (31.9) into (31.6) and (31.7), respectively, we obtain, omitting the indices of the functions $R(r)$ and $\Phi(\varphi)$,

$$(\xi^2 R'' + \xi R' \pm \xi^2 R)/R = -\Phi''/\Phi.$$

Since the left-hand side depends only on ξ and the right-hand side depends only on φ, we have

$$\left. \begin{array}{l} (\xi^2 R'' + \xi R' \pm \xi^2 R)/R = \varepsilon^2, \\ -\Phi''/\Phi = \varepsilon^2, \end{array} \right\} \tag{31.10}$$

where ε^2 is a constant. From the equation (31.10) we have

$$\Phi = \sin(\varepsilon \varphi + \alpha),$$

where α is a constant. Since the value of a function should not be changed after adding an angle which is an integer multiple of 2π to φ, ε can only be equal to an integer n. Thus, we have the following equations for the function R:

$$\xi^2 R'' + \xi R' - (n^2 \pm \xi^2) R = 0. \tag{31.11}$$

Each of the equations (31.11) represents the Bessel equation, hence,

$$R_{1n} + R_{2n} = C_{1n} J_n(\xi) + C_{3n} Y_n(\xi) + C_{2n} I_n(\xi) + C_{4n} K_n(\xi).$$

The form of the oscillations can be represented in the following form

$$w_n(r, \varphi) = [C_{1n} J_n(\xi) + C_{3n} Y_n(\xi) + C_{2n} I_n(\xi) + C_{4n} K_n(\xi)] \sin(n\varphi + \alpha_n).$$

The constants C and α as well as the angular frequency w, on which λ, hence, and $\xi = \lambda r$ depend, should be determined.

First, consider in great detail the most important special case: the axially symmetric oscillations of a circular plate. The form of the oscillations will have the form

$$w = C_1 J_0(\xi) + C_3 Y_0(\xi) + C_2 I_0(\xi) + C_4 K_0(\xi).$$

If the plate is solid and there is no support at the center of the plate, then the deflection at $\xi = 0$ should be finite. The expansions of the functions $Y_0(\xi)$ and $K_0(\xi)$ into power series imply that these functions are infinite as $\xi \to 0$. However, this fact does not imply that $C_3 = 0$ and $C_4 = 0$. In the vicinity of the point $\xi = 0$, these functions can be represented in the following form[1]

$$Y_0(\xi) = \frac{2}{\pi}\left(1 - \frac{\xi^2}{4}\right)\ln\xi + \frac{2\gamma}{\pi} + \dots, \tag{31.12}$$

$$K_0(\xi) = -\left(1 + \frac{\xi^2}{4}\right)\ln\xi - \gamma + \dots. \tag{31.13}$$

The ellipses denote the omitted terms which as well as their derivatives which appear in the result of the fulfilled differentiation tend to zero as $\xi \to 0$. Hence, after omitting these terms, the results of the calculation will not be changed. Let us set $C_4 = 2C_3/\pi$ and form the sum of the second and the fourth terms

$$\tilde{w}_0 = C_3[Y_0(\xi) + 2K_0(\xi)/\pi]. \tag{31.14}$$

For ξ close to zero, this sum has the form

$$\tilde{w}_0 = -C_3\xi^2 \ln\xi/\pi + \dots.$$

Thus, this part of the solution also tends to zero as $\xi \to 0$. Let us show that from the mechanical viewpoint the solution (31.14) contains a singularity of the type of a concentrated load.

For this purpose we find the transversal force along a circumference of a small radius ξ, calculate the sum of the transversal forces along this circumference and pass to the limit as $\xi \to 0$. Obviously, the conditions of the equilibrium of the part of the plate which is bounded by the circumference of the radius ξ/λ_1, imply that this limit is equal to the concentrated load P, acting on the plate (Fig. 12), taken with the opposite sign. Thus, in order to determine the constant, we need to introduce the expression (31.14) into the following equation:

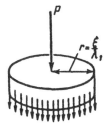

Figure 12

$$\lim_{\xi \to 0}\left[-D\lambda_1^3 \frac{d}{d\xi}\left(\frac{d^2 w_0}{d\xi^2} + \frac{1}{\xi}\frac{dw_0}{d\xi}\right)\right]2\pi\frac{\xi}{\lambda_1} = -P. \tag{31.15}$$

When making the calculation, it is useful to use the relations

$$\frac{d^2 Y_0(\xi)}{d\xi^2} + \frac{1}{\xi}\frac{dY_0(\xi)}{d\xi} = -Y_0(\xi), \tag{31.16}$$

$$\frac{d^2 K_0(\xi)}{d\xi^2} + \frac{1}{\xi}\frac{dK_0(\xi)}{d\xi} = K_0(\xi). \tag{31.17}$$

---

[1] See pages 11, 12.

After introducing the expression (31.14) into the equation (31.15) and using the equations (31.16) and (31.17), we obtain

$$\lim_{\xi \to 0} \left\{ -DC_3 \lambda_1^2 \left[-Y_0'(\xi) + \frac{2}{\pi} K_0'(\xi) \right] 2\pi \xi \right\} = -P. \qquad (31.18)$$

Using the expressions (31.12) and (31.13), after differentiating and passing to the limit, for the determination of C_3 we obtain the relation $-8DC_3 \lambda_1^3 = P$, hence $C_3 = -P/(8D\lambda_1^2)$.

Thus, the expression (31.14) can be written in the following form:

$$\tilde{w}_0 = -\frac{P}{8D\lambda_1^2} \left[Y_0(\xi) + \frac{2}{\pi} K_0(\xi) \right]. \qquad (31.19)$$

If the plate has a support at the center, then the function $w(\xi)$ should be written in the form

$$w = C_1 J_0(\xi) + C_2 I_0(\xi) - \frac{P}{8D\lambda_1^2} \left[Y_0(\xi) + \frac{2}{\pi} K_0(\xi) \right], \qquad (31.20)$$

where P is the support reaction.

In the case, when there is neither a support, nor a concentrated load at the center, we have

$$w(\xi) = C_1 J_0(\xi) + C_2 I_0(\xi).$$

Since the plate is fixed along the contour, we have $w = 0$ and $dw/d\xi = 0$ at $\xi = \beta = \lambda b$.

From the boundary conditions we obtain the following equations which are homogeneous with respect to C_1 and C_2:

$$C_1 J_0(\beta) + C_2 I_0(\beta) = 0,$$
$$C_1 J_0'(\beta) + C_2 I_0'(\beta) = 0.$$

This system has a non-trivial solution only if the determinant composed of the coefficients of the indeterminates is equal to zero.

From this fact we obtain the following equation:

$$J_0(\beta) I_0'(\beta) - J_0'(\beta) I_0(\beta) = 0. \qquad (31.21)$$

Obviously, if the general case of the oscillations is considered, then with each form, which has n nodal diameters, the following expression is associated:

$$w_n(\xi, \varphi) = [C_{1n} J_n(\xi) + C_{2n} I_n(\xi)] \sin(n\varphi + \alpha).$$

In this case we have

$$J_n(\beta) I_n'(\beta) - J_n'(\beta) I_n(\beta) = 0. \qquad (31.22)$$

The roots of the equations (31.21) and (31.22) define the values of β. Knowing these values, we can easily determine the frequencies of the oscillations.

If β_{in} denotes the root of the transcendental equation (31.22), where i is the ordinal number of the root, then

$$\omega_{in} = \sqrt{D\lambda_{in}^4/(\gamma h)} = \beta_{in}^2 \sqrt{D/(\gamma h)} \Big/ b^2.$$

The roots of the equation (31.22) are given in different reference books.

One can easily consider the oscillations of a simply supported plate. In this case the boundary conditions for $r = b$ have the form: $w = 0$,

$$\frac{\partial^2 w}{\partial r^2} + \frac{\sigma}{r}\frac{\partial w}{\partial r} + \frac{\sigma}{r^2}\frac{\partial^2 w}{\partial \varphi^2} = 0,$$

where σ is Poisson's ratio. The second of these boundary conditions can be written in the following form:

$$\frac{\partial^2 w}{\partial r^2} + \frac{\sigma}{r}\frac{\partial w}{\partial r} + \frac{\sigma}{r^2}\frac{\partial^2 w}{\partial \varphi^2} = \nabla^2 w + (\sigma - 1)\frac{1}{r}\frac{\partial w}{\partial r} = 0. \qquad (31.23)$$

Since

$$\nabla^2[J_n(\lambda_n r)\cos n\varphi] = -\lambda_n^2 J_n(\lambda_n r)\cos n\varphi,$$
$$\nabla^2[I_n(\lambda_n r)\cos n\varphi] = \lambda_n^2 I_n(\lambda_n r)\cos n\varphi,$$

from (31.23) we have

$$-\lambda_{in}[C_1 J_n(\lambda_{in}b) - C_2 I_n(\lambda_{in}b)] - (1-\sigma)\left\{[]C_1 J_n'(\lambda_{in}b) + C_2 I_n'(\lambda_{in}b)]/b\right\} = 0.$$

Considering this equation together with the equation

$$C_1 J_n(\lambda_{in}b) - C_2 I_n(\lambda_{in}b) = 0,$$

we obtain a transcendental equation, which determines the frequencies of the free oscillations.

Let us consider the problem of free oscillations of a circular plate with a point support at the center. We shall use the method, which is called the *method of forces* in structural mechanics,, assuming the amplitude of the reaction of the support by an indeterminate X. We imagine the support to be omitted and its action to be replaced by the force $X\sin\omega t$. We consider the forced oscillations which are induced by this force. Suppose that the circular plate is fixed along the contour. Then, with the help of the results obtained above and using the formula (31.20), we obtain the equation of the form of the oscillations in the following form:

$$w = w_0 + w_k = C_{10}J_0(\xi) + C_{20}I_0(\xi) - \frac{X}{8D\lambda^3}\left[Y_0(\xi) + \frac{2}{\pi}K_0(\xi)\right],$$

where $\lambda^4 = \omega^2\gamma h/D$, and A_0 and B_0 are determined by the boundary conditions on the contour. Let us write the formula for the amplitude of the deflection at the center of the plate

$$w_0(0) = C_{10}J_0(0) + C_{20}I_0(0) - \frac{X}{8D\lambda^3}\left[Y_0(\xi) + \frac{2}{\pi}K_0(\xi)\right]\Big|_{\xi\to 0}. \qquad (31.24)$$

Since the plate has a support at the center, $w(0) = 0$. We simplify the expression (31.24), taking into account that $J_0(0) = 1$, $I_0(0) = 1$ and that, as has been shown, the sum $\frac{2}{\pi}K_0(\xi) + Y_0(\xi)$ vanishes at $\xi = 0$. Using these facts, we rewrite (31.24) and from the boundary conditions for the fixed edge we find C_{10} and C_{20}. After canceling out the factor $X/(8D\lambda^2)$ from the expression obtained, we reduce the expression to the form

$$\frac{I_1(\beta) + Y_0(\beta) + I_0(\beta)Y_1(\beta) + 2/(\pi\beta)}{J_0(\beta)I_1(\beta) + J_1(\beta)I_1(\beta)}$$

$$+ \frac{2[J_1(\beta)K_0(\beta) - J_0(\beta)K_1(\beta)]/\pi + 2/(\pi\beta)}{J_0(\beta)I_1(\beta) + J_1(\beta)I_0(\beta)} = 0.$$

Finally, we deduce from this the following transcendental equation:

$$I_1(\beta)Y_0(\beta) + I_0(\beta)Y_1(\beta) + \frac{2}{\pi}J_1(\beta)K_0(\beta) - \frac{2}{\pi}J_0(\beta)K_1(\beta) + \frac{4}{\pi\beta} = 0. \quad (31.25)$$

Having found the roots β_i of this transcendental equation, we also determine the frequencies of the eigenoscillations

$$\omega_i = \frac{\beta_i^2}{b^2}\sqrt{\frac{D}{\gamma h}},$$

whose forms are such that the reactions at the support arise there.

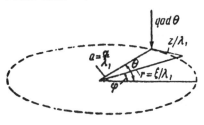

Figure 13

Let us consider the problem of the forced oscillations of a circular plate fixed along the contour, which are stipulated by a loading which is uniformly distributed on a circumference of the radius a which is concentric with the contour. Let q denote the amplitude of the loading $q \sin pt$.

Let us represent the oscillations in the form of a sum of a *basic solution* w_0 and a *compensating solution* w_k.

The basic solution can be sought as the solution of the problem of loading a plate which is not bounded, i.e. an infinite plate. Of course, the basic solution, as a rule, cannot also satisfy the boundary conditions, if the problem on the plates bounded in the plane are considered.

In order to satisfy these conditions, the so-called compensating solution is introduced.

From the above-said it is clear that the compensating solution must satisfy the homogeneous differential equation of the problem (hence, it must contain no singularities in the domain occupied by the plate), and together with the basic solution must satisfy the boundary conditions of the problem.

Suppose, for instance, the deflection and the slopes on the contour of the plate should be equal to zero. The basic solution does not satisfy this condition; therefore, non-zero deflections w_0 and slopes of the deflected plate surface φ_0 corresponding to the basic solution arise on the contour of the plate. The problem can be treated as solved, if we managed to find a compensating solution, which satisfies the conditions $w_k = -w_0$ and $\frac{\partial w_k}{\partial r} = -\varphi_0$ on the contour of the plate. Then the sum of the basic solution and the compensating one gives us the required solution of the problem formulated.

First, we consider the search for the basic solution in detail. For this purpose, we imagine dividing the loading into a series of elementary forces and sum the result of the actions of these elementary loadings. Let us take a point with the coordinates (a, θ) on the circumference, on which the loading is acting. The elementary loading corresponding to a segment of the arc of the length $a\,d\theta$ is equal to

$$qa\,d\theta = qa\,d\theta/\lambda_1, \quad \lambda_1 = \sqrt[4]{p^2\gamma h/D}. \quad (31.26)$$

Let us find w at the point with the coordinates $r = \xi_1/\lambda_1$, φ (Fig. 13). For this purpose we use the formula (31.19), replacing P by the expression (31.26) and ξ by the reduced distance from the point of application of the elementary force to the point of the plate under consideration, which is equal to

$$z = \sqrt{\alpha^2 + \xi^2 - 2\alpha\xi\cos(\theta - \varphi)},$$

and then we integrate the expression obtained. Thus, we have

$$w_0 = -\int_0^{2\pi} \frac{q\alpha}{8D\lambda_1^3} \left[Y_0(z) + \frac{2}{\pi}K_0(z) \right] d\theta$$

$$= -\frac{q\alpha}{8D\lambda_1^3} \left[\int_0^{2\pi} Y_0(\sqrt{\alpha^2 + \xi^2 - 2\alpha\xi\cos(\theta - \varphi)}) \, d\theta \right.$$

$$\left. + \frac{2}{\pi} \int_0^{2\pi} K_0(\sqrt{\alpha^2 + \xi^2 - 2\alpha\xi\cos(\theta - \varphi)}) \, d\theta \right]. \qquad (31.27)$$

In order to calculate the integrals in the expression (31.27), we use the addition formulae for the cylindrical functions, which in the case considered have the form:
for $\xi \le \alpha$

$$Y_0(\sqrt{\alpha^2 + \xi^2 - 2\alpha\xi\cos(\theta - \varphi)}) = 2\sum_0^{\infty}{}' J_n(\xi)Y_n(\alpha)\cos n(\theta - \varphi), \qquad (31.28)$$

$$K_0(\sqrt{\alpha^2 + \xi^2 - 2\alpha\xi\cos(\theta - \varphi)}) = 2\sum_0^{\infty}{}' I_n(\xi)K_n(\alpha)\cos n(\theta - \varphi); \qquad (31.29)$$

for $\xi \ge \alpha$

$$Y_0(\sqrt{\alpha^2 + \xi^2 - 2\alpha\xi\cos(\theta - \varphi)}) = 2\sum_0^{\infty}{}' J_n(\alpha)Y_n(\xi)\cos n(\theta - \varphi), \qquad (31.30)$$

$$K_0(\sqrt{\alpha^2 + \xi^2 - 2\alpha\xi\cos(\theta - \varphi)}) = 2\sum_0^{\infty}{}' I_n(\alpha)K_n(\xi)\cos n(\theta - \varphi). \qquad (31.31)$$

After introducing the relations (31.28)–(31.31) into (31.27) and integrating, we note that all the terms of the series, which contain a cosine, vanish after the integration. Thus, the solution only contains the result of the integration of the constant term; therefore,
for $\xi \le \alpha$

$$w_0 = -\frac{\pi q\alpha}{4D\lambda_1^3} \left[J_0(\xi)Y_0(\alpha) + \frac{2}{\pi}I_0(\xi)K_0(\alpha) \right], \qquad (31.32)$$

for $\xi \ge \alpha$

$$w_0 = -\frac{\pi q\alpha}{4D\lambda_1^3} \left[Y_0(\xi)J_0(\alpha) + \frac{2}{\pi}K_0(\xi)I_0(\alpha) \right]. \qquad (31.33)$$

Adding the compensating solution to the basic one, we obtain the solution of the problem formulated

$$w = w_0 + A_0 J_0(\xi) + B_0 I_0(\xi).$$

The boundary conditions have the form

$$w_0(\beta) + w_k(\beta) = 0,$$
$$w'_0(\beta) + w'_k(\beta) = 0.$$

We now determine A_0 and B_0 from the boundary conditions. For the fixed edge we have the following equations:

$$A_0 J_0(\beta) + B_0 I_0(\beta) - \frac{\pi q \alpha}{4D\lambda_1^3}\left[Y_0(\beta)J_0(\alpha) + \frac{2}{\pi}K_0(\beta)I_0(\alpha)\right] = 0,$$

$$-A_0 J_1(\beta) + B_0 I_1(\beta) + \frac{\pi q \alpha}{4D\lambda_1^3}\left[Y_1(\beta)J_0(\alpha) + \frac{2}{\pi}K_1(\beta)I_0(\alpha)\right] = 0.$$

Solving these equations and using the Wronskian[2], we obtain

$$A_0 = \frac{\pi q \alpha}{4D\lambda_1^3}\,\frac{J_0(\alpha)[I_1(\beta)Y_0(\beta) + I_0(\beta)Y_1(\beta)] + 2I_0(\alpha)/(\pi\beta)}{I_0(\beta)J_1(\beta) + I_1(\beta)J_0(\beta)},$$

$$B_0 = \frac{q\alpha}{2D\lambda_1^3}\,\frac{J_0(\alpha)/\beta + I_0(\alpha)[J_1(\beta)K_0(\beta) - J_0(\beta)K_1(\beta)]}{I_0(\beta)J_1(\beta) + I_1(\beta)J_0(\beta)}.$$

Let us note that a variation of the boundary condition only changes the equations for the determination of the constants A_0 and B_0 which are present in the compensating solution.

Now consider the forced oscillations stipulated by the loading $q_1 \sin pt$ which is uniformly distributed on the ring whose internal reduced radius is equal to α_1 and where external one is equal to α_2.

In order to determine the basic solution we divide the ring into concentric elementary rings. The α denote the reduced internal radius of the elementary ring and $\alpha + d\alpha$ denote its reduced external radius. The loading per unit length of the ring is equal to $q = q_1\, d\alpha/\lambda_1$, where the dimension of q_1 is kg/cm^2. The solution corresponding to the action of this elementary loading is given by the formulae (31.32) and (31.33), in which we should replace q by $q_1\, d\alpha/\lambda_1$. By integrating these expressions with respect to α from α_1 to α_2, we obtain the solution of the problem formulated. If $\xi \le \alpha_1$, i.e. the cross-section lies in the internal circle which is free of the external loading, then we should integrate (31.32), therefore,

$$w_0 = -\frac{\pi q_1}{4D\lambda_1^4}\int_{\alpha_1}^{\alpha_2}\left[J_0(\xi)Y_0(\alpha) + \frac{2}{\pi}I_0(\xi)K_0(\alpha)\right]\alpha\,d\alpha. \qquad (31.34)$$

Using the formulae

$$\int^x xY_0(x)\,dx = xY_1(x),$$

$$\int^x xK_0(x)\,dx = -xK_1(x),$$

we can rewrite (31.34) in the following form:

$$w_0 = -\frac{\pi q_1}{4D\lambda_1^4}\left\{[\alpha_2 Y_1(\alpha_2) - \alpha_1 Y_1(\alpha_1)]J_0(\xi) - \frac{2}{\pi}[\alpha_2 K_1(\alpha_2) - \alpha_1 K_1(\alpha_1)]I_0(\xi)\right\}.$$

---

[2]See page 20.

For $\xi \geq \alpha_2$ we must integrate (31.33) and should use the formulae

$$\int^x x J_0(x)\, dx = x J_1(x),$$

$$\int^x x I_0(x)\, dx = x I_1(x).$$

Thus, for $\xi \geq \alpha_2$ we have

$$w_0 = -\frac{\pi q_1}{4D\lambda_1^4} \int_{\alpha_1}^{\alpha_2} \left[J_0(\alpha) Y_0(\xi) + \frac{2}{\pi} I_0(\alpha) K_0(\xi) \right] \alpha\, d\alpha$$

$$= -\frac{\pi q_1}{4D\lambda_1^4} \Big\{ [\alpha_2 J_1(\alpha_2) - \alpha_1 J_1(\alpha_1)] Y_0(\xi)$$

$$+ \frac{2}{\pi} [\alpha_2 I_1(\alpha_2) - \alpha_1 I_1(\alpha_1)] K_0(\xi) \Big\}.$$

If the considered point of the plate lies in the domain occupied by the loading, i.e. $\alpha_2 \geq \xi \geq \alpha_1$, then in order to solve the problem for $\alpha_1 \leq \alpha \leq \xi$ we should integrate (31.33) and for $\xi \leq \alpha \leq \alpha_2$ we should integrate (31.32), after replacing there, as before, q by $q_1\, d\alpha/\lambda_1$. In this case we obtain the following result for $\alpha_2 \geq \xi \geq \alpha_1$:

$$w_0 = -\frac{\pi q_1}{4D\lambda_1^4} \Big\{ [\xi J_1(\xi) - \alpha_1 J_1(\alpha_1)] Y_0(\xi)$$

$$+ \frac{2}{\pi} [\xi I_1(\xi) - \alpha_1 I_1(\alpha_1)] K_0(\xi) + [\alpha_2 Y_1(\alpha_2) - \xi Y_1(\xi)] J_0(\xi)$$

$$- \frac{2}{\pi} [\alpha_2 K_1(\alpha_2) - \xi K_1(\xi)] I_0(\xi) \Big\}.$$

Rearranging some terms in this formula, we obtain

$$w_0 = -\frac{\pi q_1}{4D\lambda_1^4} \Big\{ \xi [J_1(\xi) Y_0(\xi) - Y_1(\xi) J_0(\xi)]$$

$$+ \frac{2\xi}{\pi} [I_1(\xi) K_0(\xi) + I_0(\xi) K_1(\xi)] - \alpha_1 J_1(\alpha_1) Y_0(\xi)$$

$$- \frac{2}{\pi} \alpha_1 I_1(\alpha_1) K_0(\xi) + \alpha_2 Y_1(\alpha_2) J_0(\xi) - \frac{2}{\pi} \alpha_2 K_1(\alpha_2) I_0(\xi) \Big\}.$$

The terms in the square brackets can be simplified, by using the Wronski determinant. After this simplification we obtain

$$w_0 = -\frac{\pi q_1}{4D\lambda_1^4} \Big[\frac{4}{\pi} - \alpha_1 J_1(\alpha_1) Y_0(\xi) - \frac{2}{\pi} \alpha_1 I_1(\alpha_1) K_0(\xi)$$

$$+ \alpha_2 Y_1(\alpha_2) J_0(\xi) - \frac{2}{\pi} \alpha_2 K_1(\alpha_2) I_0(\xi) \Big].$$

After summing the basic solution obtained with the compensating one, we have

$$w = w_0 + w_k = w_0 + A_0 J_0(\xi) + B_0 I_0(\xi),$$

where A_0 and B_0 are determined by the boundary conditions.

Now consider the problem on the forced oscillations of a circular plate, on which a concentrated force $P \sin pt$ is acting, which is applied at a point A whose reduced distance from the center of the plate is equal to $\alpha = \lambda_1 z$. Let θ denote the angle between the straight line OA and the fixed radius. We write the basic solution which, in fact, has been determined above. For this purpose we replace the argument ξ by z in the formula (31.15)

$$w_0 = -\frac{P}{8D\lambda_1^2}\left[Y_0(z) + \frac{2}{\pi}K_0(z)\right],\qquad(31.35)$$

where $z = \sqrt{\alpha^2 + \xi^2 - 2\alpha\xi\cos(\theta - \varphi)}$. With the help of the addition formulae we rewrite (31.35) for $\xi \geq \alpha$ in the form

$$w_0 = -\frac{P}{4D\lambda_1^2}\sum_{n=0}^{\infty}{}'\left[Y_n(\xi)J_n(\alpha) + \frac{2}{\pi}K_n(\xi)I_n(\alpha)\right]\cos n(\theta - \varphi).\qquad(31.36)$$

We represent the compensating solution w_k in the form

$$w_k = \sum_{n=0}^{\infty}{}'[A_n J_n(\xi) + B_n I_n(\xi)]\cos n(\theta - \varphi).\qquad(31.37)$$

The coefficients A_n and B_n can be found from the conditions on the contour of the plate. Suppose that the plate is fixed along the contour whose reduced radius is equal to β. Then for $\xi = \beta$ we have

$$\frac{\partial w}{\partial \xi} = 0,\quad w = 0,\qquad(31.38)$$

where $w = w_0 + w_k$.

After replacing w in the conditions (31.38) by the sums (31.36) and (31.37) and making the coefficients of the cosines of equal arguments equal, we obtain the following equations with respect to the unknown A_n and B_n:

$$-\frac{P}{4D\lambda_1^2}\left[Y_n(\beta)J_n(\alpha) + \frac{2}{\pi}K_n(\beta)I_n(\alpha)\right] + A_n J_n(\beta) + B_n I_n(\beta) = 0,$$

$$-\frac{P}{4D\lambda_1^2}\left[Y_n'(\beta)J_n(\alpha) + \frac{2}{\pi}K_n'(\beta)I_n(\alpha)\right] + A_n J_n'(\beta) + B_n I_n'(\beta) = 0.$$

Solving this system of equations, after simple transformations associated with using of the Wronskian, we obtain

$$A_n = \frac{P}{4D\lambda_1^2}\frac{J_n(\alpha)\left[-Y_n(\beta)I_n'(\beta) + Y_n'(\beta)I_n(\beta)\right] - 2I_n(\alpha)/(\pi\beta)}{I_n(\beta)J_n'(\beta) - J_n(\beta)I_n'(\beta)},\qquad(31.39)$$

$$B_n = \frac{P}{4D\lambda_1^2}\frac{2I_n(\alpha)\left[J_n'(\beta)K_n(\beta) - J_n(\beta)K_n'(\beta)\right]/\pi - 2J_n(\alpha)/(\pi\beta)}{I_n(\beta)J_n'(\beta) - J_n(\beta)I_n'(\beta)}.\qquad(31.40)$$

Let us give a numerical example. Consider forced oscillations of a circular plate fixed along the contour and loaded with a force $P\sin pt$ applied at the distance $a = 0.5b$ from the center, where b is the radius of the contour (Fig. 14).

We take the following main data, assuming the plate made of reinforced concrete: the modulus of elasticity is $E = 2\cdot 10^6\ t/m^2$, Poisson's ratio is $\sigma = 1/6$, the specific weight is $\gamma_1 = 2.4\ t/m^3$, the density is $\gamma = \gamma_1/g$, the frequency of the forced oscillations is $p = 480\cdot 2\pi/60 \approx 10\ 1/sec$, the thickness of the plate is $h = 0.2\ m$, the radius of the plate is $b = 3\ m$.

Let us determine the flexural rigidity

$$D = Eh^3/[12(1 - \sigma^2)] = 2\cdot 10^6\cdot(0.2)^3/[12(1 - (1/6)^2)] = 1370\ tm.$$

Calculate

$$\lambda_1 = \sqrt[4]{\gamma_1 h p^2/(Dg)} = \sqrt[4]{2.4 \cdot 0.2 \cdot 50^2/(1370 \cdot 9.81)} = 0.69 \ 1/m,$$

$$\beta = \lambda_1 b = 0.69 \cdot 3 \sim 2,$$

$$\alpha = \lambda_1 a \sim 1,$$

the factor $P/8D\lambda_1^2$ is assumed equal to 1 in the calculation.

The calculation is fulfilled by the formulae (31.35), (31.37), (31.39), and (31.40). We are interested in the process of obtaining of the solution as well as in the convergence rate of the series. Therefore, below we give the y-coordinate of the elastic surface (Fig. 15) as well as the summary table of the calculation.

The first column of Table 31 contains the coordinates of the points for which the deflection was computed; w_0 denotes the basic solution; w_{k_n} denotes the nth term of the series (31.37) which represents the compensating solution, where A_n and B_n are computed according to the formulae (31.39)–(31.40). The intermediate obvious results, obtained by substituting numerical values into the formulae, are omitted. The factor $P/8D\lambda_1^2$ is also omitted in the table.

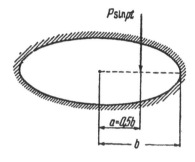

Figure 14

Let us note that a small error occurs for the points of the contour, i.e. for $\xi = 2$, for the considered method of obtaining of the solution. One can see from

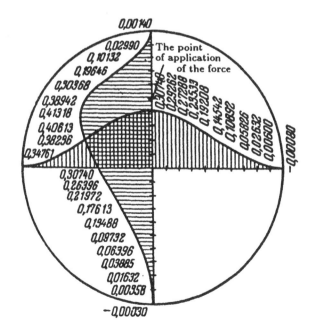

Figure 15

point (ξ, φ)	w_0	w_{k_0}	w_{k_1}	w_{k_2}	w_{k_3}	w_{k_4}	w
$(0.2; 0)$	-0.2731	0.6603	-0.0185	-0.00126	-0.00000385	0.00000156	0.34761
$(0.4; 0)$	-0.1860	0.6504	-0.0358	-0.00491	-0.000290	-0.0000204	0.38236
$(0.6; 0)$	-0.1035	0.6345	-0.0507	-0.0107	-0.000950	-0.0000850	0.40613
$(0.8; 0)$	-0.0350	0.6136	-0.0621	-0.0182	-0.00215	-0.000258	0.41318
$(1.2; 0)$	0	0.5892	-0.0688	-0.0266	-0.0039	-0.000590	0.38942
$(1.2; 0)$	-0.0350	0.5627	-0.0696	-0.0352	-0.00612	-0.00109	0.30368
$(1.4; 0)$	-0.1035	0.5363	-0.0638	-0.0429	-0.00970	-0.00177	0.19646
$(1.6; 0)$	-0.1860	0.5123	-0.0503	-0.0486	-0.0111	-0.00249	0.10132
$(1.8; 0)$	-0.2731	0.4934	-0.0282	-0.0510	-0.0129	-0.00310	0.02990
$(2.0; 0)$	-0.3563	0.48224	$+0.00304$	-0.04890	-0.01330	-0.00311	0.00140
$(0.2; \frac{\pi}{2})$	-0.3642	0.6603	0	0.00126	0	0.00000156	0.29262
$(0.4; \frac{\pi}{2})$	-0.3873	0.6504	0	0.00491	0	0.0000204	0.27288
$(0.6; \frac{\pi}{2})$	-0.4204	0.6345	0	0.0107	0	-0.0000850	0.23533
$(0.8; \frac{\pi}{2})$	-0.4574	0.6136	0	0.0182	0	-0.000258	0.19208
$(1.2; \frac{\pi}{2})$	-0.4958	0.5892	0	0.0266	0	-0.000590	0.14542
$(1.2; \frac{\pi}{2})$	-0.5220	0.5627	0	0.0352	0	-0.00109	0.10892
$(1.4; \frac{\pi}{2})$	-0.5603	0.5363	0	0.0429	0	-0.00177	0.05826
$(1.6; \frac{\pi}{2})$	-0.5782	0.5123	0	0.0486	0	-0.00249	0.02632
$(1.8; \frac{\pi}{2})$	-0.5832	0.4934	0	0.0510	0	-0.00310	0.00600
$(2.0; \frac{\pi}{2})$	-0.5745	0.4823	0	0.0489	0	-0.00310	0.00080
$(0.2; \pi)$	-0.4309	0.6603	0.0185	-0.00126	0.0000385	0.00000156	0.26393
$(0.4; \pi)$	-0.4930	0.6504	0.0358	-0.00491	0.000290	-0.0000204	0.21972
$(0.6; \pi)$	-0.5401	0.6345	0.0507	-0.0107	0.000950	-0.0000850	0.17613
$(0.8; \pi)$	-0.5703	0.6136	0.0621	-0.0182	0.00215	-0.000258	0.13488
$(1.2; \pi)$	-0.5829	0.5892	0.0688	-0.0266	0.0039	-0.000590	0.09732
$(1.2; \pi)$	-0.5776	0.5627	0.0696	-0.0352	0.00612	-0.00109	0.06396
$(1.4; \pi)$	-0.5551	0.5363	0.0638	-0.0429	0.00970	-0.00177	0.03886
$(1.6; \pi)$	-0.5166	0.5123	0.0503	-0.0486	0.0111	-0.00249	0.01632
$(1.8; \pi)$	-0.4638	0.4934	0.0282	-0.0510	0.0129	-0.00310	0.00358
$(2.0; \pi)$	-0.3990	0.4823	-0.0031	-0.0489	0.0133	-0.00310	0.00030
$(0; 0)$	-0.3563	0.6637	0.0000	-0.0000	-0.0000	0.0000	0.3974

[1] When computing the deflection w, the terms w_{k_1}, w_{k_2}, w_{k_3}, w_{k_4}, should be doubled; in the notation of the points, the first number shows the reduced distance from the center, and the second one shows the angle in radians.

the table that we have obtained a non-zero deflection on the contour; however, its greatest value is approximately $\frac{1}{4}\%$ of the maximal deflection, which is at the point $(0.8; 0)$.

It should be noted that the function w becomes infinite for

$$I_n(\beta) J_n'(\beta) - I_n'(\beta) J_n(\beta) = 0. \qquad (31.41)$$

Thus, we have the phenomenon of the resonance corresponding to the frequency of the forced oscillations

$$p = \sqrt{\frac{D}{\gamma h}} \frac{\beta_n^2}{b^2},$$

where the β_n are the roots of the equation (31.41).

We now pass to consideration of the free oscillations of a plate with several additional masses $(n = 1, 2, 3, \ldots, l)$, which are concentrated at the points α_n, θ_n. After reasoning similar to those used in Section 4, one can obtain the following frequency equation:

$$\begin{vmatrix} \omega^2 M_1 w(\alpha_1, \theta_1; \alpha_1, \theta_1; \omega) - 1 & \cdots & \omega^2 M_l w(\alpha_1, \theta_1; \alpha_l, \theta_l; \omega) \\ \omega^2 M_1 w(\alpha_2, \theta_2; \alpha_1, \theta_1; \omega) & \cdots & \omega^2 M_l w(\alpha_2, \theta_2; \alpha_l, \theta_l; \omega) \\ \vdots & \ddots & \vdots \\ \omega^2 M_1 w(\alpha_l, \theta_l; \alpha_1, \theta_1; \omega) & \cdots & \omega^2 M_l w(\alpha_l, \theta_l; \alpha_l, \theta_l; \omega) - 1 \end{vmatrix} = 0. \quad (31.42)$$

The frequency equation also has the same form after replacing the additional masses by elastic supports of the point type. In this case, we must set $M_n = -\dfrac{C_n}{\omega^2}$ in the frequency equation, where C_n is the stiffness of the support.

The frequency equation for a plate which has an additional mass M at the point $(\alpha, 0)$, has the following form:

$$\omega^2 w(\alpha, 0; \alpha, 0) = \frac{1}{M}. \quad (31.43)$$

Replacing w here by the expression obtained at the beginning of this section, when we considered the problem on the forced oscillations, we can obtain the appropriate frequency equation. Making the right-hand side of this equation equal to zero, we obtain the frequency equation for the case, when the additional mass is infinitely large, or, which is equivalent, for a plate with one stiff support of the point type.

2. Consider a girderless floor, which represents a plate of reinforced concrete, supported by columns located on a square or rectangular frame. The columns are enlarged at the point of contact with the plate and form capitals. As a first approximation, the girderless floor can be considered as a plate with supports of the point type.

First, consider the problem on the forced oscillations of a circular plate with supports of the point type, which is subject to the action of a concentrated force, applied at the center and varied according to the harmonic rule. We set $\dfrac{P}{8D\lambda_1^2} = 1$ and give the corresponding reduced distances in Fig. 16. First, we find the deflections at points 1 and 2 of the plates loaded by unit forces as is shown in Fig. 16.

Let us find w_{1a} and $w_{2a}{}^3$. For this purpose we represent the solution as a sum of the basic solution and the compensating one

$$w = A_0 J_0(\xi) + B_0 I_0(\xi) - \frac{P}{8D\lambda_1^2}\left[Y_0(\xi) + \frac{2}{\pi}K_0(\xi)\right], \quad (31.44)$$

where A_0 and B_0 are given by the formulae page 140.

Using these formulae, we find

$$A_0 = -0.381\,;$$
$$B_0 = -0.0181\,.$$

---

[3] Here, as below, the first index denotes the number of the point, at which the deflection is determined, and the second one denotes the group of the forces, which lead to the deflection.

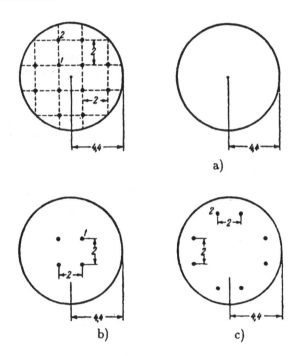

Figure 16

We find the deflection at point 1, by substituting $\xi = 1.41$ into (31.20). Note that the last term is the basic solution and the sum of the first two terms is the compensating solution

$$w_{1a} = -(0.3427 + 0.1531) - 0.381 \cdot 0.5614 - 0.0181 \cdot 1.562 = -0.738\,;$$

the numerical values of the Bessel functions present in these formulae are taken from the appropriate tables.

In order to find the deflection at point 2, we substitute $\xi = 3.16$ into (31.20) and then we find in just the same manner:

$$w_{2a} = -0.323\,.$$

Now consider the case, when the loading of the plate is fulfilled by four equal forces applied at the points 1 and varying in time according to the harmonic rule (Fig. 16b). As before, we find the deflections at the point 1 and 2 as the sums of the basic solution and the compensating one. In this case the basic solution is defined with the help of the principle of composition of forces as the sum of four solutions of the type (31.19)

$$w_0 = \sum_{n=1}^{4} \left[Y_0(\lambda_1 r_n) + \frac{2}{\pi} K_0(\lambda_1 r_n) \right],$$

where r_n is the distance from the point, at which the deflection is determined, to the corresponding force.

When determining the compensating solution w_k, we restrict ourselves only by the constant term of the series (31.20), because the coefficients A_1, B_1, A_2, B_2, A_3,

B_3 of the series vanish in the considered case from the symmetry conditions, and the influence of the term with the index $n = 4$ on the result is small.

Thus, we find that for the loading according to the scheme of Fig. 16b, the deflections at the points 1 and 2 are equal to

$$w_{1b} = w_0 + w_k = -1.624 - 0.585 = -2.209,$$
$$w_{2b} = w_0 + w_k = -0.849 - 0.059 = -0.790.$$

Just in the same way, we find that for the loading according to the scheme of Fig. 16c, the deflections at the points 1 and 2 are equal to

$$w_{1c} = 1.581,$$
$$w_{2c} = -0.351.$$

We now pass to calculation of the reaction X_1 of the support at the points 1 and X_2 at the points 2, using the method of forces. In our case the equations have the form

$$w_{1a} + X_1 w_{1b} + X_2 w_{1c} = 0,$$
$$w_{2a} + X_1 w_{2b} + X_2 w_{2c} = 0.$$

Solving these equations, we obtain

$$X_1 = -0.380,$$
$$X_2 = 0.064.$$

Thus, following the usual scheme of the method of forces, we reduce the problem on the calculation of a plate with supports to the calculation of a plate which is loaded by the given force as well as by the reactions of the supports X_1 and X_2 defined above. Then, using the principle of addition of actions, we find the deflections at the points I–VIII and I–V' (see Fig. 17), which are caused by the forces mentioned. The calculations are collected in Table 31, where w_a, w_b and w_c denote the deflections caused by the elementary forces applied as is shown in schemes a, b and c of Fig. 16 (see Table 31).

For the plate shown in Fig. 18 the problem has been solved in the

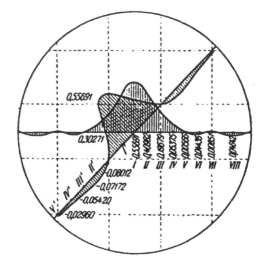

Figure 17

same way. The force X_1 in this case does almost not differ from the force calculated in the previous example, the deflections at the center differ by 1%. We will return to the problem on circular plates as the basic system and, in this connection, to the problem on girderless floors in Section 35.

Obviously, when solving the problem on forced oscillations of a girderless floor, the refinement of the numerical scheme, which allows one to represent the real

point	w_a	w_b	$X_1 w_b$	w_c	$X_2 w_c$	w
I	-0.39913	-2.95184	1.12126	0.74878	-0.16522	0.55691
II	-0.52081	-2.85958	1.08622	0.59122	-0.15559	0.40982
III	-0.67072	-2.62284	0.99643	0.34925	-0.12892	0.19679
IV	-0.74334	-2.33440	0.88740	0.16446	-0.09031	0.05375
V	-0.70939	-1.96782	0.74940	0.05586	-0.04567	-0.00566
VI	-0.57874	-1.41140	0.54014	-0.02987	-0.00575	-0.04435
VII	-0.38802	-0.97440	-0.37696	-0.00910	0.01967	$+0.00851$
VIII	-0.04294	0.08233	-0.02024	-0.07496	0.01676	-0.04642

conditions of the work of construction better, is of interest. First of all, in this case we should take into account that the reaction of the support is not applied at a point but distributed on some area. If we assume the reaction of a support distributed along a circumference, then the basic solution for such loading should be found by the formulae (31.32) and (31.33). If the loading is distributed uniformly on the area of a ring and, in a special case, on the area of a circle, then we should use the formulae (31.46), (31.51), (31.54). On the basis of Saint-Venant's principle we can, when constructing the compensating solution, as before assume the reactions to be concentrated forces, applied at the centers of the corresponding circumferences or circles. Moreover, when calculating the deflections corresponding to the basic solution, we can replace the reactions of the supports which are sufficiently distant from the point under consideration, by concentrated forces. Then, we construct the canonical equations as before.

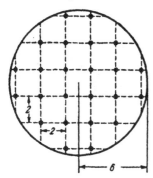

Let us note that we can find the compensating solution with higher accuracy (although, in fact, it is not needed). For this purpose we should expand the basic solution on the contour into the trigonometric series, using the addition formulae.

3. Consider the problem on oscillations of a circular plate compressed by a constant radial pressure p_0 which acts in the median plane. This problem is more general than the problems considered in previous sections. As before, we begin with considering the problem on a concentrated force. The differential equation of the deflection surface has the form

Figure 18

$$\Delta\Delta w + \frac{p_0}{2D}\Delta w - \rho\frac{p^2}{D}w = 0. \tag{31.45}$$

After passing to dimensionless coordinates, the differential equation takes the form

$$\Delta_\xi\Delta_\xi w + 2b_0\Delta_\xi w - w = 0 \tag{31.46}$$

where

$$b_0 = \frac{p_0}{2D\lambda^2}; \quad \lambda^4 = \frac{\rho p^2}{D}; \quad \xi = \lambda r.$$

Let s_1 and s_2 be the roots of the characteristic equation. They are equal to

$$s_1 = -b_0 + \sqrt{b_0^2 + 1},$$

$$s_2 = -b_0 - \sqrt{b_0^2 + 1}.$$

Let us note that the signs of s_1 and s_2 are always distinct. Denoting $\lambda\sqrt{s_1} = \lambda_1$ and $\lambda\sqrt{s_2} = i\lambda_2$, we can write the general integral of the equation (31.46) in the following form:

$$\begin{aligned}
w = &\sum J_n(\lambda_1 r)[A_n \cos n\varphi + A_n' \sin n\varphi] \\
&+ \sum I_n(\lambda_2 r)[B_n \cos n\varphi + B_n' \sin n\varphi] \\
&+ \sum Y_n(\lambda_1 r)[C_n \cos n\varphi + C_n' \sin n\varphi] \\
&+ \sum K_n(\lambda_2 r)[D_n \cos n\varphi + D_n' \sin n\varphi].
\end{aligned} \tag{31.47}$$

Let us denote $\lambda_1 r = \xi_1$, $\lambda_2 r = \xi_2$, $\lambda_1 a = \alpha_1$, $\lambda_2 a = \alpha_2$. The solution, which has a singularity of the type of a concentrated force at (a, θ), is of the form

$$w_0 = -\frac{1}{8D}\left[\frac{1}{\lambda_1^2}Y_0(z_1) + \frac{2}{\pi\lambda_2^2}K_0(z_2)\right], \tag{31.48}$$

where

$$z_1 = \lambda_1\sqrt{r^2 + a^2 - 2ar\cos(\theta - \varphi)},$$

$$z_2 = \lambda_2\sqrt{r^2 + a^2 - 2ar\cos(\theta - \varphi)}.$$

Consider the oscillations of a circular plate of the radius b, fixed along the contour, which are caused by the force $1 \cdot \sin pt$ applied at the point with the coordinates (a, θ). For this purpose we should first expand the values of the function w_0 and its normal derivative on the contour into the trigonometric series. Note that for $r \leq a$

$$w_0 = -\frac{1}{4D}\sum_{n=0}^{\infty}{}'\left[\frac{1}{\lambda_1^2}J_n(\xi_1)Y_n(\alpha_1) + \frac{2}{\pi\lambda_2^2}K_n(\alpha_2)I_n(\xi_2)\right]\cos n(\theta - \varphi), \tag{31.49}$$

for $r \geq a$

$$w_0 = -\frac{1}{4D}\sum_{n=0}^{\infty}{}'\left[\frac{1}{\lambda_1^2}J_n(\alpha_1)Y_n(\xi_1) + \frac{2}{\pi\lambda_2^2}K_n(\xi_2)I_n(\alpha_2)\right]\cos n(\theta - \varphi). \tag{31.50}$$

We represent the required solution in the form $w = w_0 + w_k$, where

$$w_k = \frac{1}{4D}\sum_{n=0}^{\infty}{}'\left[\frac{A_n}{\lambda_1}J_n(\xi_1) + \frac{B_b}{\lambda_2^2}I_n(\xi_2)\right]\cos n(\theta - \varphi). \tag{31.51}$$

The coefficients A_n and B_n are determined by the conditions on the contour. For the plates with the fixed edge these equation have the form

$$-\left[\frac{1}{\lambda_1^2}J_n(\alpha_1)Y_n(\beta_1) + \frac{2}{\pi\lambda_2^2}I_n(\alpha_2)K_n(\beta_2)\right] + \frac{A_n}{\lambda_1^2}J_n(\beta_1) + \frac{B_n}{\lambda_2^2}I_n(\beta_2) = 0, \tag{31.52}$$

$$-\left[\frac{1}{\lambda_1}J_n(\alpha_1)Y_n'(\beta_1) + \frac{2}{\pi\lambda_2}I_n(\alpha_2)K_n'(\beta_2)\right] + \frac{A_n}{\lambda_1}J_n'(\beta_1) + \frac{B_n}{\lambda_2}I_n'(\beta_2) = 0, \tag{31.53}$$

where $\lambda_1 b = \beta_1$ and $\lambda_2 b = \beta_2$.

After some simplifications connected with using the Wronskian of the Bessel equation, we obtain:

$$A_n = \frac{-\frac{2}{\lambda_2^2 \pi \beta_2} I_n(\alpha_1) + J_n(\alpha_1)\left[\frac{-1}{\lambda_1^2}Y_n(\beta_1)I_n'(\beta_2) + \frac{1}{\lambda_1\lambda_2}I_n(\beta_2)Y_n'(\beta_1)\right]}{\frac{1}{\lambda_1^2}J_n(\beta_1)I_n'(\beta_2) + \frac{1}{\lambda_1\lambda_2}J_n'(\beta_1)I_n(\beta_2)}, \qquad (31.54)$$

$$B_n = \frac{-\frac{1}{\lambda_1^2\beta_1}J_n(\alpha_1) + \frac{2}{\pi}I_n(\alpha_2)\left[\frac{1}{\lambda_1\lambda_2}K_n'(\beta_2)J_n(\beta_1) - \frac{1}{\lambda_2}J_n'(\beta_1)K_n(\beta_2)\right]}{\frac{1}{\lambda_1^2}J_n(\beta_1)I_n'(\beta_2) + \frac{1}{\lambda_1\lambda_2}J_n'(\beta_1)I_n(\beta_2)}. \qquad (31.55)$$

If the denominator of the expressions (31.54) and (31.55) vanish, then we have the phenomenon of resonance. In this case we must take into account the damping. For this purpose we should introduce the imaginary part of the determinant Δ into the denominator, in accordance with the discussion in Section 43 of Part 2. The expansion of the basic solution into a series in $\cos n(\theta - \varphi)$ allows us to obtain functions which have the property of the unit matrix.

4. The differential equation of this problem has the form

$$D\Delta\Delta w + p_n \Delta w = 0, \qquad (31.56)$$

where, as before, p_0 is the compressing radial pressure which acts in the middle plane.

One can easily show that, by introducing the dimensionless coordinate $\xi = \lambda r$, we can reduce the equation (31.56) to the system of two equations

$$\Delta_\xi w_1 + \lambda^2 w_1 = 0, \qquad (31.57)$$
$$\Delta_\xi w_2 = 0, \quad w = w_1 + w_2, \qquad (31.58)$$

where the operator is

$$\Delta_\xi = \frac{d^2}{d\xi^2} + \frac{1}{\xi}\frac{d}{d\xi}$$

and

$$\lambda^2 = \frac{p_0}{D}. \qquad (31.59)$$

In order to obtain the basic solution of the problem formulated, we need to find the integral of the system which has the singularity of the concentrated force type. As has been shown, this means that the solution w_0 of the system of equations (31.57) and (31.58) must satisfy the condition

$$\lim_{\xi \to 0}\left[-D\lambda^3 \frac{d}{d\xi}\Delta_\xi w_0\right]\cdot 2\pi\frac{\xi}{\lambda} = -P. \qquad (31.60)$$

It has been shown that the function $C_0 Y_0(\xi)$ is one of solutions of the equation (31.57). The solution of the equation (31.58) has the form

$$w = A_0 \ln(\xi) + B_0.$$

The solution of the system of equations (31.57) and (31.58) which satisfies the conditions (31.60) and $w(0) = 0$, can be represented in the form

$$w_0 = C_0 Y_0(\xi) + A_0 \ln \xi + B_0. \qquad (31.61)$$

The constants C_0, A_0 and B_0 must be chosen such that these conditions hold. One can easily show that

$$w_0 = -\frac{P}{4D\lambda^3}\left[Y_0(\xi) - \frac{2}{\pi}\ln\xi - \frac{2\gamma}{\pi}\right], \qquad (31.62)$$

where $\gamma = 1.781072\cdots = e^C$ (C is the Euler constant). Let us note that the constant term in the square brackets can be, broadly speaking, omitted, because in this case it will be automatically present in the boundary conditions, if we consider the plates bounded in the plane.

The second term of the expression (31.62), i.e. $\dfrac{P}{2\pi D\lambda^2}\ln\xi$, can be interpreted from the mechanical viewpoint as the basic solution of the problem on the membrane which is extended by the forces $D\lambda^2$ and loaded by the concentrated force P. If we denote the reduced distance from the point of application of the force to the point under consideration by

$$z = \sqrt{\xi^2 - \alpha^2 - 2\alpha\xi\cos(\theta - \varphi)},$$

then we can represent the basic solution as a series in $\cos n(\theta - \varphi)$.

Thus, it remains only to expand the function $\ln z$ into a similar series. We only mention, without making the calculations, that such an expansion can be easily obtained, if, using the remark made above concerning the static interpretation of the expression $\dfrac{P}{2\pi D\lambda^2}\ln z$, we expand the concentrated force into a series in $\cos n(\theta - \varphi)$ and find the solution (i.e. the deflection of the membrane), which corresponds to every term of this series.

Finally, we obtain in this way:

for $\xi \leq \alpha$

$$w_0 = w_\mathrm{I} = -\frac{P}{4D\lambda^2}\left[J_0(\xi)Y_0(\alpha) + 2\sum_{n=1}^{\infty}J_n(\xi)Y_n(\alpha)\cos n(\theta - \varphi)\right.$$

$$\left. -\frac{2}{\pi}\ln\alpha - \frac{2}{\pi}\gamma + \frac{2}{\pi}\sum_{n=1}^{\infty}\frac{\xi^n}{n\alpha^n}\cos n(\theta - \varphi)\right], \quad (31.63)$$

for $\xi \geq \alpha$

$$w_0 = w_\mathrm{II} = -\frac{P}{4D\lambda^2}\left[J_0(\alpha)Y_0(\xi) + 2\sum_{n=1}^{\infty}J_n(\alpha)Y_n(\xi)\cos n(\theta - \varphi)\right.$$

$$\left. -\frac{2}{\pi}\ln\xi - \frac{2}{\pi}\gamma + \frac{2}{\pi}\sum_{n=1}^{\infty}\frac{\alpha^n}{n\xi^n}\cos n(\theta - \varphi)\right]. \quad (31.64)$$

One can easily construct the functions which have the property of the unit matrix. For this purpose we should apply the method given in Section 37, which consists in constructing the functions of the direct direction by formulae of the type

$$w_\mathrm{II} - w_\mathrm{I}.$$

After making the calculations described in Section 32, we obtain the formula for the fundamental function in the following form:

$$Y_4 = -\frac{\pi\alpha}{2}\left[J_n(\alpha)Y_n(\xi) - Y_n(\alpha)J_n(\xi) + \frac{2}{\pi}\frac{\alpha^n}{n\xi^n} - \frac{2}{\pi}\frac{\xi^n}{n\alpha^n}\right]. \qquad (31.65)$$

Consider several examples connected with using the formulae (31.63) and (31.64).

Consider the stability of a circular plate of a constant thickness, which is fixed along the contour and has a stiff ring support (Fig. 19). We will calculate the smallest critical force, by considering the losses of stability which are symmetric with respect to the center of the form. As before, let p_0 be the intensivity of the radial pressure, D be the flexural rigidity, $\lambda = \sqrt{\dfrac{p_0}{D}}$, $\xi = \lambda r$. The solution, which has the singularity corresponding to the action of a loading perpendicular to the middle plane and uniformly distributed along the circumference at $\xi = \lambda a = \alpha$, has the following form, as can be seen from (31.68) and (31.69):

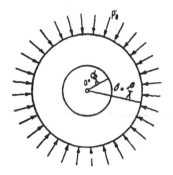

Figure 19

for $\xi \leq \alpha$

$$w_0 = J_0(\xi)Y_0(\alpha) - \frac{2}{\pi}\ln\alpha, \tag{31.66}$$

for $\xi \geq \alpha$

$$w_0 = J_0(\alpha)Y_0(\xi) - \frac{2}{\pi}\ln\xi. \tag{31.67}$$

The boundary conditions for $\xi = \lambda b = \beta$ have the form

$$w = 0, \qquad \frac{dw}{d\xi} = 0. \tag{31.68}$$

The solution, which satisfies the boundary conditions and has the required singularity at $\xi = \alpha$, can be written in the following form:

$$w = w_0 + A_0 J_0(\xi) + B_0. \tag{31.69}$$

Using the conditions (31.68), we find the coefficients A_0 and B_0:

$$\left.\begin{aligned} A_0 &= -\frac{J_0(\alpha)Y_1(\beta) + \frac{2}{\pi\beta}}{J_1(\beta)}, \\[2mm] B_0 &= \frac{2}{\pi}\ln\beta + \frac{2}{\pi\beta}\frac{J_0(\beta) - J_0(\alpha)}{J_1(\beta)}. \end{aligned}\right\} \tag{31.70}$$

Taking into account the fact that at $\xi = \alpha$ we have

$$w = 0, \tag{31.71}$$

we obtain the following transcendental equation for the determination of the critical loading:

$$J_0(\alpha)Y_0(\alpha) + \frac{2}{\pi}\ln\frac{b}{a} + \frac{1}{J_1(\beta)}\left\{\frac{2}{\pi\beta}[J_0(\beta) - 2J_0(\alpha)] - J_0^2(\alpha)Y_1(\beta)\right\} = 0. \tag{31.72}$$

If we set $\alpha = 0$ in equation (31.72), then the transcendental equation obtained gives the solution to the problem of the stability of a plate which has a support at the center and is fixed along the contour. This problem has been considered by Nadai. If $\alpha = 1$, then we obtain the problem on a circular plate which has been considered by A.N. Dinnik.

Let us give the values of the smallest root of the equation (31.72) for different values of the ratio $\frac{a}{b}$[4]:

$$\frac{a}{b} = 0.1; \quad 0.2; \quad 0.3; \quad 0.4; \quad 0.5; \quad 0.6; \quad 0.7; \quad 0.8; \quad 0.9; \quad 1.0;$$

$$\beta_k = 6.77; \quad 5.96; \quad 6.99; \quad 6.66; \quad 6.07; \quad 5.47; \quad 4.95; \quad 4.51; \quad 5.15; \quad 3.83.$$

The critical loading p_k is defined by the formula

$$p_k = \frac{\beta_k^2 D}{b^2}. \tag{31.73}$$

Consider the problem on stability of a plate which is simply supported along the contour and has an intermediate ring support.

This problem differs from the preceding one only in one of the boundary conditions. The condition of equality to zero of the slopes on the contour should be replaced by the condition that the intensivity of the bending moments G_1 on the contour is equal to zero, i.e.
for $\xi = \beta$,

$$w = 0,$$

$$G_1 = -D\lambda^2 \left[\sigma \Delta w + (1 - \sigma) \frac{\partial^2 w}{\partial \xi^2} \right] = 0, \tag{31.74}$$

where σ is Poisson's ratio.

These conditions lead to the following expressions for the constants A_0 and B_0:

$$A_0 = \frac{J_0(\alpha)[(1 - \sigma)Y_1(\beta) - \beta Y_0(\beta)] + \frac{2}{\pi \beta}(1 - \sigma)}{\beta J_0(\beta) - (1 - \sigma)J_1(\beta)}, \tag{31.75}$$

$$B_0 = \frac{2}{\pi} \ln \beta - J_0(\alpha)Y_0(\beta) - A_0 J_0(\beta). \tag{31.76}$$

Making the deflection at $\xi = \alpha$ equal to zero, we obtain the transcendental equation

$$J_0(\alpha)[Y_0(\alpha) - Y_0(\beta)] + \frac{2}{\pi} \ln \frac{b}{a}$$

$$+ [J_0(\alpha) - J_0(\beta)] \frac{J_0(\alpha)[(1 - \sigma)Y_1(\beta) - \beta Y_0(\beta)] + \frac{2}{\pi \beta}(1 - \sigma)}{\beta J_0(\beta) - (1 - \sigma)J_1(\beta)} = 0. \tag{31.77}$$

Let us give the obtained values of the least root of this equation for different values of the ratio a/b for $\sigma = 1/3$:

$$\frac{a}{b} = 0.2; \quad 0.4; \quad 0.6; \quad 0.8; \quad 1.0;$$

$$\beta_k = 4.79; \quad 5.13; \quad 5.18; \quad 4.42; \quad 3.19.$$

We now pass to investigating the stability of plates which have point supports constraining the vertical displacement. The reactions at such supports can be reduced to concentrated forces.

Consider a plate with a fixed edge. As before, we divide the solution into two parts: the basic solution and the compensating one. We set $\xi = \beta$ in the basic

---

[4]As far as we know, other authors did not make the calculations for the ratio a/b different from 0 and 1.

solution (31.64). In this case we obtain that the deflection on the edge of the plate, which corresponds to the basic solution, is equal to

$$w_0(\beta) = \frac{-P}{4D\lambda^2}\left[J_0(\alpha)Y_0(\beta) + 2\sum_{n=1}^{\infty} J_n(\alpha)Y_n(\beta)\cos n(\theta - \varphi) - \frac{2}{\pi}\ln\beta \right.$$

$$\left. - \frac{2}{\pi}\gamma + \frac{2}{\pi}\sum_{n=1}^{\infty}\frac{\alpha^n}{n\beta^n}\cos n(\theta - \varphi)\right]. \quad (31.78)$$

Differentiating (31.780 with respect to ξ and setting $\xi = \beta$, one can obtain the values of the slopes on the edge of the plate, which corresponds to the basic solution

$$\left(\frac{dw}{d\xi}\right)_{\xi=\beta} = -\frac{P}{4D\lambda^2}\left[J_0(\alpha)Y_0'(\beta) + 2\sum_{n=1}^{\infty} J_n(\alpha)Y_n'(\beta)\cos n(\theta - \varphi) - \frac{2}{\pi\beta} \right.$$

$$\left. - \frac{2}{\pi}\sum_{n=1}^{\infty}\frac{\alpha^n}{\beta^{n+1}}\cos n(\theta - \varphi)\right]. \quad (31.79)$$

We seek the compensating solution in the form of the series

$$w_k = \sum_{n=0}^{\infty}[A_n J_n(\xi) + B_n\xi^n]\cos n(\theta - \varphi). \quad (31.80)$$

Then we have the following equations for the determination of the coefficients A_n and B_n for $n \neq 0$:

$$\frac{-P}{2D\lambda^2}\left[J_n(\alpha)Y_n(\beta) + \frac{\alpha^n}{\pi n\beta^n}\right] + A_n J_n(\beta) + B_n\beta^n = 0, \quad (31.81)$$

$$\frac{-P}{2D\lambda^2}\left[J_n(\alpha)Y_n'(\beta) - \frac{\alpha^n}{\pi\beta^{n+1}}\right] + A_n J_n'(\beta) + n B_n\beta^{n-1} = 0; \quad (31.82)$$

hence,

$$A_n = \frac{P}{2D\lambda^2}\cdot\frac{J_n(\alpha)Y_{n+1}(\beta) + \frac{2\alpha^n}{\pi\beta^{n+1}}}{J_{n+1}(\beta)}, \quad (31.83)$$

$$B_n = \frac{P}{2D\lambda^2}\cdot\frac{2J_n(\alpha) - J_{n-1}(\beta)\frac{\alpha^n}{n\beta^{n-1}}}{\pi\beta^{n+1}J_{n+1}(\beta)}. \quad (31.84)$$

For $n = 0$ the constants A_0 and B_0 can be obtained from the formulae (31.70), after introducing the additional factor $-\frac{P}{4D\lambda^2}$ into their right-hand side.

Thus, we have

$$w = w_0 + w_k = \frac{-P}{4D\lambda^2}\left[Y_0(r) - \frac{2}{\pi}\ln r - \frac{2\gamma}{\pi}\right] + w_k. \quad (31.85)$$

For a circular plate with a stiff point support, we obtain the critical loading, by making the deflection at the support equal to zero

$$w(\alpha, \theta; \alpha, \theta) = 0, \quad (31.86)$$

or

$$\sum_{n=0}^{\infty}[A_n J_n(\alpha) + B_n\alpha^n] = 0, \quad (31.87)$$

because

$$w_0(0) = 0.$$

Suppose that a circular plate fixed along the contour and loaded by a radial pressure has l supports of the point type. We take off the supports, by replacing their action by the forces. The critical loading can be obtained, by vanishing the determinant of the system of equations which express the fact that the displacements at the supports are equal to zero

$$\begin{vmatrix} w(\alpha_1, \theta_1; \alpha_1, \theta_1; p) & \cdots & w(\alpha_1, \theta_1; \alpha_l, \theta_l; p) \\ w(\alpha_2, \theta_2; \alpha_1, \theta_1; p) & \cdots & w(\alpha_2, \theta_2; \alpha_l, \theta_l; p) \\ \vdots & \ddots & \vdots \\ w(\alpha_l, \theta_l; \alpha_1, \theta_1; p) & \cdots & w(\alpha_l, \theta_l; \alpha_l, \theta_l; p) \end{vmatrix} = 0. \tag{31.88}$$

32. Equilibrium of a circular plate lying on an elastic foundation. Axially symmetrical deformation

In this section we consider the problems on a circular plate, which lies on an elastic Winkler-type foundation, when the reactive pressure p at a point is proportional to the deflection of the plate at the same point:

$$p = k_0 w.$$

The coefficient of proportionality k_0 is called the *bedding constant of the foundation material*. Moreover, we suppose that radial forces are applied to the plate, compressing or stretching it.

Such problems arise, when calculating a plate which is compressed or stretched due to the temperature effect as well as when calculating prestressed coatings.

The differential equation of the deflection surface of a plate which lies on a Winkler-type foundation has the following form, if we take into account the influence of the radial forces:

$$D\nabla^2\nabla^2 w - p_0\nabla^2 w + k_0 w = q, \tag{32.1}$$

where D is the flexural rigidity, k_0 is the bedding constant of the foundation material, q is the intensivity of the external loading, p_0 are the radial forces which are assumed positive in the case of stretching. In this section we only consider the axially symmetrical problem; therefore, the Laplace operator has the form $\nabla^2 = \dfrac{d^2}{dr^2} + \dfrac{1}{r}\dfrac{d}{dr}$. We denote $l = \sqrt[4]{D/k_0}$, $\xi = r/l$ and rewrite (32.1) in the following form:

$$\left(\frac{d^2}{d\xi^2} + \frac{1}{\xi}\frac{d}{d\xi}\right)^2 w - 2b_0\left(\frac{d^2}{d\xi^2} + \frac{1}{\xi}\frac{d}{d\xi}\right)w + w = \frac{ql^4}{D},$$

where

$$b_0 = p_0 l^2/(2D).$$

Consider the homogeneous differential equation

$$\left(\frac{d^2}{d\xi^2} + \frac{1}{\xi}\frac{d}{d\xi}\right)^2 w - 2b_0\left(\frac{d^2}{d\xi^2} + \frac{1}{\xi}\frac{d}{d\xi}\right)w + w = 0. \tag{32.2}$$

We shall seek a solution of this equation in the form

$$w = AZ_0\left(\sqrt{s}\xi\right), \tag{32.3}$$

where Z_0 is the cylindrical function with zero index. We substitute expression (32.3) into equation (32.2) and use the relation

$$\frac{d^2 Z_0\left(\sqrt{s}\xi\right)}{d\xi^2} + \frac{1}{\xi}\frac{d Z_0\left(\sqrt{s}\xi\right)}{d\xi} = -s Z_0\left(\sqrt{s}\xi\right).$$

After cancelling by $A Z_0\left(\sqrt{s}\xi\right)$, we obtain the characteristic equation

$$s^2 + 2b_0 s + 1 = 0. \tag{32.4}$$

Let s_1 and s_2 denote the roots of this equation. Suppose that $s_1 \neq s_2$. Then the solution of the differential equation (32.2) can be written in the following form:

$$w = A_1^* J_0\left(\xi\sqrt{s_1}\right) + A_2^* Y_0\left(\xi\sqrt{s_1}\right) + A_3^* J_0\left(\xi\sqrt{s_2}\right) + A_4^* Y_0\left(\xi\sqrt{s_2}\right). \tag{32.5}$$

Consider in more detail the roots of the characteristic equation (32.4), which are equal to $s_{1,2} = -b_0 \pm \sqrt{b_0^2 - 1}$. Assume that $|b_0| < 1$. Denoting

$$\xi\sqrt{s} = \rho e^{\pm i\varphi}, \quad \xi \equiv \rho, \quad \varphi = \left[\arctan\left(-b_0/\sqrt{1 - b_0^2}\right)\right]/2,$$

we rewrite (32.5) in the following form:

$$w = A_1^* J_0\left(\rho e^{i\varphi}\right) + A_2^* H_0^{(1)}\left(\rho e^{i\varphi}\right) + A_3^* J_0\left(\rho e^{-i\varphi}\right) + A_4^* H_0^{(2)}\left(\rho e^{-i\varphi}\right). \tag{32.6}$$

Let us use the notation

$$J_0\left(\rho e^{\pm i\varphi}\right) = \bar{u}_0(\rho) \pm i\bar{v}_0(\rho),$$
$$H_0^{(1)}\left(\rho e^{+i\varphi}\right) = \bar{f}_0(\rho) + i\bar{g}_0(\rho),$$

and set

$$A_1 = A_1^* + A_3^*,$$
$$A_2 = i(A_1^* - A_3^*),$$
$$A_3 = A_2^* + A_4^*,$$
$$A_4 = i(A_2^* - A_4^*).$$

Then (32.6) takes the form

$$w = A_1 \bar{u}_0(\rho) + A_2 \bar{v}_0(\rho) + A_3 \bar{f}_0(\rho) + A_4 \bar{g}_0(\rho). \tag{32.7}$$

When stretching the plate, we have $b_0 > 0$ and $\varphi > \pi/4$; when compressing it, we have $b_0 < 0$ and $\varphi < \pi/4$. If $b_0 = 1$, then the characteristic equation has multiple roots.

Consider the problem on a plate of an infinite radius which is loaded by a concentrated force. We take the point at which the force is applied as the origin.

If the characteristic equation has complex roots, then we should set A_1 and A_2 in the solution (32.7) of the homogeneous differential equation equal to zero, because the linearly independent functions u_0 and v_0 also become infinite as $\rho \to \infty$. The functions f_0 and g_0 which are present in this solution tend to zero at infinity. In order to find the coefficients of these functions, we should consider the conditions at the origin.

There are two such conditions. The first one asserts that the deflections of the plate are bounded. The second condition has a static character and consists in the fact that the sum of transversal forces which act on a circumference whose center is at the origin, tends to a given force, if the reduced radius of the circumference tends to zero. Thus, the second condition can, just as in the previous section, be written in the form

$$-\frac{D}{l^3}\frac{d}{d\rho}\nabla^2 w = -\frac{P}{2\pi l\rho},$$

$$\nabla^2 = \frac{d^2}{d\rho^2} + \frac{1}{\rho}\frac{d}{d\rho} = \Delta_0.$$

We give the expansions of the real and imaginary parts of the Hankel function of a complex argument, keeping only the terms which are essential for the calculations connected with passing to the limit as $\rho \to 0$:

$$\bar{g}_0(\rho) \sim \frac{2}{\pi}\left(\ln\frac{\rho}{2} + C\right) + \ldots,$$

$$\bar{f}_0(\rho) \sim \frac{1}{2\pi}\rho^2 \ln\rho\sin 2\varphi + 1 - \frac{2\varphi}{\pi} + \ldots,$$

where C is the Euler constant.

The first of these formulae shows that the coefficient of $g_0(\rho)$ must be assumed to be equal to zero, because the solution should not become infinite as $\rho \to 0$. Thus, $w = A_3\bar{f}_0(\rho)$. One can easily see[5], that as $\rho \to 0$

$$\frac{d}{d\rho}\Delta_0\bar{f}_0 = \frac{2}{\pi\rho}\sin 2\varphi.$$

Hence, $A_3 = Pl^2/(4D\sin 2\varphi)$; if $\varphi = \pi/4$, then $A_3 = Pl^2/(4D)$.

Consider an unbounded plate which is loaded by forces uniformly distributed along the circumference of a radius a; the reduced radius of this circumference is $\alpha = a/l$.

In order to obtain the solution of this problem, we use the equation of the deflection surface of the plate which is loaded by a concentrated force P. Consider the element of the arc of the circumference, which has the length $ds = a\,d\theta$. The force, which acts on this element, is equal to $q\,ds$. Let us find the deflection which is caused by this force at a point B whose polar coordinates will be denoted by (r, θ_1). The distance between the point at which the elementary force is applied and the point at which we calculate the deflection, is equal to $\sqrt{r^2 + a^2 - 2ar\cos(\theta - \theta_1)}$, and the deflection caused by the elementary loading is equal to

$$dw = \frac{qal^2}{4D\sin 2\varphi}\Re H_0^{(1)}\left(\frac{e^{i\varphi}}{l}\sqrt{r^2 + a^2 - 2ar\cos(\theta - \theta_1)}\right)d\theta.$$

In order to find the deflection caused by the given loading, we should integrate the expression obtained

$$w = \int_0^{2\pi}\frac{qal^2}{4D\sin 2\varphi}\Re H_0^{(1)}\left(\frac{e^{i\varphi}}{l}\sqrt{r^2 + a^2 - 2ar\cos(\theta - \theta_1)}\right)d\theta.$$

In order to calculate this integral, we should, as before, use the addition formulae for the cylindrical functions. These formulae have the form for $a \leq r$

$$H_0^{(1)}\left(\frac{1}{l}\sqrt{r^2 + a^2 - 2ar\cos(\theta - \theta_1)}\right) = 2{\sum}' J_n(a/l)H_n^{(1)}(r/l)\cos n(\theta - \theta_1),$$

_____

[5] See pages 14, 16.

for $a \geq r$

$$H_0^{(1)} \left(\frac{1}{l}\sqrt{r^2 + a^2 - 2ar\cos(\theta - \theta_1)} \right) = 2\sum' H_n^{(1)}(a/l)J_n(r/l)\cos n(\theta - \theta_1).$$

After integrating and separating the real and imaginary parts, we obtain the solution of the problem in the following form:

for $r \leq a$

$$w = w_\mathrm{I} = \frac{\pi q \alpha l^3}{2D\sin 2\varphi}[\bar{f}_0(\alpha)\bar{u}_0(\rho) - \bar{g}_0(\alpha)\bar{v}_0(\rho)]; \tag{32.8}$$

for $r \geq a$

$$w = w_\mathrm{II} = \frac{\pi q \alpha l^3}{2D\sin 2\varphi}[\bar{u}_0(\alpha)\bar{f}_0(\rho) - \bar{v}_0(\alpha)\bar{g}_0(\rho)], \tag{32.9}$$

where $\rho = r/l$, $\alpha = a/l$.

As a check shows, the functions $w(r)$, $w'(r)$ and $\Delta_0 w(r)$ are continuous at $r = a$. Hence, the moments G_1 and G_2 are also continuous; the function $\Delta_0 w'$ is discontinuous with the break q/D; hence, the transversal force has the required break equal to q.

The solution of the problem on a plate loaded along the circumference of the radius a by the moments q_m, whose vectors are directed along the tangent to the circumference, can be obtained by applying two loadings, equal in value and directed in opposite directions, to the circumferences, whose reduced radii are equal to α and $\alpha + \varepsilon$, setting $q_m = q\varepsilon$ and calculating the limit to which the solution tends as $\varepsilon \to 0$. The equation of the deflection surface, obtained in this way, has the form for $\rho \leq a$

$$w = \frac{\pi q_m \alpha l^2}{2D\sin 2\varphi}[\bar{f}_0'(\alpha)\bar{u}_0(\rho) - \bar{g}_0'(\alpha)\bar{v}_0(\rho)];$$

for $\rho \geq a$

$$w = \frac{\pi q_m \alpha l^2}{2D\sin 2\varphi}[\bar{u}_0'(\alpha)\bar{f}_0(\rho) - \bar{v}_0'(\alpha)\bar{g}_0(\rho)].$$

This solution together with $w'(\rho)$ and $\Delta_0 w'(\rho)$ remain continuous at $\rho = \alpha$. The function $\Delta_0 w(r)$ and the radial moment G_1 are discontinuous.

Let us consider the problem on loading a plate by the angular deformation θ_0. The solution, having a break in the slope of the given value θ_0 at $\rho = \alpha$ should remain continuous at $\rho = \alpha$ together with the moment G_1 and the transversal force N_1.

This solution has the form:

for $\alpha \geq \rho$

$$w = \frac{\pi \alpha \theta_0 l}{2\sin 2\varphi}\left\{ \left[\bar{f}_0(\alpha)\cos 2\varphi + \bar{g}_0(\alpha)\sin 2\varphi - \frac{1-\sigma}{\alpha}\bar{f}_0'(\alpha) \right] \bar{u}_0(\rho) \right.$$
$$\left. + \left[-\bar{g}_0(\alpha)\cos 2\varphi + \bar{f}_0(\alpha)\sin 2\varphi + \frac{1-\sigma}{\alpha}\bar{g}_0'(\alpha) \right] \bar{v}_0(\rho) \right\},$$

for $\alpha \leq \rho$

$$w = \frac{\pi \alpha \theta_0 l}{2 \sin 2\varphi} \left\{ \left[\bar{u}_0(\alpha) \cos 2\varphi + \bar{v}_0(\alpha) \sin 2\varphi - \frac{1-\sigma}{\alpha} \bar{u}_0'(\alpha) \right] \bar{f}_0(\rho) \right.$$

$$\left. + \left[-\bar{v}_0(\alpha) \cos 2\varphi + \bar{u}_0(\alpha) \sin 2\varphi + \frac{1-\sigma}{\alpha} \bar{v}_0'(\alpha) \right] \bar{g}_0(\rho) \right\}.$$

For $\alpha = \rho$, $\Delta_0 w$ as well as the circular moment G_2 are discontinuous.

5. The problem on the loading of a plate by the deflection f has the following solution:

for $\alpha > \rho$

$$w = -\frac{\pi \alpha f}{2} \left\{ [\bar{g}_0'(\alpha) + \bar{f}_0'(\alpha) \cotan 2\varphi] \bar{u}_0(\rho) + [\bar{f}_0'(\alpha) - \bar{g}_0'(\alpha) \cotan 2\varphi] \bar{v}_0(\rho) \right\} ;$$

for $\alpha < \rho$

$$w = -\frac{\pi \alpha f}{2} \left\{ [\bar{v}_0'(\alpha) + \bar{u}_0'(\alpha) \cotan 2\varphi] \bar{f}_0(\rho) + [\bar{u}_0'(\alpha) - \bar{v}_0(\alpha) \cotan 2\varphi] \bar{g}_0(\rho) \right\}.$$

For $\alpha = \rho$ this solution has a break of the value f. The first derivative, the bending moment G_1 and the transversal force N_1 remain continuous.

In conclusion we consider the problem of the calculation of a plate of an infinite radius, which is subjected to the loading q_0 uniformly distributed along the circle of the reduced radius β (Fig. 20).

Figure 20

In order to obtain the solution, we use the principle of addition of the actions, by integrating formulae (32.8) and (32.9). Thus, we have

$$\left. \begin{array}{ll} \text{for } \rho \leq \beta & w = \displaystyle\int_0^\rho w_{II}\, d\alpha + \int_\rho^\beta w_I\, d\alpha, \\[4mm] \text{for } \rho \geq \beta & w = \displaystyle\int_0^\beta w_{II}\, d\alpha, \end{array} \right\} \qquad (32.10)$$

where w_I and w_{II} are expressed by the formulae (32.8) and (32.9), respectively, in which the value q should be replaced by the value $l q_0$, taking into account that the dimension of q_0 is kg/cm^2.

Introducing the expressions for w_I and w_{II} into (32.10), we obtain

$$w = \frac{\pi q_0}{2 k_0 \sin 2\varphi} \left\{ \int_0^\rho \alpha [\bar{u}_0(\alpha) \bar{f}_0(\rho) - \bar{v}_0(\alpha) \bar{g}_0(\rho)]\, d\alpha \right.$$

$$\left. + \int_\rho^\beta \alpha [\bar{f}_0(\alpha) \bar{u}_0(\rho) - \bar{g}_0(\alpha) \bar{v}_0(\rho)]\, d\alpha \right\}.$$

In order to calculate these integrals, we use the elementary formula

$$\frac{d}{dz}[z Z_1(z)] = z Z_0(z),$$

where Z denotes an arbitrary cylindrical function. By integrating this equality, we obtain

$$\int z Z_0(z)\, dz = z Z_1(z). \qquad (32.11)$$

Setting $z = \rho e^{i\varphi}$, $Z_0 = J_0$ and taking into account the obvious equalities

$$z = \rho(\cos\varphi + i\sin\varphi);$$
$$J_0 = \bar{u}_0 + i\bar{v}_0;$$
$$dz = d\rho(\cos\varphi + i\sin\varphi),$$

we rewrite (32.11) in the following form:

$$\int z J_0(z)\, dz = \int \rho(\cos\varphi + i\sin\varphi)^2 [\bar{u}_0(\rho) + i\bar{v}_0(\rho)]\, d\rho$$
$$= (\cos\varphi + i\sin\varphi)^2 \int \rho[\bar{u}_0(\rho) + i\bar{v}_0(\rho)]\, d\rho.$$

On the other hand, setting

$$J_1(\rho e^{i\varphi}) = \bar{u}_1(\rho) + i\bar{v}_1(\rho)$$

and

$$H_1^{(1)}(\rho e^{i\varphi}) = \bar{f}_1(\rho) + i\bar{g}_1(\rho),$$

we obtain

$$z J_1(z) = \rho(\cos\varphi + i\sin\varphi)(\bar{u}_1 + i\bar{v}_1).$$

Thus, we have

$$\rho(\cos\varphi + i\sin\varphi)(\bar{u}_1 + i\bar{v}_1) = (\cos\varphi + i\sin\varphi)^2 \int \rho[\bar{u}_0(\rho) + i\bar{v}_0(\rho)]\, d\rho.$$

Cancelling by $\cos\varphi + i\sin\varphi$ and separating the real and imaginary parts, we obtain

$$\rho\bar{u}_1 = \cos\varphi \int \rho\bar{u}_0(\rho)\, d\rho - \sin\varphi \int \rho\bar{v}_0(\rho)\, d\rho,$$
$$\rho\bar{v}_1 = \sin\varphi \int \rho\bar{u}_0(\rho)\, d\rho + \cos\varphi \int \rho\bar{v}_0(\rho)\, d\rho;$$

therefore,

$$\int \rho\bar{u}_0(\rho)\, d\rho = \rho(\bar{u}_1 \cos\varphi + \bar{v}_1 \sin\varphi),$$
$$\int \rho\bar{v}_0(\rho)\, d\rho = \rho(\bar{v}_1 \cos\varphi - \bar{u}_1 \sin\varphi).$$

Without giving the obvious calculations, we write

$$\int \rho\bar{f}_0(\rho)\, d\rho = \rho(\bar{f}_1 \cos\varphi + \bar{g}_1 \sin\varphi),$$
$$\int \rho\bar{g}_0(\rho)\, d\rho = \rho(-\bar{f}_1 \sin\varphi + \bar{g}_1 \cos\varphi).$$

Using these formulae and noting that $\bar{u}_1(0) = 0$, $\bar{v}_1(0) = 0$, we transform (32.10), applying the Wronski determinant for necessary simplifications. As a result,

we obtain
for $\rho \leq \beta$

$$w = \frac{\pi q_0 \beta}{2k_0 \sin 2\varphi} \left\{ \frac{2 \sin 2\varphi}{\pi \beta} + [\bar{f}_1(\beta) \cos \varphi + \bar{g}_1(\beta) \sin \varphi] \bar{u}_0(\rho) \right.$$

$$\left. - [\bar{g}_1(\beta) \cos \varphi - \bar{f}_1(\beta) \sin \varphi] \bar{v}_0(\rho) \right\}. \quad (32.12)$$

Just in the same way we obtain:
for $\rho \geq \beta$

$$w = \frac{\pi q_0 \beta}{2k_0 \sin 2\varphi} \left\{ +[\bar{u}_1(\beta) \cos \varphi + \bar{v}_1(\beta) \sin \varphi] \bar{f}_0(\rho) \right.$$

$$\left. - [\bar{v}_1(\beta) \cos \varphi - \bar{u}_1(\beta) \sin \varphi] \bar{g}_0(\rho) \right\}. \quad (32.13)$$

For $\rho = 0$, i.e. at the center of the plate we have

$$w = \frac{\pi q_0 \beta}{2k_0 \sin 2\varphi} \left[\frac{2}{\pi \beta} \sin 2\varphi + \bar{f}_1(\beta) \cos \varphi + \bar{g}_1(\beta) \sin \varphi \right].$$

The bending moment at the center of the plate can be obtained by differentiating the expression (32.12). However, we use another approach, by calculating the moment at the center for the loading on the circumference and integrating the result obtained.

For loading by the force q uniformly distributed along the circumference of the reduced radius α, we have:

$$\text{for } \rho \leq \alpha \qquad w = \frac{\pi q l^3 \alpha}{2D \sin 2\varphi} \left[\bar{f}_0(\alpha) \bar{u}_0(\rho) - \bar{g}_0(\alpha) \bar{v}_0(\rho) \right].$$

The bending moment G_1 is expressed by the following formula:

$$G_1 = -\frac{D}{l^2} \left[\Delta w - (1 - \sigma) \frac{1}{\rho} \frac{dw}{d\rho} \right].$$

Let us give the formulae of the differentiation

$$\Delta \bar{u}_0 = -\bar{u}_0(\rho) \cos 2\varphi + \bar{v}_0(\rho) \sin 2\varphi, \left. \right\}$$
$$\Delta \bar{v}_0 = -\bar{u}_0(\rho) \sin 2\varphi - \bar{v}_0(\rho) \cos 2\varphi, \right\}$$

hence,

$$G_1 = \frac{-\pi q l \alpha}{2 \sin 2\varphi} \left\{ \left[-\bar{u}_0(\rho) \cos 2\varphi + \bar{v}_0(\rho) \sin 2\varphi - (1 - \sigma) \frac{\bar{u}_0'(\rho)}{\rho} \right] \bar{f}_0(\alpha) \right.$$

$$\left. + \left[\bar{u}_0(\rho) \sin 2\varphi + \bar{v}_0(\rho) \cos 2\varphi + (1 - \sigma) \frac{\bar{v}_0'(\rho)}{\rho} \right] \bar{g}_0(\alpha) \right\}. \quad (32.14)$$

Setting $\rho = 0$, we can determine the moment at the center of the circumference from the expression (32.14). In this case we should take into account that

$$\bar{u}_0(0) = 1,$$
$$\bar{v}_0(0) = 0.$$

Taking into account the equality

$$\frac{dJ_0(\rho e^{i\varphi})}{e^{i\varphi} d\rho} = -J_1(\rho e^{i\varphi}),$$

we obtain that as $\rho \to 0$

$$\bar{u}_0'(\rho) + i\bar{v}_0'(\rho) \approx -\frac{\rho}{2}(\cos 2\varphi + i\sin 2\varphi)$$

or

$$\bar{u}_0'(\rho) \approx -\frac{\rho}{2}\cos 2\varphi,$$

$$\bar{v}_0'(\rho) \approx -\frac{\rho}{2}\sin 2\varphi,$$

hence,

$$\frac{\bar{u}_0'(\rho)}{\rho} \approx -\frac{\cos 2\varphi}{2};$$

$$\frac{\bar{v}_0'(\rho)}{\rho} \approx -\frac{\sin 2\varphi}{2}.$$

Thus, the bending moment at the center of the circle is equal to

$$G_1(0) = \frac{\pi q l \alpha}{2}\left[\frac{1+\sigma}{2}\bar{f}_0(\alpha)\cotan 2\varphi - \frac{1+\sigma}{2}\bar{g}_0(\alpha)\right]$$

$$= \frac{-\pi q_0 l \alpha}{4}(1+\sigma)[\bar{g}_0(\alpha) - \bar{f}_0(\alpha)\cotan 2\varphi]. \qquad (32.15)$$

In order to obtain the bending moment at the center of the circle of the reduced radius β for the loading which is uniformly distributed on the area of the circle, we should integrate the expression (32.15) with respect to α from 0 to β. After integrating, we obtain

$$G_1^*(0) = \frac{-\pi q_0 l^2 \beta}{4}(1+\sigma)\{-[\cotan 2\varphi \cos \varphi + \sin \varphi]\bar{f}_1(\beta)$$

$$+ [\cos \varphi - \cotan 2\varphi \sin \varphi]\bar{g}_1(\beta)\}, \qquad (32.16)$$

where q_0 denotes the loading per the unit of the area.

33. Equilibrium of a circular plate lying on an elastic foundation. Non-axially symmetrical deformation

In this section, in contrast to the previous one, we consider a plate which lies on an elastic Winkler-type foundation which is not loaded, however, by axial forces. Therefore, p_0 should be assumed equal to zero, and we should take $\varphi = \pi/4$ in the argument of $\rho e^{i\varphi}$. In the notation for the real and imaginary parts of the Bessel functions we should omit the bars over the letters u, v, f and g in this case.

It follows from Section 32 that the solution of the problem on an unbounded plate which is loaded by a concentrated force P has the form

$$w = P l^2 f_0(\xi)/(4D).$$

By integrating this expression, which is the basic influence function for $P = 1$, we obtain the basic solutions for some special problems.

Consider a plate which is loaded by the forces $q \cos n\theta$ distributed along the circumference whose reduced radius is denoted by α.

Using the principle of addition of the actions, we represent the deflection at a point with the coordinates ξ, φ in the following form:

$$w = \frac{qal^3}{4D} \int_0^{2\pi} f_0 \left(\sqrt{\alpha^2 + \xi^2 - 2\alpha\xi \cos(\theta - \varphi)} \right) \cos n\theta \, d\theta.$$

In order to compute this integral, we use, as before, the addition formula for the cylindrical functions

$$Z_0 \left(\sqrt{\alpha^2 + \xi^2 - 2\alpha\xi \cos(\theta - \varphi)} \right) = 2 \sum_{n=0}^{\infty} {}' J_n(\alpha) Z_n(\xi) \cos n(\theta - \varphi), \qquad (33.1)$$

where the sign $'$ means that we should introduce the coefficient $1/2$ for $n = 0$. This formula holds for $\alpha \le \xi$. If $\alpha \ge \xi$, then we should interchange α and ξ in the right-hand side of (33.1). Setting

$$Z_0 = H_0^{(1)} \left(\sqrt{i} \sqrt{\alpha^2 + \xi^2 - 2\alpha\xi \cos(\theta - \varphi)} \right),$$

after integrating and taking the real part, we obtain
for $\xi \le \alpha$

$$w = w_{\mathrm{I}} = \pi\alpha q l^3 [u_n(\xi) f_n(\alpha) - v_n(\xi) g_n(\alpha)] \cos n\varphi/(2D), \qquad (33.2)$$

for $\xi \ge \alpha$

$$w = w_{\mathrm{II}} = \pi\alpha q l^3 [u_n(\alpha) f_n(\xi) - v_n(\alpha) g_n(\xi)] \cos n\varphi/(2D). \qquad (33.3)$$

Setting $n = 0$ in the expressions (33.2) and (33.3), we obtain the formulae for the axially symmetric problem.

Using the Wronskian of the Bessel equation, one can show that the solutions obtained satisfy the conditions of the conjunction at $\xi = \alpha$.

By integrating the expressions (33.2) and (33.3), we can solve the problem on an unbounded plate, which is loaded according the law $\Psi(\xi) \cos n\theta$, where $\Psi(\xi)$ is a given function.

If a loading $q(\varphi)$ distributed along the circumference of the reduced radius α acts on the plate, then, expanding this loading into the Fourier series, we have

$$q = \sum_{n=0}^{\infty} (a_n \cos n\theta + b_n \sin n\theta)$$

and, using the formulae (33.2) and (33.3), we obtain the solution
for $\alpha \le \xi$

$$w = \frac{\pi\alpha l^3}{2D} \left\{ \sum_{n=0}^{\infty} a_n [u_n(\alpha) f_n(\xi) - v_n(\alpha) g_n(\xi)] \sin n\varphi \right.$$

$$\left. + \sum_{n=0}^{\infty} b_n [u_n(\alpha) f_n(\xi) - v_n(\alpha) g_n(\xi)] \cos n\varphi \right\}; \qquad (33.4)$$

for $\alpha \geq \xi$

$$w = \frac{\pi \alpha l^3}{2D}\left\{ \sum_{n=0}^{\infty} a_n[u_n(\xi)f_n(\alpha) - v_n(\xi)g_n(\alpha)]\sin n\varphi \right.$$

$$\left. + \sum_{n=0}^{\infty} b_n[u_n(\xi)f_n(\alpha) - v_n(\xi)g_n(\alpha)]\cos n\varphi \right\}. \tag{33.5}$$

With the help of the formulae (33.4) and (33.5) it is convenient to construct the basic solution for the loading represented in the form of the series

$$q = \sum_{n=0}^{\infty} \Psi_n(\xi)[a_n \cos n\theta + b_n \sin n\theta].$$

If the loading applied to the plate can be represented in the form

$$q = \sum_{n=1}^{\infty} \xi^n(a_n \cos n\theta + b_n \sin n\theta) \tag{33.6}$$

or

$$q = \sum_{n=1}^{\infty} \xi^{-n}(a_n \cos n\theta + b_n \sin n\theta), \tag{33.7}$$

where n is a positive integer, then the integration is especially simple[6].
Remember the formulae of differentiation of the Bessel functions

$$\left. \begin{array}{c} \dfrac{d}{dz}[z^n J_n(z)] = z^n J_{n-1}(z), \\[2mm] \dfrac{d}{dz}[z^n H_n^{(1)}(z)] = z^n H_{n-1}^{(1)}(z). \end{array} \right\}$$

Setting $z = \xi\sqrt{i}$, integrating and dividing the real and imaginary parts, we obtain the following formulae:

$$\int \xi^n u_{n-1}(\xi)\, d\xi = \xi^n[u_n(\xi) + v_n(\xi)]/\sqrt{2},$$

$$\int \xi^n v_{n-1}(\xi)\, d\xi = -\xi^n[u_n(\xi) - v_n(\xi)]/\sqrt{2},$$

$$\int \xi^n f_{n-1}(\xi)\, d\xi = \xi^n[f_n(\xi) + g_n(\xi)]/\sqrt{2},$$

$$\int \xi^n g_{n-1}(\xi)\, d\xi = -\xi^n[f_n(\xi) - g_n(\xi)]/\sqrt{2}.$$

The solution for the loading which is varied according to the rule (33.6) and is distributed on the area of a circular ring, can be obtained after integrating the expressions (33.2) and (33.3). For this purpose we should firstly replace the loading q in these expressions by the elementary loading $q_0\alpha^n\, dz = q_0\alpha^n l\, d\alpha$, where q_0 has the dimension kg/cm^2. Let us note that in this case the factor $l^4/D = 1/k_0$ appears. Then we integrate these expressions from $\alpha = \alpha_1$ to $\alpha = \alpha_2$.

---

[6] If the exponent of ξ in (33.6) and (33.7) is different from $\pm n$, then the solution can be obtained via the Lommel functions of a complex argument.

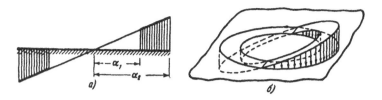

Figure 21

For $\xi \leq \alpha_1 < \alpha_2$ the solution has the form

$$w = \int_{\alpha_1}^{\alpha_2} \frac{\pi q_0 \alpha^{n+1}}{2k_0} [u_n(\xi) f_n(\alpha) - v_n(\xi) g_n(\alpha)] \cos n\varphi \, d\alpha$$

$$= \frac{\pi q_0}{2k_0\sqrt{2}} (\{\alpha_2^{n+1}[f_{n+1}(\alpha_2) + g_{n+1}(\alpha_2)]$$
$$- \alpha_1^{n+1}[f_{n+1}(\alpha_1) + g_{n+1}(\alpha_1)]\} u_n(\xi)$$
$$+ \{\alpha_2^{n+1}[f_{n+1}(\alpha_2) - g_{n+1}(\alpha_2)]$$
$$- \alpha_1^{n+1}[f_{n+1}(\alpha_1) - g_{n+1}(\alpha_1)]\} v_n(\xi)) \cos n\varphi;$$

For $\xi \geq \alpha_2 > \alpha_1$

$$w = \frac{\pi q_0}{2k_0\sqrt{2}} (\{\alpha_2^{n+1}[u_{n+1}(\alpha_2) + v_{n+1}(\alpha_2)]$$
$$- \alpha_1^{n+1}[u_{n+1}(\alpha_1) + v_{n+1}(\alpha_1)]\} f_n(\xi)$$
$$+ \{\alpha_2^{n+1}[u_{n+1}(\alpha_2) - v_{n+1}(\alpha_2)]$$
$$- \alpha_1^{n+1}[u_{n+1}(\alpha_1) - v_{n+1}(\alpha_1)]\} g_n(\xi)) \cos n\varphi.$$

For $\alpha_2 \geq \xi \geq \alpha_1$ the solution can be written in the form

$$w = \frac{\pi q_0}{2k_0\sqrt{2}} (\{\alpha_2^{n+1}[f_{n+1}(\alpha_2) + g_{n+1}(\alpha_2)] - \xi^{n+1}[f_{n+1}(\xi) + g_{n+1}(\xi)]\} u_n(\xi)$$
$$+ \{\alpha_2^{n+1}[f_{n+1}(\alpha_2) - g_{n+1}(\alpha_2)] - \xi^{n+1}[f_{n+1}(\xi) - g_{n+1}(\xi)]\} v_n(\xi)$$
$$+ \{\xi^{n+1}[u_{n+1}(\xi) + v_{n+1}(\xi)] - \alpha_1^{n+1}[u_{n+1}(\alpha_1) + v_{n+1}(\alpha_1)]\} f_n(\xi)$$
$$+ \{\xi^{n+1}[u_{n+1}(\xi) - v_{n+1}(\xi)] - \alpha_1^{n+1}[u_{n+1}(\alpha_1) - v_{n+1}(\alpha_1)]\} g_n(\xi)) \cos n\varphi$$

$$= \frac{\pi q_0}{2k_0\sqrt{2}} \{\alpha_2^{n+1}[f_{n+1}(\alpha_2) + g_{n+1}(\alpha_2)] u_n(\xi)$$
$$+ \alpha_2^{n+1}[f_{n+1}(\alpha_2) - g_{n+1}(\alpha_2)] v_n(\xi)$$
$$- \alpha_1^{n+1}[u_{n+1}(\alpha_1) + v_{n+1}(\alpha_1)] f_n(\xi)$$
$$- \alpha_1^{n+1}[u_{n+1}(\alpha_1) - v_{n+1}(\alpha_1)] g_n(\xi) + 2\xi^n \sqrt{2}/\pi\} \cos n\varphi.$$

These formulae give the expression of the nth term of the series, which represents the deflection for the loading given by (33.6).

The case $n = 1$, i.e. the loading by the forces varying according to the rule $q(\xi)\cos\theta$ (Fig. 21a,b), has an independent practical value. Such a loading can appear when calculating the foundations, on which circular cylinders rest which are subjected to the action of horizontal forces; with such a loading the circumferences with the center at origin remain plane, rotating one with respect to another.

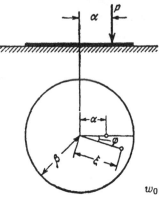

Figure 22

The work [21] contains tables of the functions, the use of which enables one to obtain numerical results for the case $n = 1$.

Consider the problem of the calculation of a circular plate with the free edge, which is loaded by a concentrated force P. Let β denote the reduced radius of the plate, and α denote the reduced distance from the point at which the concentrated force is applied to the center of the circle. The basic solution for the loading by a concentrated force has the form

$$w_0 = \frac{Pl^2}{4D} f_0 \left(\sqrt{\alpha^2 + \xi^2 - 2\alpha\xi \cos\varphi} \right), \tag{33.8}$$

where ξ, φ are the polar coordinates of the point considered (Fig. 22). In this case φ is calculated from the radius which passes through the point at which the force is applied.

In order to obtain the compensating solution, we first, using the addition formula, write the basic solution w_0 for $\xi > \alpha$ in the form of the series

$$w_0 = \frac{Pl^2}{4D} f_0 \sum{}' [u_n(\alpha) f_n(\xi) - v_n(\alpha) g_n(\xi)] \cos n\varphi. \tag{33.9}$$

On the circumference of the radius β, when deforming the plate on the surface w_0, the bending moments arise, which can be calculated

$$G_1 = -\frac{P}{2} \sum_{n=0}^{\infty}{}' \left\{ u_n(\alpha) \left[g_n(\beta) - \frac{1-\sigma}{\beta} \left(f_n'(\beta) - \frac{n^2 f_n(\beta)}{\beta} \right) \right] \right.$$
$$\left. + v_n(\alpha) \left[f_n(\beta) + \frac{1-\sigma}{\beta} \left(g_n'(\beta) - \frac{n^2 g_n(\beta)}{\beta} \right) \right] \right\} \cos n\varphi. \tag{33.10}$$

The reduced transversal forces will be defined with the help of the formulae

$$N_1 - \frac{\partial H_1}{\partial s} = -\frac{P}{2l} \sum{}' \left\{ u_n(\alpha) \left[g_n'(\beta) - \frac{(1-\sigma)n^2}{\beta^2} \left(f_n'(\beta) - \frac{f_n(\beta)}{\beta} \right) \right] \right.$$
$$\left. + v_n(\alpha) \left[f_n'(\beta) + \frac{(1-\sigma)n^2}{\beta^2} \left(g_n'(\beta) - \frac{g_n(\beta)}{\beta} \right) \right] \right\} \cos n\varphi. \tag{33.11}$$

We seek the compensating solution in the form of a series

$$w_k = \frac{Pl^2}{2D} \sum_{n=0}^{\infty}{}' [a_n u_n(\xi) + b_n v_n(\xi)] \cos n\varphi. \tag{33.12}$$

In order to determine the unknown coefficients a_n and b_n we use the conditions on the contour. Since the plate has the free edge, for $\xi = \beta$ the bending moment and the reduced transversal force are equal to zero.

We have determined the values of the moments and transversal forces on the contour, which appear as a result of the deformation described by the basic solution. We now find the moments and the transversal forces caused by the compensating solution.

Substituting the expression (33.5) into formula (32.10) and setting $\xi = \beta$, we obtain:

$$G_1^* = -\frac{P}{2} \sum_{n=0}^{\infty}{}' \left\{ a_n \left[v_n(\beta) - \frac{1-\sigma}{\beta} \left(u_n'(\beta) - \frac{n^2 u_n(\beta)}{\beta} \right) \right] \right.$$

$$\left. - b_n \left[u_n(\beta) + \frac{1-\sigma}{\beta} \left(v_n'(\beta) - \frac{n^2 v_n(\beta)}{\beta} \right) \right] \right\} \cos n\varphi. \qquad (33.13)$$

Just in the same way we obtain

$$N_1^* - \frac{\partial H_1^*}{\partial s} = -\frac{P}{2l} \sum{}' \left\{ a_n \left[v_n'(\beta) - \frac{n^2(1-\sigma)}{\beta^2} \left(u_n'(\beta) - \frac{u_n(\beta)}{\beta} \right) \right] \right.$$

$$\left. - b_n \left[u_n'(\beta) + \frac{n^2(1-\sigma)}{\beta^2} \left(v_n'(\beta) - \frac{v_n(\beta)}{\beta} \right) \right] \right\} \cos n\varphi. \qquad (33.14)$$

It follows from the boundary conditions that at $\xi = \beta$ we have

$$G_1 + G_1^* = 0;$$

$$N_1 - \frac{\partial H_1}{\partial s} + N_1^* - \frac{\partial H_1^*}{\partial s} = 0. \qquad (33.15)$$

Making the coefficients of the cosines with equal n equal and solving the systems of the equations, which are obtained in this case, we find:

$$a_n = \frac{A_n u_n^{[Q]}(\beta) - B_n u_n^{[M]}(\beta)}{u_n^{[M]}(\beta) v_n^{[Q]}(\beta) - v_n^{[M]}(\beta) u_n^{[Q]}(\beta)}, \qquad (33.16)$$

$$b_n = \frac{-A_n v_n^{[Q]}(\beta) + B_n v_n^{[M]}(\beta)}{u_n^{[M]}(\beta) v_n^{[Q]}(\beta) - v_n^{[M]}(\beta) u_n^{[Q]}(\beta)}, \qquad (33.17)$$

where

$$A_n = u_n(\alpha) g_n^{[M]}(\beta) - v_n(\alpha) f_n^{[M]}(\beta),$$

$$B_n = u_n(\alpha) g_n^{[Q]}(\beta) - v_n(\alpha) f_n^{[Q]}(\beta),$$

$$u_n^{[M]}(\beta) = -u_n(\beta) - \frac{1-\sigma}{\beta} \left[v_n'(\beta) - \frac{n^2}{\beta} v_n(\beta) \right],$$

$$v_n^{[M]}(\beta) = v_n(\beta) - \frac{1-\sigma}{\beta} \left[u_n'(\beta) - \frac{n^2}{\beta} u_n(\beta) \right],$$

$$f_n^{[M]}(\beta) = -f_n(\beta) - \frac{1-\sigma}{\beta} \left[g_n'(\beta) - \frac{n^2}{\beta} g_n(\beta) \right],$$

$$g_n^{[M]}(\beta) = g_n(\beta) - \frac{1-\sigma}{\beta} \left[f_n'(\beta) - \frac{n^2}{\beta} f_n(\beta) \right],$$

$$u_n^{[Q]}(\beta) = -u_n'(\beta) - (1-\sigma)\frac{n^2}{\beta^2} \left[v_n'(\beta) - \frac{v_n(\beta)}{\beta} \right],$$

$$v_n^{[Q]}(\beta) = v_n'(\beta) - (1-\sigma)\frac{n^2}{\beta^2} \left[u_n'(\beta) - \frac{u_n(\beta)}{\beta} \right],$$

$$f_n^{[Q]}(\beta) = -f_n'(\beta) - (1-\sigma)\frac{n^2}{\beta^2} \left[g_n'(\beta) - \frac{g_n(\beta)}{\beta} \right],$$

$$g_n^{[Q]}(\beta) = g_n'(\beta) - (1-\sigma)\frac{n^2}{\beta^2} \left[f_n'(\beta) - \frac{f_n(\beta)}{\beta} \right].$$

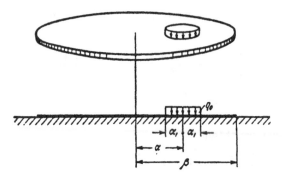

Figure 23

Now consider the problem of the calculation of a circular plate with the free edge, which is loaded by a loading q_0 uniformly distributed on the circle of the radius α_1 whose center is located on the distance α from the center of the plate (Fig. 23). We assume that $\alpha + \alpha_1 < \beta$, i.e. the whole circle lies inside the contour of the plate.

Consider a point with the coordinates (ξ, φ). Its distance from the center of the circle, on which the loading acts, is equal to

$$z = \sqrt{\alpha^2 + \xi^2 - 2\alpha\xi \cos\varphi}.$$

The basic solution represents an deflection surface of an unbounded plate which is loaded by the loading uniformly distributed on the circle. From the formulae of Chapter II we see that
for $z \leq \alpha_1$

$$w_0 = \frac{q_0 \pi \alpha_1}{2k_0} \left[-\frac{2}{\pi \alpha_1} + g_0'(\alpha_1) u_0(z) + f_0'(\alpha_1) v_0(z) \right], \qquad (33.18)$$

and for $z \geq \alpha_1$

$$w_0 = \frac{q_0 \pi \alpha_1}{2k_0} \left[v_0'(\alpha_1) f_0(z) + u_0'(\alpha_1) g_0(z) \right]. \qquad (33.19)$$

In order to obtain the compensating solution, we should first represent the basic solution as a series in $\cos n\varphi$.

The basic solution can be considered as the sum of two terms, the first of which contains $f_0(z)$ and the second one contains $g_0(z)$.

We have used the expansion of $f_0(z)$ many times. Therefore, we now consider only the representation of the function $g_0(z)$ in the form of a series. By definition

$$g_0(z) = \Im H_0^{(1)}(z\sqrt{i}).$$

After replacing z in the addition formula by its value and setting $\xi \leq \alpha$, we have

$$H_0^{(1)}\left(\sqrt{i}\sqrt{\alpha^2 + \xi^2 - 2\alpha\xi \cos\varphi}\right) = 2\sum_{n=0}^{\infty}{}' J_n(\alpha\sqrt{i}) H_n^{(1)}(\xi\sqrt{i}) \cos n\varphi. \qquad (33.20)$$

Introducing in accordance with Section 1 the notation

$$H_n^{(1)}(\xi\sqrt{i}) = f_n(\xi) + ig_n(\xi),$$
$$J_n(\xi\sqrt{i}) = u_n(\xi) + iv_n(\xi),$$

and separating the imaginary part, we obtain

$$g_0(z) = 2\sum_{n=0}^{\infty}{}'[v_n(\alpha)f_n(\xi) + u_n(\alpha)g_n(\xi)]\cos n\varphi.$$

Thus, for $\xi \geq \alpha$ the basic solution can be represented in the form of the following series:

$$w_0 = \frac{\pi q_0 \alpha_1}{k_0}\Big\{v_0'(\alpha_1)\sum{}'[u_n(\alpha)f_n(\xi) - v_n(\alpha)g_n(\xi)]\cos n\varphi$$

$$+u_0'(\alpha_1)\sum{}'[v_n(\alpha)f_n(\xi) + u_n(\alpha)g_n(\xi)]\cos n\varphi\Big\}. \qquad (33.21)$$

The bending moments and the reduced transversal forces on the contour of the plate for the deformation which is described by the basic solution, have the form

$$G_1^* = -\frac{\pi q_0 D \alpha_1}{k_0 l^2}\Big\{v_0'(\alpha_1)\sum{}'[u_n(\alpha)g_n^{[M]}(\beta) - v_n(\alpha)f_n^{[M]}(\beta)]\cos n\varphi$$

$$+ u_0'(\alpha_1)\sum{}'[v_n(\alpha)g_n^{[M]}(\beta) + u_n(\alpha)f_n^{[M]}(\beta)]\cos n\varphi\Big\}. \qquad (33.22)$$

$$N_1 - \frac{\partial H_1}{\partial s} = -\frac{\pi q_0 D \alpha_1}{k_0 l^3}\Big\{v_0'(\alpha_1)\sum{}'[u_n(\alpha)g_n^{[Q]}(\beta) - v_n(\alpha)f_n^{[Q]}(\beta)]\cos n\varphi$$

$$+ u_0'(\alpha_1)\sum{}'[v_n(\alpha)g_n^{[Q]}(\beta) + u_n(\alpha)f_n^{[Q]}(\beta)]\cos n\varphi\Big\}. \qquad (33.23)$$

As before, we seek the compensating solution in the form

$$w_k = \frac{Pl^2}{2D}\sum{}'[a_n u_n(\xi) + b_n v_n(\xi)]\cos n\varphi,$$

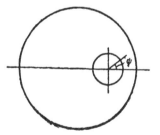

where P is the total loading which is equal to $q_0 \pi \alpha_1^2 l^2$.

Then by formulae (32.10) and (32.13) we obtain the expressions for the bending moment G_{1k} and the transversal force N_{1k} which correspond to the compensating solution. The coefficients a_n and b_n can easily be found by solving the system of equations

$$G_1^* + G_{1k} = 0,$$
$$N_1^* + N_{1k} = 0.$$

The problem can now be treated as solved.

Let us note that the result obtained can easily be generalized to the case of an arbitrary loading, which is distributed on the circle of the radius α_1 and is symmetrical with respect to the center of the circle. The reason is that for any such loading, the basic solution has for $z > \alpha_1$ the form

Figure 24

$$w_0 = A f_0(z) + B g_0(z).$$

Thus, in this case we should only replace the values $\dfrac{q_0\pi\alpha_1}{2k_0}v_0'(\alpha_1)$ and $\dfrac{q_0\pi\alpha_1}{2k_0}u_0'(\alpha_1)$ by A and B, respectively.

Consider the problem for the special case, when the loading is varied according to the rule

$$q = az\cos\psi$$

(the notation is given in Fig. 24).

The basic solution has the form

$$w_0 = [Af_1(z) + Bg_1(z)]\cos\psi,$$

where A and B are defined by the formulae of this section. For $\alpha \geq \xi$ the following formula holds

$$Z_1\left(\sqrt{\alpha^2 + \xi^2 - 2\alpha\xi\cos\varphi}\right)\cos\psi = 2\sum_{m=0}^{\infty}{}' Z_{m+1}(\alpha)J_m(\xi)\cos m\varphi.$$

Replacing α and ξ by $\alpha\sqrt{i}$ and $\xi\sqrt{i}$, respectively, setting $Z_1 = H_1^{(1)}(\xi\sqrt{i})$ and separating the real and imaginary parts, we obtain

$$\Re Z_1\cos\psi = 2\sum_{m=0}^{\infty}{}'[f_{m+1}(\alpha)u_m(\xi) - g_{m+1}(\alpha)v_m(\xi)]\cos m\varphi,$$

$$\Im Z_1\cos\psi = 2\sum_{m=0}^{\infty}{}'[f_{m+1}(\alpha)v_m(\xi) + g_{m+1}(\alpha)u_m(\xi)]\cos m\varphi.$$

Using these expansions, we can easily obtain the compensating solution.

34. Calculation of a circular conic shell under the action of axially symmetrical loadings and non-uniform heating

Consider the calculation of a circular conic shell of constant thickness, which is under the action of axially symmetrical forces and non-uniform heating. For the sake of simplifying the calculations, we apply the method which is known in structural mechanics as the method of initial parameters.

This method is developed in detail in connection with problems on the bending of a beam, on the bending of a beam on an elastic Winkler-type foundation and on the oscillations of a beam, which can be reduced to the solution of differential equations of the fourth order

$$\frac{d^4y}{dx^4} = f(x),$$

$$\frac{d^4y}{dx^4} + y = f_1(x),$$

$$\frac{d^4y}{dx^4} - y = f_2(x).$$

Its foundation is presented, for instance, in the books by A.N. Krylov [25, 26]. It is known that the essential feature of this method is the special choice of the fundamental system of the solutions Y_1, Y_2, Y_3, Y_4, which satisfy the following conditions at $x = 0$:

	Y_1	Y_2	Y_3	Y_4
Y	1	0	0	0
Y'	0	1	0	0
Y''	0	0	1	0
Y'''	0	0	0	1

This allows us to simplify the fulfillment of the conditions of conjunction in the process of finding a particular solution for discontinuous right-hand side and of determination the constants present in the expression for the general solution of the homogeneous equation. The application of this method to the problems which are described by a system of two Bessel equations, has been presented in [18]–[20].

It is most fruitful to apply the method of the initial parameters to shells of a small length or which have a large angle at the top. The basis of the method of initial parameters in connection with the problem on a conic shell has been considered in the work [20].

Let us first give the main formulae of the calculations.

$$T_1 = B\left[\frac{du}{dr} + \sigma\left(\frac{u}{r} - \frac{w}{r\tan\gamma}\right)\right], \tag{34.1}$$

$$T_2 = B\left[\sigma\frac{du}{dr} + \frac{u - w/\tan\gamma}{r}\right], \tag{34.2}$$

$$G_1 = -D\left(\frac{d^2w}{dr^2} + \sigma\frac{dw}{r\,dr}\right), \tag{34.3}$$

$$G_1 = -D\left(\sigma\frac{d^2w}{dr^2} + \frac{dw}{r\,dr}\right), \tag{34.4}$$

$$N = -T_1\cotan\gamma. \tag{34.5}$$

In accordance with Fig. 25, T_1, G_1 and N_1 here denote, respectively, the normal force, the bending moment and the transversal force in the cross-sections normal to the generator; and T_2 and G_2 are the normal force and the bending moment in the cross-sections along the generator. Let u and w denote the components of the displacement along the generator and normal to the surface; r denote the distance from the point considered of the median surface to the top of the cone, 2γ denote the angle at the top of the cone. Let D denote the flexural rigidity of the shell, and, finally, $B = Eh/(1 - \sigma^2)$.

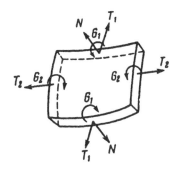

Figure 25

It is known that from the equations of equilibrium and of consistency, as well as from (34.1)–(34.5) (see, for instance, [36]), one can obtain the differential equations

$$\frac{d^2\xi}{d\zeta^2} + \frac{1}{\zeta}\frac{d\xi}{d\zeta} - \frac{4\xi}{\zeta^2} + \frac{24}{h^2}\eta = 0, \tag{34.6}$$

$$\frac{d^2\eta}{d\zeta^2} + \frac{1}{\zeta}\frac{d\eta}{d\zeta} - \frac{4\eta}{\zeta^2} - \frac{2(1-\sigma^2)}{\tan^2\gamma}\xi = 0, \tag{34.7}$$

where $\xi = dw/dr$, $\eta = Nr/B$, $\zeta = \sqrt{2}r$.

Setting $k^4 = 48(1-\sigma^2)/(h^2\tan^2\gamma)$ and passing to the argument $x = k\zeta$, we obtain the equations

$$\frac{d^2\xi}{dx^2} + \frac{1}{x}\frac{d\xi}{dx} - \frac{4\xi}{x^2} + \frac{2\sqrt{3}\tan\gamma}{h\sqrt{1-\sigma^2}}\eta = 0,$$

$$\frac{d^2\eta}{dx^2} + \frac{1}{x}\frac{d\eta}{dx} - \frac{4\eta}{x^2} - \frac{h\sqrt{1-\sigma^2}}{2\sqrt{3}\tan\gamma}\xi = 0,$$

which, after introducing the operator $L = \dfrac{d^2}{dx^2} + \dfrac{1}{x}\dfrac{d}{dx} - \dfrac{4}{x^2}$, can be concisely written in the following form:

$$L(\xi) + \frac{2\sqrt{3}\tan\gamma}{h\sqrt{1-\sigma^2}}\eta = 0, \tag{34.8}$$

$$L(\eta) - \frac{h\sqrt{1-\sigma^2}}{2\sqrt{3}\tan\gamma}\xi = 0. \tag{34.9}$$

Eliminating the indeterminate η from the equations (34.8) and (34.9), we obtain the equation

$$L^2(\xi) + \xi = 0. \tag{34.10}$$

The integral of this equation can be written in the following form (see Section 13, Part I):

$$\xi = C_1 J_2(x\sqrt{i}) + C_2 J_2(x\sqrt{-i}) + C_3 Y_2(x\sqrt{i}) + C_4 Y_2(x\sqrt{-i}),$$

where J_2 is the cylindrical function of the first kind of the second order, and Y_2 is the cylindrical function of the second kind of the second order.

We represent ξ and η as a linear combination of the four functions $Y_1(x,a)$, $Y_2(x,a)$, $Y_3(x,a)$ $Y_4(x,a)$, which satisfy equation (34.10) as well as some relations written below, which simplify the fulfillment of the conjunction conditions. These functions are formed from the Bessel function $J_2(x\sqrt{i}) = u_2(x) + iv_2(x)$ and the Hankel function of the first kind $H_2^{(1)}(x\sqrt{i}) = f_2(x) + ig_2(x)$. For this purpose we set that

for $x < a$

$$Y_n(x,a) = 0 \qquad (n = 1, 2, 3, 4),$$

for $x \geq a$

$$
\left.
\begin{aligned}
Y_1(x, a) &= \frac{\pi a}{2}\left[-u_2'(a)g_2(x) + f_2'(a)v_2(x) - v_2'(a)f_2(x) + g_2'(a)u_2(x)\right], \\
Y_2(x, a) &= \frac{\pi a}{2}\left[-u_2(a)g_2(x) + f_2(a)v_2(x) - v_2(a)f_2(x) + g_2(a)u_2(x)\right], \\
Y_3(x, a) &= \frac{\pi a}{2}\left[u_2'(a)f_2(x) - f_2'(a)u_2(x) - v_2'(a)g_2(x) + g_2'(a)v_2(x)\right], \\
Y_4(x, a) &= \frac{\pi a}{2}\left[u_2(a)f_2(x) - f_2(a)u_2(x) - v_2(a)g_2(x) + g_2(a)v_2(x)\right].
\end{aligned}
\right\}
\tag{34.11}
$$

The functions $Y_n(x, a)$ are chosen such that they satisfy at $x = a$ the following conditions:

n	Y_n	Y_n'	$-L(Y_n)$	$-L'(Y_n)$
1	1	0	0	0
2	0	1	0	0
3	0	0	1	0
4	0	0	0	1

$$\tag{34.12}$$

There exist the following dependences between the functions $u_2(x)$, $v_2(x)$, $f_2(x)$, and $g_2(x)$:

$$
L(u_2) = v_2, \qquad L(v_2) = -u_2, \qquad L(f_2) = g_2, \qquad L(g_2) = -f_2, \tag{34.13}
$$

$$
u_2(x)f_2'(x) - v_2(x)g_2'(x) - u_2'(x)f_2(x) + v_2'(x)g_2(x) = 0, \tag{34.14}
$$

$$
v_2(x)f_2'(x) + u_2(x)g_2'(x) - u_2'(x)g_2(x) - v_2'(x)f_2(x) = 2/(\pi x). \tag{34.15}
$$

The formulae (34.13) can easily be checked directly. The formulae (34.14) and (34.15) are obtained from consideration of the Wronski determinant.

There also exist the differential dependences between the functions Y_1, Y_2, Y_3, and Y_4:

$$
L(Y_1) = Y_3, \qquad L(Y_2) = Y_4, \qquad L(Y_3) = -Y_1, \qquad L(Y_4) = -Y_2.
$$

We now write the solutions of equations (34.8) and (34.9) in the following form:

$$
\xi = A_1 Y_1 + A_2 Y_2 + A_3 Y_3 + A_4 Y_4, \tag{34.16}
$$

$$
\eta = -\frac{h\sqrt{1 - \sigma^2}}{2\sqrt{3}\tan\gamma}(A_1 Y_3 + A_2 Y_4 - A_3 Y_1 - A_4 Y_2). \tag{34.17}
$$

Knowing η, we can obtain the expressions for the forces

$$
N = B\frac{\eta}{r} = \frac{Eh^2 k^2}{x^2\sqrt{3(1 - \sigma^2)}\tan\gamma}(A_1 Y_3 + A_2 Y_4 - A_3 Y_1 - A_4 Y_2), \tag{34.18}
$$

$$
T_1 = -\frac{1}{\cotan\gamma}N = \frac{Eh^2 k^2}{x^2\sqrt{3(1 - \sigma^2)}}(A_1 Y_3 + A_2 Y_4 - A_3 Y_1 - A_4 Y_2), \tag{34.19}
$$

$$
T_2 = \frac{d}{dr}(T_1 r) = \frac{Eh^2 k^2}{2x\sqrt{3(1 - \sigma^2)}}(A_1 Y_3' + A_2 Y_4' - A_3 Y_1' - A_4 Y_2'). \tag{34.20}
$$

The bending moments should be determined from the formulae (34.3) and (34.4) which, with the help of the notation for ξ and x given above, take the form

$$G_1 = -\frac{Dk^2}{x}\left(\frac{d\xi}{dx} + \frac{2\sigma}{x}\xi\right),$$

$$G_2 = -\frac{Dk^2}{x}\left(\sigma\frac{d\xi}{dx} + \frac{2}{x}\xi\right).$$

Hence,

$$G_1 = -\frac{Eh^2}{x\sqrt{3(1-\sigma^2)}\tan\gamma}\left[A_1\left(Y_1' + \frac{2\sigma}{x}Y_1\right) + A_2\left(Y_2' + \frac{2\sigma}{x}Y_2\right)\right.$$
$$\left. + A_3\left(Y_3' + \frac{2\sigma}{x}Y_3\right) + A_4\left(Y_4' + \frac{2\sigma}{x}Y_4\right)\right]; \qquad (34.21)$$

$$G_2 = -\frac{Eh^2}{x\sqrt{3(1-\sigma^2)}\tan\gamma}\left[A_1\left(\sigma Y_1' + \frac{2}{x}Y_1\right) + A_2\left(\sigma Y_2' + \frac{2}{x}Y_2\right)\right.$$
$$\left. + A_3\left(\sigma Y_3' + \frac{2}{x}Y_3\right) + A_4\left(\sigma Y_4' + \frac{2}{x}Y_4\right)\right]. \qquad (34.22)$$

Thus, all the internal forces can be determined, if we know one of the forces N or T_1. In the latter case, we find N and hence η by formula (34.5), then by comparing (34.16) and (34.17) we determine ξ as well as the bending moments.

The displacements u and v can be defined from the equalities (34.1), (34.2) and (34.19) and (34.20).

However, for our aims we do not need to find these displacements, because below we need to determine the absolute elongations of the radii of parallel circumferences which are obtained as a result of a section of the conic shell by the planes which are perpendicular to its axis. We call these elongations *transversal displacements* and we denote them by the letter δ. In order to determine them we can use the formula

$$\delta = u\sin\gamma - w\cos\gamma. \qquad (34.23)$$

Equality (34.23) implies that

$$\frac{\delta}{r\sin\gamma} = \frac{u}{r} - \frac{w}{r\tan\gamma}.$$

According to formulae (34.1) and (34.2) we obtain from this equality that

$$\delta = \frac{r\sin\gamma(T_2 - \sigma T_1)}{B(1-\sigma 2)}. \qquad (34.24)$$

From formulae (34.1) and (34.2) we also obtain

$$\frac{du}{dr} = \frac{T_1 - \sigma T_2}{B(1-\sigma^2)}. \qquad (34.25)$$

Let us give some examples of the use of the method of initial parameters when calculating a conic shell under the action of different loadings.

1. First, consider a conical shell (Fig. 26) whose smaller base is clamped, and a radial force Q_1 uniformly distributed is applied to the greater base.

The boundary conditions have the following form:

for $r = r_B$, $\dfrac{dw}{dr} = 0$, $\delta = 0$,

for $r = r_C$, $T_1 = Q_1 \sin \gamma$, $G_1 = 0$.

Figure 26

The solution of the problem consists in the determination of the constants A_1, A_2, A_3 and A_4. Suppose that the origin is taken on the clamped end. i.e. $a = k\sqrt{2r_B}$. Since at $x = a$ we have $Y_1 = 1$, $Y_2 = Y_3 = Y_4 = 0$, formula (34.16) and the condition of the vanishing of the slope in the cross-section B imply that $A_1 = 0$.

From the second condition on the cross-section B we obtain that $T_2 - \sigma T_1 = 0$ at $x = a$. At the same time, according to (34.19), (34.20) and (34.12), we have at $x = a$

$$T_1 = \frac{Eh^2 k^2}{a^2 \sqrt{3(1 - \sigma^2)}}(-A_3),$$

$$T_2 = \frac{Eh^2 k^2}{2a \sqrt{3(1 - \sigma^2)}}(-A_4);$$

hence, $A_4 = 2\sigma A_3/a$.

Thus, the principal unknown functions ξ and η take the form

$$\xi = A_2 Y_2 + A_3(Y_3 + 2\sigma Y_4/a),$$

$$\eta = -\frac{h\sqrt{1 - \sigma^2}}{2\sqrt{3}\tan\gamma}\left[A_2 Y_4 - A_3\left(Y_1 + \frac{2\sigma}{a}Y_2\right)\right].$$

In order to determine the constants A_2 and A_3, which are until now unknown, we use the conditions in the cross-section C. Setting $b = k\sqrt{2r_C}$, we obtain the following equations:

$$\frac{Eh^2 k^2}{b^2 \sqrt{3(1 - \sigma^2)}}\left[A_2 Y_4(a, b) - A_3\left\{Y_1(a, b) + \frac{2\sigma}{a}Y_2(a, b)\right\}\right] = Q_1 \sin\gamma, \qquad (34.26)$$

$$\frac{Eh^2}{b\sqrt{3(1 - \sigma^2)}\tan\gamma}\left[A_2\left\{Y_2'(a, b) + \frac{2\sigma}{b}Y_2(a, b)\right\}\right.$$

$$\left. + A_3\left\{Y_3'(a, b) + \frac{2\sigma}{b}Y_3(a, b) + \frac{2\sigma}{2}Y_4'(a, b) + \frac{4\sigma^2}{ab}Y_4(a, b)\right\}\right] = 0. \quad (34.27)$$

By solving (34.26) and (34.27), we find A_2 and A_3. If we do not use the method of initial parameters, then in order to determine the arbitrary constants we should solve a system which consists of four equations.

Consider other cases of fixing the cross-section B. If B is simply supported, then we have $G_1 = 0$ and $\delta = 0$ at $x = a$.

The first condition implies that $2\sigma A_1/a + A_2 = 0$, and the second condition implies, as is shown in the previous example, that $A_4 = 2\sigma A_3/a$.

If we have supporting fastening in the cross-section B, which prevents only the turning, then we have $\xi = 0$ and $\eta = 0$ at $x = a$, hence, $A_1 = 0$ and $A_3 = 0$.

At last, if the shell is free in the cross-section B, then we have $G_1 = 0$ and $\eta = 0$ at $x = a$, hence, $2\sigma A_1/a + A_2 = 0$ and $A_3 = 0$.

2. We pass now to the calculation of a shell loaded by radial forces which act in the planes of transversal cross-sections and are uniformly distributed on the cross-section (Fig. 27).

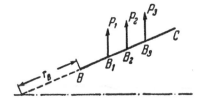

Let P_1, P_2, P_3 denote the intensity of these forces, B_1, B_2, B_3 denote the cross-section in which these forces are applied, and a_1, a_2, a_3 be the corresponding values of the argument x.

Suppose that the smaller base of the cone is

Figure 27

fixed against the turning. Consider the interval $a_1 > x > a$. For the sake of brevity, we denote the value for T_1 on this interval by $T_{1,\mathrm{I}}$

$$T_{1,\mathrm{I}} = \frac{Eh^2k^2}{x^2\sqrt{3(1-\sigma^2)}}\,[A_2Y_4(a,x) - A_4Y_2(a,x)]. \tag{34.28}$$

Let us go to the second interval, where $a_2 > x > a_1$. The normal force T_1 on this interval will be denoted by $T_{1,\mathrm{II}}$. We add such a solution to $T_{1,\mathrm{I}}$ which gives the required break of the transversal force and of the force $T_{1,\mathrm{I}}$ without spoiling the continuity of the bending moment G_1 and the displacements u and ξ.

We try to satisfy these requirements, by adding the expression $A_3^{*}Y_1(x,a_1)$ to (34.28), where a_1 denotes the reduced distance from the top of the cone to the cross-section B_1, i.e. $a_1 = k\sqrt{2r_{B_1}}$.

From expressions (34.18)–(34.20) and (34.11) we see that in this case we have breaks in N, T_1 and T_2; G_1 and G_2 remain continuous, as well as ξ. However, the discontinuity of T_2 results in the discontinuity of the displacement δ. In order to evade this effect, we seek an additional solution in the following form:

$$T_{1,\mathrm{II}} - T_{1,\mathrm{I}} = A_3^{*}Y_1(x,a_1) + A_4^{*}Y_2(x,a_1).$$

and in accordance with (34.20) we obtain

$$T_{2,\mathrm{II}} - T_{2,\mathrm{I}} = \frac{x}{2}\,[A_3^{*}Y_1'(x,a_1) + A_4^{*}Y_2'(x,a_1)].$$

We now require that $T_{2,\mathrm{II}} - T_{2,\mathrm{I}} - \sigma(T_{1,\mathrm{II}} - T_{1,\mathrm{I}}) = 0$ at $x = a_1$. This implies $A_1^{*} = 2\sigma A_3^{*}/a_1$ and

$$T_{1,\mathrm{II}} - T_{1,\mathrm{I}} = A_3^{*}[Y_1(x,a_1) + 2\sigma Y_4(x,a_1)/a_1].$$

In order to determine A_3^{*} we note that $T_{2,\mathrm{II}} - T_{2,\mathrm{I}} = P_1\sin\gamma$ at $x = a_1$. Hence $A_3^{*} = P_1\sin\gamma$.

Thus, we finally obtain

$$T_{1,\mathrm{II}} = T_{1,\mathrm{I}} + P_1\sin\gamma\left[Y_1(x,a_1) + \frac{2\sigma}{a_1}Y_4(x,a_1)/a_1\right].$$

For the next interval $a_3 > x > a_2$ we have

$$T_{1,\mathrm{III}} = T_{1,\mathrm{II}} + P_2\sin\gamma\left[Y_1(x,a_2) + \frac{2\sigma}{a_2}Y_4(x,a_2)/a_2\right].$$

For the last interval $x > a_3$ we have

$$T_{1,\mathrm{IV}} = T_{1,\mathrm{III}} + P_3\sin\gamma\left[Y_1(x,a_3) + \frac{2\sigma}{a_3}Y_4(x,a_3)/a_3\right].$$

Obviously, the arbitrary constants A_2 and A_4 which are present in expression (34.28) can be determined from the boundary conditions in the cross-section C of the shell.

With the help of formulae (34.20), (34.21), (34.22), (34.24), and (34.25) we can immediately obtain all the forces and displacements.

We now pass to the construction of solutions which are discontinuous in G_1. Let a moment whose intensity is equal to G_1^* be applied to the cross-section B_1 of the shell (Fig. 28). Suppose that the moment is uniformly distributed on the cross-section. Let the index I denote the forces and the displacements for $x < a_1$, i.e. to the left from the cross-section B_1, and the index II denote the same for $x > a_1$. As before, we seek an

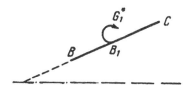

Figure 28

additional solution, which is equal to the difference of the forces or displacements with the indices II and I in the form of a linear combination of the functions $A_n^* Y_n(x, a_1)$.

In order to find the constants A_n^* we use the conditions (34.12). Since ξ and η must be continuous in the cross-section B_1, the conditions (34.16) and (34.18) imply that $A_1^* = 0$, $A_3^* = 0$, and, finally, from the condition of continuity of the transversal displacement we find, using (34.19), (34.20), and (34.24), that $A^* = 0$.

Thus, it only remains to find A_2. For this purpose we use formula (34.21), which implies the following equation

$$G_1^* = -\frac{Eh^2 A_2^*}{a_1 \sqrt{3(1 - \sigma^2)}\tan \gamma}.$$

Therefore,

$$A_2^* = -\frac{G_1^* \sqrt{3(1 - \sigma^2)}a_1 \tan \gamma}{Eh^2}. \tag{34.29}$$

Hence, if $T_1 = T_{1,\mathrm{I}}$ for $x < a_1$, then, taking into account (34.19), we obtain for $x > a_1$

$$T_{1,\mathrm{II}} = T_{1,\mathrm{I}} + \frac{Eh^2 k^2}{x^2 \sqrt{3(1 - \sigma^2)}} A_2^* Y_4(x, a_1),$$

or, using (34.29), we have

$$T_{1,\mathrm{II}} = T_{1,\mathrm{I}} - G_1^* \frac{a_1 k^2}{x^2} \tan \gamma \cdot Y_4(x, a_1).$$

One can construct solutions which are discontinuous in the slope or in the transversal displacement in the same way. These solutions are important in connection with the calculation of non-uniform heating.

3. Let us briefly consider the problem on non-uniform heating of a conic shell. We assume that the temperature t is constant with respect to the width of the shell and varies along the generator. In this case we should modify the right-hand sides of formulae (34.1) and (34.2) so that they express the elongations which are caused only by the stresses.

For this purpose we should replace the total elongations in formulae (34.1) and (34.2), which are equal to $\varepsilon_1 = \dfrac{du}{dr}$ and $\varepsilon_2 = \dfrac{u}{r} - \dfrac{w}{r \tan \gamma}$, respectively, by $\varepsilon_1^* = \varepsilon_1 - \alpha t$

and $\varepsilon_2^* = \varepsilon_2 - \alpha t$, where α is the coefficient of the temperature lengthening. After this substitution the normal forces will be expressed via the displacements in the following way:

$$T_1 = B\left[\frac{du}{dr} + \sigma\left(\frac{u}{r} - \frac{w}{r\tan\gamma}\right)\right] - B(1+\sigma)\alpha t, \qquad (34.30)$$

$$T_2 = B\left[\sigma\frac{du}{dr} + \frac{uw/\tan\gamma}{r}\right] - B(1+\sigma)\alpha t. \qquad (34.31)$$

This modification of the expressions for the normal forces results in a right-hand side arising in equation (34.7), and equation (34.6) remains without changes.

Actually, equation (34.6) is a consequence of the equilibrium conditions and formulae (34.3) and (34.4), which were not changed after taking into account the temperature.

Equation (34.7) has been obtained from the consistency condition, in which the relative elongations are replaced by the expressions given by formulae (34.1) and (34.2). It is natural that as a result of the replacement of formulae (34.1) and (34.2) by (34.30) and (34.31) the right-hand side appears.

Thus, instead of equations (34.8) and (34.9) we have the following system of equations:

$$\left.\begin{array}{l} L(\xi) + \dfrac{2\sqrt{3}\tan\gamma}{h\sqrt{1-\sigma^2}}\eta = 0, \\[4mm] L(\eta) - \dfrac{h\sqrt{1-\sigma^2}}{2\sqrt{3}\tan\gamma}\xi = r\dfrac{h\sqrt{1-\sigma^2}}{2\sqrt{3}}\dfrac{d(\alpha t)}{dr}. \end{array}\right\}$$

After eliminating η, we obtain the inhomogeneous equation

$$L^2(\xi) + \xi = -r\frac{d(\alpha t)}{dr}\tan\gamma.$$

It is convenient to solve this equation, by using the method of initial parameters, particularly, if the right-hand side is discontinuous.

The method of initial parameters can also be used in all the problems of the three previous sections. In this case, one can show (see, for instance, [20]) that the particular solution of the equations obtained is determined, by applying the method of initial parameters, up to a constant factor, as an integral of the form

$$\int_0^\xi Z_4(\xi, x)q(x)\,dx,$$

where q is the right-hand side and Z_4 is the fourth function of the method of initial parameters.

4. Consider a conic shell whose end B is fastened against the turning and the end C is free, Let the shell be heated so that the temperature is a step function. For the sake of determinacy we assume that the temperature on the intervals BB_1, B_1B_2, B_2B_3, B_3C (Fig. 29) is equal to t_1, t_2, t_3, t_4, respectively.

In this case the forces G_1, T_1, N and the slope should be continuous at the ends of the intervals. The transversal displacements should also be continuous. These displacements are defined by the formula

$$\delta = r\sin\gamma\frac{T_2 - \sigma T_1}{B(1-\sigma^2)} + \alpha tr\sin\gamma, \qquad (34.32)$$

where the first term represents the elastic part of the displacement, and the second part can be called the *free temperature displacement*.

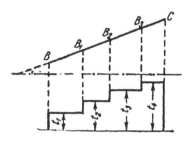

Since t has breaks at the ends of the intervals, it is necessary for the continuity of the transversal displacement that the first term of (34.32) also has a break. All these condition can be satisfied, by using the functions which have a break only in T_2. We can verify by direct checking that the required discontinuous solution has the form (34.12) and (34.17), where we should set $A_1 = A_2 = A_3 = 0$ and replace the argument a by a_1.

We now pass to the immediate determination of the forces and displacements which appear in the shell under consideration.

Figure 29

On the first interval, i.e. for $x < a_1$ the normal forces are expressed in the following way:

$$T_{1,\mathrm{I}} = \frac{Eh^2k^2}{x^2\sqrt{3(1-\sigma^2)}}[A_2Y_4(a,x) - A_4Y_2(a,x)], \tag{34.33}$$

$$T_{2,\mathrm{I}} = \frac{Eh^2k^2}{2x\sqrt{3(1-\sigma^2)}}[A_2Y_4'(a,x) - A_4Y_2'(a,x)], \tag{34.34}$$

and the transversal displacement has the form

$$\delta_1 = \frac{h\sin\gamma}{2\sqrt{3(1-\sigma^2)}}\left[A_2\left\{Y_4'(a,x) - \frac{\sigma x}{2}Y_4(a,x)\right\}\right.$$
$$\left. - A_4\left\{Y_2'(a,x) - \frac{\sigma x}{2}Y_2(a,x)\right\}\right] + \alpha t_1 r\sin\gamma. \tag{34.35}$$

On the second interval we seek the solution in the form

$$T_{1,\mathrm{II}} = T_{1,\mathrm{I}} - \frac{Eh^2k^2}{x^2\sqrt{3(1-\sigma^2)}}A_4^*Y_2(a_1,x), \tag{34.36}$$

$$T_{2,\mathrm{II}} = T_{2,\mathrm{I}} - \frac{Eh^2k^2}{2x\sqrt{3(1-\sigma^2)}}A_4^*Y_2'(a_1,x); \tag{34.37}$$

then the transversal displacement can be expressed in the following form:

$$\delta_2 = \delta_1 - \frac{h\sin\gamma}{2\sqrt{3(1-\sigma^2)}}A_4^*\left[Y_2'(a_1,x) - \frac{\sigma x}{2}Y_2(a_1,x)\right] + \alpha r(t_2 - t_1)\sin\gamma. \tag{34.38}$$

Since the transversal displacement in the cross-section B_1 is continuous, setting $x = a_1$ in (34.38) we obtain the equation, which allows to determine A_4^*:

$$-\frac{h\sin\gamma}{2\sqrt{3(1-\sigma^2)}}A_4^* + \alpha(t_2 - t_1)\frac{a_1^2}{2k^2}\sin\gamma = 0. \tag{34.39}$$

Hence

$$A_4^* = \frac{\alpha a_1^2(t_2 - t_1)\sqrt{3(1-\sigma^2)}}{hk^2}. \tag{34.40}$$

In the same way we construct the solution for the third and fourth intervals. For the sake of brevity, we only write the expression for the force $T_{1,\text{III}}$:

$$T_{1,\text{III}} = T_{1,\text{II}} - \frac{Eha_2^2\alpha(t_3 - t_2)}{x^2}Y_2(a_2, x). \qquad (34.41)$$

On the fourth interval we have

$$T_{1,\text{IV}} = T_{1,\text{III}} - \frac{Eha_3^2\alpha(t_4 - t_3)}{x^2}Y_2(a_3, x). \qquad (34.42)$$

As before, A_2 and A_4 can be determined from the boundary conditions on the right edge of the shell.

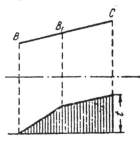

Figure 30

5. Let us now consider a shell in which the temperature along the generator varies according to the rule of a polygonal line (Fig. 30). In this case the right-hand side of the equation has the form

$$\frac{h\sqrt{1-\sigma^2}}{2\sqrt{3}}rm_n,$$

where $m_n = \dfrac{d(\alpha t)}{dr}$ is the angular coefficient of the curve of variation of the temperature, which is constant on each part of the polygonal line.

The question here arises concerning the solution with a break in the slope.

Suppose that we have obtained a solution for the first part as the sum of the general integral and a particular solution. When passing to the second interval, the conditions of continuity of the forces N_1, T_1, G_1 and of the transversal displacement should hold. Because of the break in the angular coefficient of the curve of variation of the temperature, the continuity of the slope only remains in this case, if the elastic part of the slope has a break of the same value but of the opposite sign. This solution can be obtained, by setting in (34.18) and (34.19)

$$A_3 = A_4 = 0; \qquad A_2 = -\frac{2\sigma}{a_1}; \qquad A_1 = -\frac{2\sigma}{a_1}A_1^*$$

and taking the point of discontinuity B_1 as the origin, i.e. introducing a_1 instead of a.

The constant A_1^* can be found from the condition of continuity of the slope. For $x = a_1$ the slope obtained from the additional solution, is equal to A_1^*. On the other hand, the variation of the difference between the slope and its elastic part for $x = a_1$ is equal to $\alpha r_{B_1}(m_2 - m_1)\tan\gamma$, hence,

$$A_1^* = \frac{a_1^2(m_2 - m_1)\alpha}{2k^2}\tan\gamma. \qquad (34.43)$$

Let us write the expression for the normal force on the second interval, using (34.43) and (34.19):

$$T_{1,\text{II}} = T_{1,\text{I}} + \frac{Eh^2a_1^2}{2x^2\sqrt{3(1-\sigma^2)}}(m_2 - m_1)\left[Y_3(a_1, x) - \frac{2\sigma}{a_1}Y_4(a_1, x)\right]. \qquad (34.44)$$

6. If the length of the generator is large and the angle γ is small, then it may be fruitful to apply the method of compensating loadings. In this case, which is more important than the one above considered, we should use the basic influence

functions in order to construct the solution. There exists a close connection between the influence functions and the fundamental functions, which is emphasized in the previous chapters.

Therefore, we only give for example the equations of the four basic influence functions. First, we write the equation of the influence function $\Gamma_4(x, a)$, which is discontinuous in $\dfrac{dL}{dx}$.

We impose the condition that the influence function remains finite at $x = 0$, does not become infinite at $x = \infty$ and satisfies equation (34.10) as well as the condition

$$\left[\frac{dL(\Gamma_4)}{dx}\right]_{x=a+0} - \left[\frac{dL(\Gamma_4)}{dx}\right]_{x=a-0} = 1. \tag{34.45}$$

Obviously, this influence function is connected with the fundamental function Y_4. It has the following form:
for $x \leq a$

$$\Gamma_4(x, a) = \frac{\pi a}{2}[-f_2(a)u_2(x) + g_2(a)v_2(x)], \tag{34.46}$$

for $x \geq a$

$$\Gamma_4(x, a) = \frac{\pi a}{2}[-u_2(a)f_2(x) + v_2(a)g_2(x)]. \tag{34.47}$$

We give, without details, the formulae for calculations of the three remaining influence functions.

The influence function Γ_3 which gives the unit break in the operator L at $x = a$, has the form:
for $x \leq a$

$$\Gamma_3(x, a) = \frac{\pi a}{2}[-f_2'(a)u_2(x) + g_2'(a)v_2(x)], \tag{34.48}$$

for $x \geq a$

$$\Gamma_3(x, a) = \frac{\pi a}{2}[-u_2'(a)f_2(x) + v_2'(a)g_2(x)]. \tag{34.49}$$

The influence function Γ_2 which gives the break in the first derivative at $x = a$, has the form:
for $x \leq a$

$$\Gamma_2(x, a) = -\frac{\pi a}{2}[g_2(a)u_2(x) + f_2(a)v_2(x)], \tag{34.50}$$

for $x \geq a$

$$\Gamma_2(x, a) = -\frac{\pi a}{2}[v_2(a)f_2(x) + u_2(a)g_2(x)]. \tag{34.51}$$

Finally, the function Γ_1 which is discontinuous at $x = a$ is:
for $x < a$

$$\Gamma_1(x, a) = -\frac{\pi a}{2}[g_2'(a)u_2(x) + f_2'(a)v_2(x)], \tag{34.52}$$

for $x > a$

$$\Gamma_1(x, a) = -\frac{\pi a}{2}[v_2'(a)f_2(x) + u_2'(a)g_2(x)]. \tag{34.53}$$

Having these influence functions, one can easily obtain the so called main solution, which has all the required breaks and quickly decreases a long way from the place where the loading is applied. Since the forces in shells of this type decreases

sufficiently quickly, one can mainly confine oneself by considering only the main solution. In the cases when the influence of the ends cannot be neglected, we should
add a linear combination of the integrals of the homogeneous equation influence of
the to the main solution. Generally, the influence of the conditions at each of the
ends can be taken into account separately. The estimate of the influence at the
internal end, i.e. for the points located nearest to the top of the cone is reduced to
the addition of the solution

$$C_1 f_2(x) + C_2 g_2(x);$$

the influence of the external end is reduced to the addition of the solution

$$C_3 u_2(x) + C_4 v_2(x).$$

If C_1 and C_2 cannot be determined independently of C_3 and C_4, then we should
solve four equations with four indeterminates.

In this case the method of initial parameters becomes much more effective than
the method of compensating loadings; therefore, one should use the formulae given
at the beginning of this section here.

Let us make some additional remarks concerning the calculation of a conic shell
under the action of an axially symmetrical loading or a heating. All these problem
have much in common with the calculation of a circular plate which lies on an
elastic foundation and is subjected to the action of the loading $q(\xi) \cos 2\theta$. In fact,
this problem was studied in the third chapter. In the cases when the loading is
varied according to the exponential rule, so that the corresponding integrals cannot
be expressed via the cylindrical functions, they can be expressed via the Lommel
functions, just as was done in the problems on rods which will be considered in the
fifth chapter.

At the present time, asymptotic integration is widely used in problems concerning shells. Similar results, which however allow a wide range variation in the
accuracy of the approximate formulae, can be obtained, by taking the explicit solutions of the problems considered here, and then using the asymptotic formulae.
Note that in some cases, for comparatively small, more exact "medium" values of
the argument, there are reasons to think that the asymptotic formulae do not give
the accuracy required for practical calculations.

Using the asymptotic formulae for the cylindrical functions and substituting
the corresponding expressions into the influence functions, we obtain the solutions
required. In connection with the fact that the tables of the Bessel functions for the
argument $x\sqrt{i}$ contain only the Bessel functions of order zero and one, we give the
recurrence relations for the determination of the functions u_2, f_2, v_2, g_2 as well as
the asymptotic formulae for the calculation of the functions for the values of the
argument, which lay outside the tabulations.

First we give the asymptotic formulae.

Let

$$u_n(x) + iv_n(x) = J_n(x\sqrt{i}),$$
$$f_n(x) + ig_n(x) = H_n^{(1)}(x\sqrt{i});$$

as $x \to 0$ we have

$$u_0 = a_0 \cos \delta, \qquad v_0 = b_0 \sin \delta, \qquad u_1 = a_0 \sin \delta, \qquad v_1 = a_0 \cos \delta,$$

where

$$a_0 = \frac{1}{\sqrt{2\pi x}} e^{x/\sqrt{2}}, \qquad \delta = \frac{x}{\sqrt{2}} - \frac{\pi}{8};$$

as $x \to \infty$ we have

$$f_0 = b_0 \sin \sigma, \qquad g_0 = -b_0 \cos \sigma, \qquad f_1 = -b_0 \cos \sigma, \qquad g_1 = -b_0 \sin \sigma,$$

where

$$b_0 = \sqrt{\frac{2}{\pi x}} e^{-x/\sqrt{2}}, \qquad \sigma = \frac{x}{\sqrt{2}} + \frac{\pi}{8}.$$

With the help of the recurrence relation

$$Z_{n+1} = \frac{2n}{z} Z_n - Z_{n-1}$$

we can write that for $z = x\sqrt{i}$ we have

$$u_2 + iv_2 = \frac{2}{x\sqrt{i}}(u_1 + iv_1) - (u_0 + iv_0).$$

Using the equality

$$\sqrt{i} = \frac{1+i}{\sqrt{2}}$$

and separating the real and imaginary parts, we obtain

$$u_2 = \frac{\sqrt{2}}{x}(u_1 + v_1) - u_0,$$

$$v_2 = \frac{\sqrt{2}}{x}(v_1 - u_1) - v_0.$$

By analogy we write

$$f_2 = \frac{\sqrt{2}}{x}(f_1 + g_1) - f_0,$$

$$g_2 = \frac{\sqrt{2}}{x}(g_1 - f_1) - g_0.$$

With the help of the formula for differentiation

$$\frac{dZ_n}{dz} = -\frac{n}{z} Z_n + Z_{n-1},$$

substituting $z = x\sqrt{i}$ and setting $n = 2$, we obtain

$$u_2' = -\frac{2}{x} u_2 + \frac{1}{\sqrt{2}}(u_1 - v_1),$$

$$v_2' = -\frac{2}{x} v_2 + \frac{1}{\sqrt{2}}(u_1 + v_1),$$

$$f_2' = -\frac{2}{x} f_2 + \frac{1}{\sqrt{2}}(f_1 - g_1),$$

$$g_2' = -\frac{2}{x} g_2 + \frac{1}{\sqrt{2}}(f_1 + g_1).$$

35. The method of compensating loadings in problems on membranes and plates

In this section we consider some problems of the theory of membranes and plates which can be solved with the help of the method of compensating loadings. In this case the solution is divided into the sum of two solutions: the basic one and the compensating one. When constructing both solutions, we consider an extended domain, which is generally unbounded, instead of the given domain. It has been noted that the basic solution satisfies the inhomogeneous differential equation and the boundary conditions for the extended domain.

The compensating solution must satisfy the homogeneous differential equation and, together with the basic solution, the boundary conditions. The compensating solution is treated as a result of the action of some loadings which are chosen in a special way and are applied to the plate or the membrane, which occupies the extended domain. Obviously, the compensating loadings must be applied to the part of the extended domain which is outside of the given domain.

The choice of the line or the system of lines along which the compensating loadings are distributed gives many opportunities in order to use the simplification suggested by the character of the problem under consideration. Moreover, the arbitrariness in the choice of the type of the loadings, namely, the possibility to choose the loading as a force, a moment or breaks in deflections and in slopes also extends the possibilities of constructing effective methods of calculation.

Practically, in order to solve the problem we should, depending on the type of the boundary conditions, make the total deflections or the slopes or make the forces on the contour equal to zero or to a given function. We obtain either an integral equation or a system of integral Fredholm equation of the first kind in this way. By solving these equations, we find the compensating loading, and then we find the total solution of the problem. The integral equations mentioned should be solved approximately. In one way or another, they are replaced by a system of linear equations with respect to the parameters, via which the compensating loadings can be expressed.

In the problems considered here, the basic influence functions are represented via Bessel functions; therefore, all kernels of the integral equations obtained are also expressed via these functions.

The calculations presented in this section show the effectiveness of the method applied.

1. Consider the problem on oscillations of a membrane. Suppose that the membrane has a constant thickness h and the special weight γ_1 and is stretched by a constant force T_0. The differential equation of free oscillations of the membrane has the form

$$T_0 \nabla^2 w = \frac{\gamma_1 h}{g} \frac{\partial^2 W}{\partial t^2}, \qquad (35.1)$$

where $\nabla^2 = \frac{\partial^2}{\partial x^2} + \frac{\partial^2}{\partial y^2}$.

We denote $\gamma_1 h/(T_0 g) = 1/c^2$ and seek the solution of equation (35.1) in the form

$$W(x, y; t) = w(x, y) \sin(\omega t),$$

where ω is the circular[7] frequency of the free oscillations, and the function $w(x,y)$ satisfies the differential equation

$$\nabla^2 w + \lambda^2 w = 0, \tag{35.2}$$

where $\lambda = \omega/c$.

It is known that a lot of problems of the mathematical physics can be reduced to the integration of this equation.

In order to see the effectiveness of the method of compensating loadings, we first apply this equation to problems which have an explicit solution.

Let us compute the frequencies and the forms of the free oscillations of the membrane. In this case the basic solution vanishes identically, and the boundary conditions are homogeneous. We represent the solution of the problem in the form

$$w = \int_C q(\sigma) K(x,y;\xi,\eta)\,d\sigma,$$

where C is the line along which the compensating loading is distributed (this line will be called singular), σ is the arc coordinate on this line; x, y are the coordinates of the considered point of the membrane; ξ, η are the coordinates of a point of the singular line; the function K represents the basic influence function. In the case considered it expresses the form of the oscillations of an unbounded membrane which are caused by the unit force $1 \cdot \sin \omega t$. Up to a constant factor the function K represents a Bessel function of the second kind with zero index of the real argument $r = \lambda\sqrt{(x-\xi)^2 + (y-\eta)^2}$, where r is the reduced distance between the point of the singular line and the considered point of the membrane. This can easily be checked, by making calculations similar to those which are necessary for obtaining formula (31.14). Since the deflections on the contour of the membrane are equal to zero, for any point of the contour we have

$$\int_C q(\sigma) Y_0(r)\,d\sigma = 0. \tag{35.3}$$

If we apply the moments instead of the forces, then the kernels represent Bessel functions of the first order multiplied by the cosines of the corresponding angles. The problem is reduced to the determination of the values of the parameter λ for which equation (35.3) has a non-trivial solution. The integral in the calculations will be approximately replaced by the sum, and the problem will be reduced to the determination of the root of a system of transcendental equations.

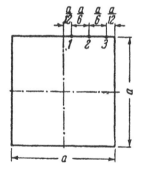

Figure 31

As a numerical example, let us consider the problem on free oscillations of a square membrane which is pulled on a rigid contour. By the way we note that this problem, in fact, coincides with the problem of the determination of the critical pressure which is uniformly distributed along the simply supported contour of a square plate as well as with the problem on free oscillations of the same plate.

Let us find the fundamental frequency of the oscillations. It is known that this problem has an explicit solution. We will show that, using the polar coordinate

---

[7]For the sake of brevity the word "circular" will sometimes be omitted.

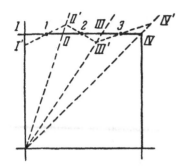

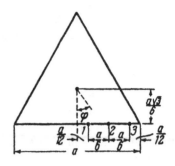

Figure 32   Figure 33

system and starting from the method of compensating loadings, we can obtain an approximate solution which has a sufficiently high accuracy from the practical viewpoint comparatively easily. Moreover, we will illustrate the possibility of applying different versions of computing the first eigenvalue.

Let a singular line be a circumference. Suppose that the compensating loadings $q(\sigma)$ are represented as a series which from the symmetry conditions has the form

$$q(\sigma) = \sum_{n=0}^{\infty} B_{4n} \cos 4n\varphi,$$

where φ is the polar angle whose vertex is at th e center of the square (Fig. 31).

Using the addition theorem, we can write equation (35.3) in the form

$$A_0 J_0(\lambda R_s) + A_4 J_4(\lambda R_s) \cos 4\varphi + \cdots = 0, \qquad (35.4)$$

where R_s is the polar coordinate of a point of the contour. If we keep the three terms in (35.4) and require that the boundary condition holds at the three points $(1, 2, 3)$ of the contour, which are shown in Fig. 31, then, as a result, we obtain a system of three equations. In these equations, the angles φ are given and all the values R_s can be expressed via the length of a side of the square a or via its reduced length $a' = \lambda a$. Making the determinant of the system equal to zero, we obtain

$$\begin{vmatrix} J_0(k_1 a') & J_4(k_1 a') \cos 4\varphi_1 & J_8(k_1 a') \cos 8\varphi_1 \\ J_0(k_2 a') & J_4(k_2 a') \cos 4\varphi_2 & J_8(k_2 a') \cos 8\varphi_2 \\ J_0(k_3 a') & J_4(k_3 a') \cos 4\varphi_3 & J_8(k_3 a') \cos 8\varphi_3 \end{vmatrix} = 0.$$

As we can see in Fig.31

$$2k_1 = \sqrt{1 + (1/6)^2}, \qquad 2k_2 = \sqrt{1 + (1/2)^2}, \qquad 2k_3 = \sqrt{1 + (5/6)^2},$$
$$\varphi_1 = \operatorname{arcsec}(2k_1), \qquad \varphi_2 = \operatorname{arcsec}(2k_2), \qquad \varphi_3 = \operatorname{arcsec}(2k_3).$$

Expanding the determinant and computing the smallest root of the transcendental equation obtained, we find the reduced length a'; therefore, we immediately obtain $\lambda = a'/a$ and the frequency of the eigenoscillations ω. We obtain the result, calculated with three significant digits, which coincides with the exact solution with sufficient accuracy for practical usage. The obtained solution approximately satisfies the boundary conditions and it is interesting to estimate the error which is present in the result.

The nodal line, i.e. the line along which all the deflections are equal to zero, passes through the points *1, 2, 3* as well as through the points symmetrical to these ones. In total, it has 24 points in common with the points of the contour (Fig. 32). In order to find out how much the nodal line deviates from the contour, we draw the radius-vectors through the equidistant points of the contour *I, II, III* which are situated at equal distances from the nodes of the interpolation and through the point *IV* which is the vertex of the square. Then we find the location of the points of intersection for these straight lines and the nodal line. For the exact solution, these points denoted by *I', II', III'*, and *IV'* should lie on the contour of the membrane. In order that the nodal line does not fuse with the contour, we need to increase the difference between the radius-vectors λR and the radius-vectors of the appropriate points of the contour (in Fig. 32 this difference is enlarged 25 times). Similar calculations were performed by keeping two terms of the series. In both cases a good coincidence has been obtained for the frequencies of the oscillations as well as for the coordinates of the nodal line.

Similar calculations have been made for a membrane in the form of an equilateral triangle. In this case the frequency is defined by the condition that the form of the oscillations which correspond to the basic tone, has a nodal line passing through the points *1, 2, 3* of the contour which are equidistant from each other (Fig. 33).

The solution can be approximately represented as the sum

$$w = A_0 J_0(\lambda R) + A_3 J_3(\lambda R) \cos 3\varphi + A_6 J_6(\lambda R) \cos 6\varphi. \qquad (35.5)$$

The reduced length a' of the side of the triangle can be determined from the equation

$$\begin{vmatrix} J_0(k_1^* a') & J_3(k_1^* a') \cos 3\varphi_1 & J_6(k_1^* a') \cos 6\varphi_1 \\ J_0(k_2^* a') & J_3(k_2^* a') \cos 3\varphi_2 & J_6(k_2^* a') \cos 6\varphi_2 \\ J_0(k_3^* a') & J_3(k_3^* a') \cos 3\varphi_3 & J_6(k_3^* a') \cos 6\varphi_3 \end{vmatrix} = 0.$$

Just as in the previous example, the coefficients k_n^* and the angles φ_n^* are determined from geometrical reasonings here. Knowing a', we define the frequency of the basic tone, which almost coincides with the exact solution (the exact value of the parameter is $a' = 7.25$ and the approximate one is $a' = 7.29$).

By keeping only the first two terms of expression (35.5), we obtain an error equal to 0.2%.

The approximate compensating solution can be represented as a linear combination of two Bessel functions of zero order; for this purpose we should specify the singularities which are of the type of a concentrated force at different points which lie outside the domain under consideration.

Consider forced oscillations of the membrane. In this case the frequency of the oscillations is prescribed. The integral equation which is used for constructing the compensating solution, has the form

$$\int_C q(\sigma) K(s, \sigma) \, d\sigma = -f_1(s), \qquad (35.6)$$

where $f_1(s)$ is the value of the basic solution on the contour.

Just as in the previous example, we replace the integral equation (35.6) by a system of linear equations, by assuming that the singular line is a circumference and the compensating loading can be expanded into a series in cosines. Then we determine the coefficients of the sum $\sum A_n J_n(\lambda R) \cos n\varphi$, which gives an approximate

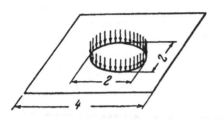

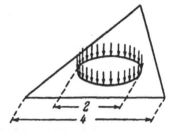

Figure 34   Figure 35

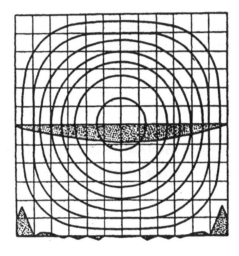

Figure 36

solution of the problem on forced oscillations of square and triangular membranes loaded with forces of intensity $q_1 \sin \omega t$ applied along the circumference as is shown in Figs. 34, 35.

The basic solution has the form

$$\text{for } \lambda R \leq \alpha, \qquad\qquad w_0 = J_0(\lambda R)Y_0(\alpha),$$
$$\text{for } \lambda R \geq \alpha, \qquad\qquad w_0 = J_0(\alpha)Y_0(\lambda R).$$

Keeping the three first terms in the compensating solution fixed, we obtain the following system of linear algebraic equations for a square membrane

$$A_0 J_0(\lambda R_m) + A_4 J_4(\lambda R_m) \cos 4\varphi_m + A_8 J_8(\lambda R_m) \cos 8\varphi_m + w_{0m} = 0,$$

where $m = 1, 2, 3$ are the indices which show the points of the contour at which the boundary conditions hold, w_{0m} are the values of the basic solution at these points.

The results of the calculations for a square membrane are given in Fig. 36, where the plan in the horizontals is shown and the graph of the function w on the contour is given.

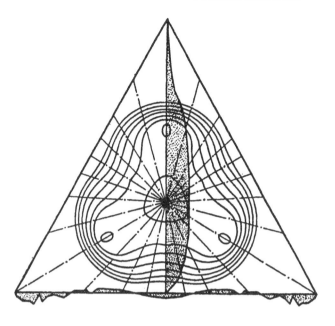

Figure 37

The results of the calculations for a membrane in the form of an equilateral triangle are presented in Fig. 37.

It should be noted that from the computational viewpoint these problems are less labour-consuming than the problems on eigenvalues which were considered above.

In fact, in all the examples considered above the solution was found by the method of point-wise interpolation.

One can easily indicate other methods of approximate numerical solution of the integral equation. We are going to consider one such method. We will seek the solution of the problem on oscillations of a membrane by the method of Trefftz. This method can be treated as one of the versions of approximating the integral in equation (35.3).

As before, we approximately represent the compensating solution as the sum

$$w_k^* = \sum_{i=1}^{m} b_i \psi_i(x, y), \tag{35.7}$$

where the $\psi_i(x, y)$ satisfy equation (35.2). Let w_k be the required solution; let us introduce the following notation for the difference $w_k - w_k^* = w_{ka}$.

Let us find the coefficients b_i such that the functional

$$F = \iint_D \left[\left(\frac{\partial w_{ka}}{\partial x} \right)^2 + \left(\frac{\partial w_{ka}}{\partial y} \right)^2 - \lambda^2 w_{ka}^2 \right] dx\, dy,$$

where D is the domain occupied by the membrane, is minimal.

In connection with the fact that w_{ka} is represented with the help of formula (35.7), the variational problem is reduced to the usual problem on the minimum whose solution leads us to the system of linear equations

$$\frac{\partial F}{\partial b_i} = 0, \qquad i = 1, \ldots, m. \tag{35.8}$$

Let us pass from the integral over the domain D occupied by the membrane to the integral along the contour C.

In this case, applying Gauss' theorem, we obtain that the system (35.8) is equivalent to the system

$$\int_C (w_k - w_k^*) \frac{\partial \psi_i}{\partial n} \, ds = 0 \tag{35.9}$$

where n is the external normal to the contour. Introducing the values w_k^* from expression (35.7 into equation (35.9), we obtain

$$\int_C w(s) \frac{\partial \psi_i}{\partial n} \, ds - \sum_{j=1}^m b_j \int_C \psi_j \frac{\partial \psi_i}{\partial n} \, ds = 0,$$

where the index $i = 1, \ldots, m$.

Let us emphasize one feature of the considered method which is very important for the practical aims and which has been demonstrated by Trefftz for problems described by the Laplace equation and by Friedrichs [65] for a larger class of problems. This method gives values of the functional considered which are too small.

In accordance with the scheme of calculations given above, let us once again consider the problem on the eigenoscillations of a square membrane.

We determine the frequency, by making the determinant of system (35.8) equal to zero for $w(s) = 0$.

If we set

$$w_k^* = b_1 J_0(\lambda R) + b_2 J_4(\lambda R) \cos 4\varphi + b_3 J_8(\lambda R) \cos 8\varphi, \tag{35.10}$$

then the equation for the frequencies takes the form

$$\begin{vmatrix} \int_{x=0}^{x=a/2} \psi_1 \frac{\partial \psi_1}{\partial n} \, dx & \int_{x=0}^{x=a/2} \psi_1 \frac{\partial \psi_2}{\partial n} \, dx & \int_{x=0}^{x=a/2} \psi_1 \frac{\partial \psi_3}{\partial n} \, dx \\ \int_{x=0}^{x=a/2} \psi_2 \frac{\partial \psi_1}{\partial n} \, dx & \int_{x=0}^{x=a/2} \psi_2 \frac{\partial \psi_2}{\partial n} \, dx & \int_{x=0}^{x=a/2} \psi_2 \frac{\partial \psi_3}{\partial n} \, dx \\ \int_{x=0}^{x=a/2} \psi_3 \frac{\partial \psi_1}{\partial n} \, dx & \int_{x=0}^{x=a/2} \psi_3 \frac{\partial \psi_2}{\partial n} \, dx & \int_{x=0}^{x=a/2} \psi_3 \frac{\partial \psi_3}{\partial n} \, dx \end{vmatrix} = 0. \tag{35.11}$$

In this equation we use the notation

$$\psi_1 = J_0(\lambda R),$$
$$\psi_2 = J_4(\lambda R) \cos 4\varphi,$$
$$\psi_3 = J_8(\lambda R) \cos 8\varphi.$$

The integrals which are present in this equation are computed approximately. As a result of the calculations, the value $\lambda a = 4.42$ was obtained.

When continuing the calculations, we restrict ourselves only to the first term of (35.10). In this case equation (35.11) takes the form

$$\int\limits_{x=0}^{x=a/2} J_0(\lambda R)J_1(\lambda R)\cos(n,R)\,dx = 0, \tag{35.12}$$

where $R = \sqrt{x^2 + a^2/4}$.

Let us collect the values of the integral which is in the left-hand side of equation (35.12) in a table. For different values of $a' = \lambda a$ we have

λa	4.1	4.2	4.3	4.4
$\displaystyle\int_0^{a/2} \psi_1\frac{\partial\psi_1}{\partial n}\,dx$	-0.053	-0.027	-0.0042	0.0175

We can see from the table that the root of the transcendental equation (35.12) is less than the root obtained from the exact solution which is equal, as has just been noted, to 4.42 and the approximation is sufficiently good.

Let us give the table of the calculations fulfilled for a membrane in the form of an equilateral triangle:

λa	7.1	7.2	7.3	7.4
$\displaystyle\int_0^{a/2} \psi_1\frac{\partial\psi_1}{\partial n}\,dx$	-0.0177	-0.0068	0.006	0.0159

The differential equation for the problem on the equilibrium of a membrane which lies on an elastic foundation has the form $\nabla^2 w - \lambda^2 w = 0$, where $\lambda = \sqrt{k_0/T}$, k_0 is the coefficient of subgrade reaction of the elastic foundation, T is the stretching of the membrane.

Suppose that the contour of the membrane cannot be displaced. In this case, the basic influence function is, up to a constant factor, the Macdonald function of order zero, which represents the solution of the problem on an unbounded membrane which lies on an elastic foundation and is loaded by a concentrated force.

2. Consider the problem on a plate which is subjected to the action of forces uniformly distributed along a contour, which lie in the plane of the plate and are directed along the normal to the contour. Moreover, the plate can be exposed to the action of any other loadings directed perpendicular to the median plane.

It is known that the differential equation of the deflection surface of the plate in this case has the form

$$D\nabla^2\nabla^2 w + p\nabla w = q, \tag{35.13}$$

where p is the intensity of the forces applied to the contour and lying in its plane, q is the loading perpendicular of the median plane, w is the deflection, D is the flexural rigidity of the plate.

Beforehand, we obtain the solution of equation (35.13) which has a singularity of the type of a concentrated force for an unbounded domain. If we assume this force to be equal to one, then this solution gives the basic influence function. The solution has the form

$$K_1^*(x, y; \xi, \eta) = -\frac{1}{4D\lambda^2}\left[Y_0(r) - \frac{2}{\pi}\ln\frac{2}{\gamma r}\right], \qquad (35.14)$$

where $r = \lambda\sqrt{(x-\xi)^2 + (y-\eta)^2}$, $\lambda = \sqrt{p/D}$.

This solution can easily be obtained by making calculations similar to those given in Section 31, when considering the problem on oscillations of the plate.

The solution which has an isolated singularity of the type of a concentrated moment, has the following form:

$$K_2^*(x, y; \xi, \eta) = \frac{1}{2D\lambda}\left[\frac{Y_1(r)}{2} + \frac{1}{\pi r}\right]\cos\varphi_1,$$

where φ_1 is the angle between the direction of the segment r and the plane of the action of the moment.

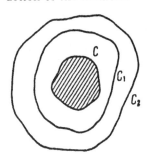

It has been noted that the basic solution can be obtained by integrating the basic influence function. This approach is especially fruitful for loadings which are distributed on a circular domain. In this case the basic solution can be represented, using the addition formulae, as well-convergent series, or, in some cases, be obtained in closed form. We will not consider this question here. Let us immediately go to constructing the compensating solution.

We write the equations of the method of compensating loadings for a plate with a clamped edge in general form. Suppose that the deflection and the angular displacement of the normal to the contour C, which are obtained from the basic solution, are represented in the form $f_1(s)$ and $f_2(s)$, respectively. We represent the compensating solution, by selecting the compensating loadings q_1 and q_2 on two non-coinciding singular lines C_1 and C_2 (Fig. 38). The intensities of these loadings can be found from the condition that the sum of the compensating solution and the basic one should satisfy the conditions on the contour, namely, the deflection and the angular displacement of the normal to the contour being equal to zero.

$$\int_{C_1} q_1(\sigma_1)K_1^*(s, \sigma_1)\,d\sigma_1 + \int_{C_2} q_2(\sigma_2)K_1^*(s, \sigma_2)\,d\sigma_2 = -f_1(s), \qquad (35.15)$$

$$\int_{C_1} q_1(\sigma_1)\frac{K_1^*(s, \sigma_1)}{dn_s}\,d\sigma_1 + \int_{C_2} q_2(\sigma_2)\frac{K_1^*(s, \sigma_2)}{dn_s}\,d\sigma_2 = -f_2(s). \qquad (35.16)$$

Here $K_1^*(s, \sigma)$ is obtained from solution (35.14) after replacing x, y in this formula by the coordinates of the point of the contour, and ξ, η by the coordinates of the point of the corresponding singular line.

Figure 38

For a plate with a free edge, the equations of the method of compensating loadings take the following form:

$$
\left.
\begin{aligned}
\int_{C_1} q_1(\sigma_1) L_1^{(M)}(s, \sigma_1)\, d\sigma_1 + \int_{C_2} q_2(\sigma_2) L_2^{(M)}(s, \sigma_2)\, d\sigma_2 = -f_3(s), \\
\int_{C_1} q_1(\sigma_1) L_1^{(Q)}(s, \sigma_1)\, d\sigma_1 + \int_{C_2} q_2(\sigma_2) L_2^{(Q)}(s, \sigma_2)\, d\sigma_2 = -f_4(s),
\end{aligned}
\right\}
\tag{35.17}
$$

where $f_3(s)$ and $f_4(s)$ are the values of the bending moments and reduced transversal forces on the contour of the plate, which are taken from the basic solution; $L^{(M)}$ and $L^{(Q)}$, respectively, are computed as the bending moments and reduced transversal forces on the contour caused by the unit force applied to a point of the singular line. In order to determine $L^{(M)}$ we should execute the operation of the calculation of the moment over the function K_1^*

$$
L^{(M)} = -D\left[\frac{\partial^2 K_1^*}{\partial n_{s_1}^2} + \frac{1}{m}\left(\frac{1}{\rho}\frac{\partial K_1^*}{\partial n_{s_1}} + \frac{\partial^2 K_1^*}{\partial s_1^2}\right)\right],
$$

where ρ is the radius of curvature of the contour, $1/m$ is Poisson's ratio. For the rectilinear segments of the contour, we should align the side of the plate with one of the axes of the Cartesian coordinates; in this case, assuming n_{s_1} directed along the x-axis, we obtain the well-known formula

$$
L^{(M)} = -D\left[\frac{\partial^2 K_1^*}{\partial x^2} + \frac{1}{m}\frac{\partial^2 K_1^*}{\partial y^2}\right].
$$

In order to calculate $L^{(Q)}$ we should execute the following operation over the function K_1^*:

$$
-D\left\{\frac{\partial}{\partial n_{s_1}}\nabla^2 + \left(1 - \frac{1}{m}\right)\frac{\partial}{\partial s_1}\left(\frac{\partial^2}{\partial n_{s_1}\,\partial s_1}\right)\right\},
$$

where n_{s_1}, s_i are fixed axes which coincide with the normal and the tangent at the point of the contour considered.

If the plate is simply supported along the contour, then the solution is given by the system of equations (35.15) and the first of equations (35.17).

It has been noted that there exist different versions of constructing the equations of the method of compensating loadings. For instance, one can prescribe the forces along one of the singular lines and the moments along the other singular line. In this case we can also make both the singular lines coincident by prescribing the forces and the moments along the same line. Lastly, one can specify the type of one loading and seek its intensity and the equation of the corresponding singular line. However, in this case the calculations become more complicated; therefore, in the examples we always prescribe the equation of the singular line.

Let us pass to the approximate solutions of problems with the help of the method of compensating loadings.

Consider the problem on the stability of a square plate clamped along the contour. Let us determine the value of the critical pressure p from the system of equations of the type (35.15) and (35.16) for $f_1 = 0$, $f_2 = 0$.

Consider the solution which has singularities on the circumference whose center is at the intersection point of the diagonals of the square and whose radius is not less than half the length of the diagonal. We expand the intensity of these singularities

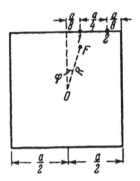

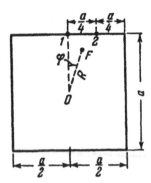

Figure 39   Figure 40

into a trigonometric series. We restrict ourselves to the two first terms, using the symmetry. As a result, the deflection at an arbitrary point F of the plate is represented by an approximate expression

$$w = A_0 J_0(\lambda R) + A_4 J_4(\lambda R) \cos 4\varphi + B_0 + B_4(\lambda R)^4 \cos 4\varphi. \qquad (35.18)$$

The notations are clear from Fig. 39.

Consider the first numerical example.

We require that the deflection and its first derivative with respect to the radius vanish at the points 1, 2 of the contour (and at the points symmetrical to these ones). This leads to the following transcendental equation:

$$\begin{vmatrix} J_0(k_1 a') & J_4(k_1 a') \cos 4\varphi_1 & 1 & (k_1 a')^4 \cos 4\varphi_1 \\ J_0(k_2 a') & J_4(k_2 a') \cos 4\varphi_2 & 1 & (k_2 a')^4 \cos 4\varphi_2 \\ -J_1(k_1 a') & J_4'(k_1 a') \cos 4\varphi_1 & 0 & 4(k_1 a')^3 \cos 4\varphi_1 \\ -J_1(k_2 a') & J_4'(k_2 a') \cos 4\varphi_2 & 0 & 4(k_2 a')^3 \cos 4\varphi_2 \end{vmatrix} = 0, \qquad (35.19)$$

where k_1, k_2, φ_1 and φ_2 can be found from elementary geometrical reasoning.

From here we find $a'/2 = \lambda a/2 = 3.58$.

Let us pass to the second example, by changing the points of interpolation and taking them according to Fig. 40. As a result of the calculations, we obtain $\lambda a/2 = 3.62$.

In the third example, the same problem will be solved, using the following approach: as before, setting w in the form (35.18), we find the critical pressure from the transcendental equation

$$\begin{vmatrix} \delta_{11} & \delta_{12} & a'/4 & \delta_{14} \\ \delta_{21} & \delta_{22} & a'/4 & \delta_{24} \\ \delta_{31} & \delta_{32} & 0 & \delta_{34} \\ \delta_{41} & \delta_{42} & 0 & \delta_{44} \end{vmatrix} = 0,$$

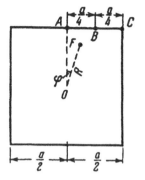

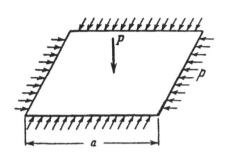

Figure 41　　　　　　　　　　　　Figure 42

$$\delta_{11} = \int\limits_{AB} J_0(\lambda R)\, dx, \qquad\qquad \delta_{12} = \int\limits_{AB} J_4(\lambda R)\cos 4\varphi\, dx,$$

$$\delta_{14} = \int\limits_{AB} (\lambda R)^4 \cos 4\varphi\, dx, \qquad\qquad \delta_{21} = \int\limits_{BC} J_0(\lambda R)\, dx,$$

$$\delta_{22} = \int\limits_{BC} J_4(\lambda R)\cos 4\varphi\, dx, \qquad\qquad \delta_{24} = \int\limits_{AB} (\lambda R)^4 \cos 4\varphi\, dx,$$

$$\delta_{31} = \int\limits_{AB} \frac{\partial}{\partial n} J_0(\lambda R)\, dx, \qquad\qquad \delta_{32} = \int\limits_{AB} \frac{\partial}{\partial n}[J_4(\lambda R)\cos 4\varphi]\, dx,$$

$$\delta_{34} = \int\limits_{AB} \frac{\partial}{\partial n}[\lambda^4 R^4 \cos 4\varphi]\, dx, \qquad\qquad \delta_{41} = \int\limits_{BC} \frac{\partial}{\partial n} J_0(\lambda R)\, dx,$$

$$\delta_{42} = \int\limits_{BC} \frac{\partial}{\partial n}[J_4(\lambda R)\cos 4\varphi]\, dx, \qquad\qquad \delta_{44} = \int\limits_{BC} \frac{\partial}{\partial n}[\lambda^4 R^4 \cos 4\varphi]\, dx,$$

which is obtained from the condition of the vanishing of all the integrals of the deflection and its normal derivative, taken along the segments AB and BC of the contour (Fig. 41). The computation gives the value $a'/2 = \lambda a/2 = 3.62$ which is close to the ones obtained before.

Thus, values of the critical pressure which are sufficiently close to each other have been obtained, as a result of the computations which have been performed by different methods.

A numerical solution of this problem has been obtained before by other authors with the help of some methods which are different from those applied in this section.

Let us introduce the quantity $\mu = \dfrac{P_{cr} a^2}{\pi^2 D}$, where a is the length of the side of the square. Note that the value a', which we have found, is connected with the critical force by the relation $P_{cr} = a'^2 D/a^2$. We now give a summary table of the values obtained by different authors[8].

---

[8]This table as well as the references are given in [20].

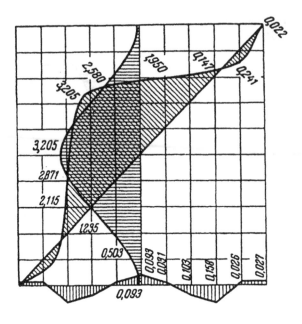

Figure 43

Taylor	Sezava	Faxen	Trefftz	Igushi	Author		
					Example 1	Example 2	Example 3
between 5.30 and 5.33	5.61	5.304	between 5.30 and 5.32	5.3036	5.18	5.31	5.31

One can see from this table that the results obtained by different authors are reasonably close (the exception is the inexact result by Sezava).

In conclusion, we present an approximate solution of the problem on the calculation of a plate which is compressed by an uniform pressure p, clamped along the contour and, moreover, loaded by a concentrated force P at the center (Fig.42). Let $P/(4D\lambda^2) = 1$, $\lambda a = 6$. Then, satisfying the boundary conditions of Example 1, i.e., the deflections and their derivatives in the radial direction vanish at the points 1, 2 (Fig. 39), we obtain a system of four equations with respect to A_0, A_4, B_0, B_4. The coefficients of the indeterminates in these equations are the appropriate elements of the rows of the determinant in the left-hand side of (35.19), and their right-hand sides are obtained with the help of formula (35.14). After solving the system of equations, we obtain the values of the deflections which are shown in Fig. 43.

Consider the problem on stability of a semicircular plate clamped along the contour and compressed by the hydrostatic pressure (Fig. 44). This problem can

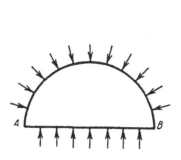

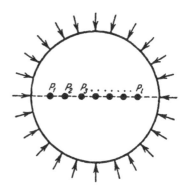

Figure 44   Figure 45

be reduced to the consideration of a question concerning the stability of a circular plate with additional supports (Fig. 45).

Let us construct a transcendental equation for the determination of the critical pressure in the plate shown in Fig. 45:

$$
\begin{vmatrix}
\delta_{11} & \delta_{12} & \cdots & \delta_{1l} \\
\delta_{21} & \delta_{22} & \cdots & \delta_{2l} \\
\vdots & \vdots & \ddots & \vdots \\
\delta_{l1} & \delta_{l2} & \cdots & \delta_{ll}
\end{vmatrix} = 0,
\tag{35.20}
$$

where δ_{ik} is the deflection under the force P_i caused by the force $P_k = 1$. The values of δ_{ik} can be determined with the help of formulae presented in the second chapter.

Since equation (35.20) corresponds to the solution of the problem which differs from the initial one only by weakening the connections against a linear shift along the straight line AB, the value of the critical pressure obtained is too small.

The initial problem can also be replaced by another problem, by considering a circular plate which is loaded by moments as is shown in Fig. 46. We obtain an equation for the determination of the critical force by the vanishing determinant (35.20), where δ_{ik} are the turns of the cross-sections at the points of the application of the moments in the directions of the action of these moments. It is useful to note that the terms along the principal diagonal should be computed by approximately replacing the concentrated moments by the forces which are distributed along circumferences of a sufficiently small radius so that the slope at the point of application of the corresponding moment becomes infinite.

Figure 46

In connection with the fact that the problem on the stability of a circular plate with supports can be solved easily enough, it is natural to use the circular plate as the basic system, i.e. take the circle (as well as a sector of the circle) as the extended domain.

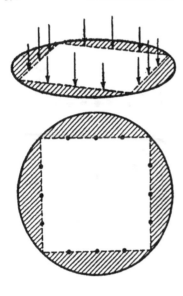

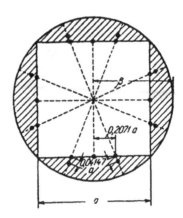

Figure 47

Figure 48

In this case, placing the supports and directing the reactive forces as is shown in Fig. 47, we can obtain an approximate solution of the problem on stability of a square plate clamped along the contour, which has been considered in this section by another method. Since this solution is obtained as a result of rejecting the connections on the contour, the approximation in the critical force is obtained from below.

Let us perform the calculation for the case, when the supports are placed according to Fig. 48. We obtain the equation for determination of the critical force by the vanishing of the determinant

$$\Delta = \begin{vmatrix} \delta_{11} & \delta_{12} & \delta_{13} \\ \delta_{21} & \delta_{22} & \delta_{23} \\ \delta_{31} & \delta_{32} & \delta_{33} \end{vmatrix},$$

where δ_{ik} are the deflections at the points of application of the forces in the circular plate which is loaded by a compressing radial pressure. The computation of these values was performed by the formulae given in Section 31. By restricting ourselves in the computation of almost all δ_{ik} (except only one) to only one term of the series, which corresponds to a rather rough approximation of the conditions on the circumference which bounds the extended domain, we obtain a value of P which is too small; the difference in the root of the transcendental equation is equal to 4.1% (the root obtained is 3.46, and its exact value is 3.61).

For a square plate which is simply supported along two adjacent sides and are clamped along the other two sides, the approximation in the compressing critical pressure from below can easily be obtained, by considering the plate shown in Fig. 49. The solution of this problem can be obtained, by using the formulae given in Section 31.

3. Let us apply the method of compensating loading to the solution of the problem on the calculation of a plate on an elastic foundation. The basic influence

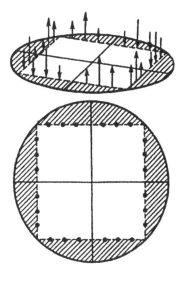

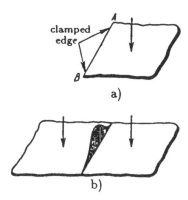

Figure 49   Figure 50

function in the considered case has the form

$$K_1^*(x, y; \xi, \eta) = \frac{1}{4D\lambda^2} f_0(r),$$

where $\lambda = \sqrt[4]{k_0/D}$, $f_0(r) = \Re H_0^{(1)}\left(r\sqrt{i}\right)$ is the real part of the Hankel function of the first kind of zero order of the complex argument $r\sqrt{i}$. Here

$$r = \lambda\sqrt{(x - \xi)^2 + (y - \eta)^2}.$$

In order to obtain the basic solution we should integrate the influence function. If the loadings are distributed over the circular domains, then we can use the formulae for calculation of an unbounded plate in order to obtain the basic solution.

Let us give the solution for a plate of an infinite dimension which is loaded by a unit concentrated moment. It has the form

$$K_2^*(r) = \frac{1}{4D\lambda\sqrt{2}}\{f_1(r) - g_1(r)\}\cos\varphi_1, \qquad (35.21)$$

where

$$f_1(r) + ig_1(r) = H_1^{(1)}\left(r\sqrt{i}\right).$$

As an example, let us write the integral equations of the method of compensating loadings for a plate with the clamped edge. Outside the domain occupied by the plate we apply the loadings $q_1(\sigma)$ and $q_2(\sigma)$ which are distributed along the curves C_1 and C_2. The necessary condition in order that the sum of the basic solution and the compensating one satisfies the boundary conditions on the contour, is

$$\int_{C_1} q_1(\sigma)K_1^*(r_1)\,d\sigma + \int_{C_2} q_2(\sigma)K_1^*(r_2)\,d\sigma = -f_s(s), \qquad (35.22)$$

$$\int_{C_1} q_1(\sigma)\frac{\partial}{\partial n_s}K_1^*(r_1)\,d\sigma + \int_{C_2} q_2(\sigma)\frac{\partial}{\partial n_s}K_1^*(r_2)\,d\sigma = -f_s(s), \qquad (35.23)$$

where r_1 is the distance from the point of the contour to a point of the singular line C_1, r_2 is the distance from the point of the contour to a point of the singular line C_2, $f_5(s)$ and $f_6(s)$ are the deflections and the slope of the normal to the contour taken from the basic solution.

Let us pass to the problem on the equilibrium of a semi-infinite plate with the clamped edge which is loaded by a unit concentrated force (Fig. 50a).

Consider the extended domain, i.e., a plate of infinite dimension. We apply, in addition to the given force, another force whose value is equal to the given one and which is located symmetrically with respect to the edge. Moreover, we apply a distributed loading along the border for the present indeterminate (Fig. 50b). As a result of the symmetry, one of the boundary conditions holds, since the slopes of the cross-sections, which are located on the straight line whihc coincides with the boundary of the semi-infinite plate are equal to zero.

When making the calculation, we replace the unknown intensity of the compensating loading by a piecewise linear function, whose ordinates are determined by the condition of vanishing deflections at the points of the straight line bounding the plate, which are located under the vertices of the polygonal line. Thus, the number of the equations is defined by the number of the points; in our case we have eight equations. Solving these equations, we obtain the values of the ordinates for the special case under consideration, which determine the compensating loading. The character of this loading is shown in Fig. 51.

Numerical checking shows that the deflections at intermediate points of the line indicated constitute less than 2% of the value of the deflection at the same point before compensation.

The plan in the horizontals of the deflection surface of the plate is presented in Fig. 52.

The same problem can be solved in another way, namely, by applying an additional inversely symmetric force P to the extended domain, i.e. to the plate of infinite dimension, (in a more general case of the loadings, an additional system of inversely symmetric loadings). In this case the deflections on the axis of the inverse symmetry vanish; this solution corresponds to the case of a semi-infinite plate which is simply supported along the rectilinear boundary.

In order for the slopes of the normal to the contour to vanish, we should apply the moments, whose vectors are directed along AB, along the straight line AB. In this case we again come to one integral equation.

Consider a square plate with a clamped edge, which is loaded by a force P concentrated at the center (Fig. 53a).

Let us note that the simplifications, which will be discussed below, can be carried over to the case of a rectangular plate and an arbitrary loading.

It is useful for us in the example being considered to introduce a loading by the forces distributed on a square frame, which is considered in the theory of girderless floor constructions. The solution of this problem is given by many authors in terms

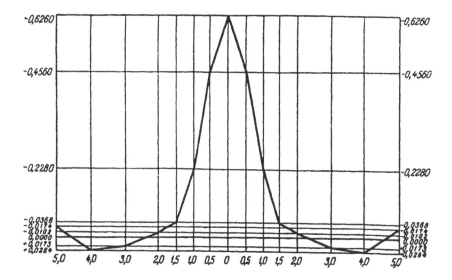

Figure 51

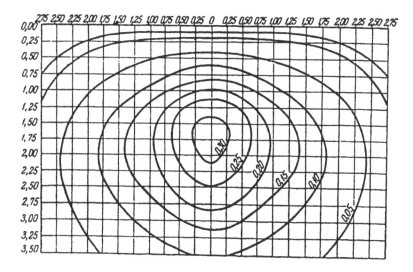

Figure 52

of Fourier series. However, for a plate on an elastic foundation it is convenient to represent the solution in the following form:

$$w = P \left[\sum_{j=0}^{\infty} \sum_{k=1}^{k=8(j+1)} K_1^*(r_k) + K_1^*(r_0) \right], \qquad (35.24)$$

where r_k is the distance of the kth force from the point considered.

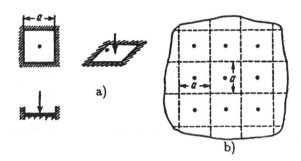

Figure 53

This formula represents the sum of the solutions for a plate of infinite dimension which is loaded by the concentrated forces applied at the nodes of a square frame. The accepted order of summing is the following one: every term of the series represents the result of the action of the loading applied to the domain which should be added to the square with the length of the side $(2j+1)a$ in order to obtain a square with the side $(2j+1)a + 2a$.

The series converges for any finite a.

In order to make sure of this fact, we find the deflection under the load. For this purpose we consider a general term of the following series:

$$\sum_{k=1}^{k=8(j+1)} f_0(r_k). \tag{35.25}$$

Using the asymptotic formula for the function $f_0(r)$ for large values of the argument

$$f_0(r) \approx \sqrt{\frac{2}{\pi r}} e^{-r/\sqrt{2}} \sin\left(\frac{r}{\sqrt{2}} + \frac{\pi}{8}\right),$$

we obtain for sufficiently large j

$$\frac{4D}{l^2} \sum_{k=1}^{k=8(j+1)} K_1^*(r_k) < 8(j+1)\sqrt{\frac{2}{\pi\alpha(j+1)}} e^{-\frac{\alpha}{\sqrt{2}}(j+1)}. \tag{35.26}$$

This inequality proves the formulated statement and proves the good convergence of the series (35.25).

Let us go to the solution of the formulated problem.

Instead of the given construction (Fig. 53a) we consider a plate of infinite dimension, which is loaded by equal forces applied at the nodes of a square frame (Fig. 54a). In this case one of the boundary conditions holds, because the slopes of the normal to the contour are equal to zero from the symmetry conditions.

In order to make the deflection on the contour vanish, we apply distributed loadings to the sides of the square (Fig. 54b). The intensity of these loadings can be determined from an integral equation; when making the computation, we replaced the compensating loading by a system of concentrated forces. One can easily see that this results in the superposition of the solutions which are expressed by a series of type (35.25).

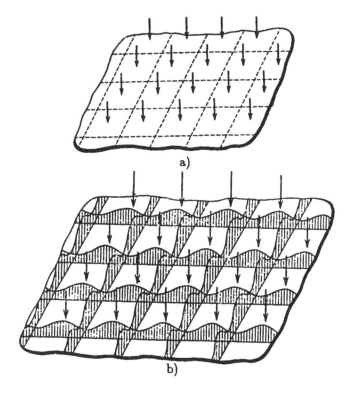

a)

b)

Figure 54

In the calculation we obtained the linear equations which express the conditions for the deflections vanishing at some points of the contour equidistant from each other. Let us note that, by summing the forces applied to the plate according to Fig. 55, we obtain the solution of the problem on the equilibrium of a plate simply supported along the four sides. This solution can be represented in the form of a series similar to (33.5). Such a plate can be considered as another basic system for the solution of the problem just considered on the equilibrium of a plate with the clamped edge. In order for the slopes to vanish, we should apply the distributed moments along the straight lines shown in Fig. 55 by dashed lines. The result of the computation is presented in Fig. 56.

Consider the equilibrium of a rectangular plate whose two opposite sides are simply supported (Fig. 57a). The plate is loaded by a concentrated force at the point of the intersection of the diagonals. We attach the plate to the strip as is shown in Fig. 57b. In this case, the boundary conditions will hold along the simply supported sides of the rectangular plate which are perpendicular to the sides of the strip. In order to make the calculation of the strip, we consider a plate of infinite dimension which is loaded by forces as is shown in Fig. 57b. The solution will be given in the form of a series similar to (35.24).

In order to satisfy the boundary conditions on the straight lines which correspond to the edges of the strip, we expand the values of the basic solution on the contour into a trigonometric series. For instance, for a strip with simply supported

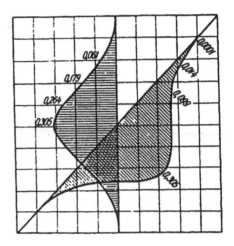

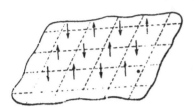

Figure 55

Figure 56

edges we should expand the values of the deflection and its normal derivative into the series. Then, applying the loadings which are varied according to the periodic rule, along straight lines parallel to the edges of the strip and located outside the domain occupied by the strip, we obtain the compensating solution.

The method which was used when considering this example, is connected with separating the singularity and leads to quickly convergent series.

Consider a plate which is bounded by two half-lines which form a right angle and is loaded by a loading distributed on the rectangular located non-symmetrically with respect to the bisectrix of the angle (Fig. 58).

For the calculation, we first complete the angle to 180° and obtain a semi-infinite plate which is loaded by the given loading. We also apply a system of concentrated forces to the semi-infinite plate and make the calculation of the semi-infinite plate with respect to the action of each of these loadings. In this case the semi-infinite plate plays the role of a specific "statically undefinable basic system".

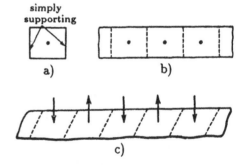

simply supporting

a)

b)

c)

Figure 57

We applied the concentrated forces at six points. They were determined by the condition that the bending moments and the transversal forces vanish at the points 1, 2, 3 (Fig. 58).

Omitting the intermediate calculations, we only present the final results.

The diagrams of the bending moments in $kg\,cm/cm$ and of transversal forces in kg/cm, which arise in the semi-infinite plate on the straight line AB corresponding to the boundary of the angle, are shown in Fig. 59a.

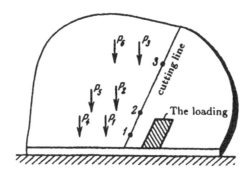

Figure 58

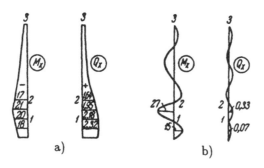

Figure 59

The same diagrams with regard to the influence of the compensating forces are shown in Fig. 59b. One can see in the figure that the transversal forces, which play an essential role in this problem when determining the moments under the loading, are sharply reduced and practically vanished. The decrease of the absolute value of the maximal bending moment in the cutting line did not occur. However, in this special case the influence of the moments which act in the cutting line, onto the value of the bending moments under the loading is rather small. Passing to the diagram of the moments with alternating signs shows that the influence of the moments on the edge on the value of the calculated moments is significantly reduced.

4. Consider some problems on bending oscillations of a plate, which are very close to the problems considered in the previous sections. As is known, the differential equation of free oscillations of a plate has the form

$$\Delta\Delta w - \lambda_1^4 w = 0, \tag{35.27}$$

where

$$\lambda_1 = \sqrt[4]{\frac{\gamma h\omega^2}{Dg}}.$$

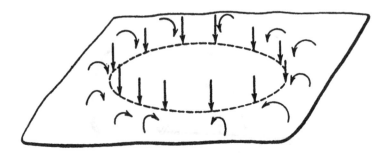

Figure 60

The solution which has a singularity of the type of a concentrated force and which satisfies the differential equation (35.27) has the form

$$K_1^*(r) = -\frac{1}{8D\lambda_1^2}\left[Y_0(r) + \frac{2}{\pi}K_0(r)\right].\qquad(35.28)$$

We will not consider the construction of the equations for the determination of the compensating loadings in detail. We only give two examples: in the first one we investigate the free oscillations of a plate, and in the second one we investigate the forced oscillations of the plate.

Consider the problem on free oscillations of a circular plate clamped along the contour. The exact solution of this problem is known. We will construct the compensating solution, by applying a system of concentrated periodically varying forces and moments to the contour, whose vectors are directed along the tangent to the contour. The disposition of the forces and the moments is shown in Fig. 60. It has been noted that the slope at the point of application of the concentrated moment becomes infinite. The deflections and the slopes at the points of application of the forces and moments should vanish. Therefore, in place of application of the concentrated moments applied at the points, we introduce the loadings $q_0 \cos \theta$, which are distributed along the circumferences of a sufficiently small radius. It is natural that when calculating we assume all the forces P and loadings q_0 to be equal. By the deflection and the slope of the normal vanishing at the indicated points, we obtain a transcendental equation with respect to the reduced radius a. In this case the smallest root of the transcendental equation is equal to 3.15, whereas the exact solution of the problem gives the root equal to 3.19.

Of course, this example is of interest only from the viewpoint of illustrating the effectiveness of the approximate solution under consideration.

Consider the problem on the forced oscillations of the girderless floor construction. We assume that the floor has equidistant point supports which are located on a square frame. The concentrated force $P \sin \omega t$ is applied to the floor at the center of the middle panel. One can see in Fig. 61 that the floor consists of 25 panels which are supported by 16 columns and the walls located along the external contour of the floor. We assume that the plate is clamped along the contour. It is clear from physical considerations that the influence of the boundary conditions on the external contour onto the work of the middle field is not essential. The fixings by point supports play the main role. Therefore, we satisfy the boundary conditions on the external contour in a rough way, by requiring that the deflection and the slope on the contour are equal to zero on average. The conditions at the point supports and at the point of application of the concentrated force will be satisfied strictly. It is clear from the symmetry conditions that the reactions at the four columns which border the middle field, are equal to each other. The reactions at the four other columns which are located on the diagonal of the square are also equal to each other, as well as the forces at all the other eight supports. Thus, we shall make the calculation for one known force at the center, setting $\dfrac{P}{8D\lambda_1^2} = 1$, and we will calculate the plate for three systems of equal unit forces, each of which combines the equal reactions at the point supports. These three schemes are presented in Fig. 62a, b, c, d. Without numerical calculations we only describe the scheme of the calculation followed by a numerical result.

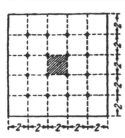

Figure 61

The solution for each of the schemes indicated will be obtained in the form of a sum of a basic solution and a compensating one. The basic solution for the first scheme is (35.21) for $\dfrac{P}{8D\lambda_1^2} = 1$. The basic solution for the remaining schemes will be obtained as a sum of such solutions, assuming the origin for each of them is at the point where the corresponding support is located. We seek the compensating solution in the form

$$w_k = A_0 J_0(\lambda_1 r) + B_0 I_0(\lambda_1 r).$$

It has been noted that we find the coefficients A_0 and B_0, by setting the average values of the deflection and the slope on the contour equal to zero.

If we have the four solutions which have been indicated, then we can easily obtain a system of canonical equations for the method of forces for the determination of the bearing reactions at the point supports; each of these equations means that the deflection over the corresponding support is equal to zero.

Then, knowing the bearing reactions and using the calculation for the unit forces, we can find the deflections with the help of the principle of addition of the actions.

In order to show that the influence of the conditions on the external contour is insignificant, we also calculate, as well as the square floor, two other girderless floors, circular in the plan, with the same reduced step of the columns (see Fig. 62e and f). One of these calculations is presented in Section 31, where a series of problem on the oscillations of circular plates is considered in detail.

points	Floor Fig. 62e	Floor Fig. 62f	Floor Fig. 62a
I	0.55691	0.55284	0.55164
II	0.40982	0.40841	0.40756
III	0.19679	0.19433	0.19981

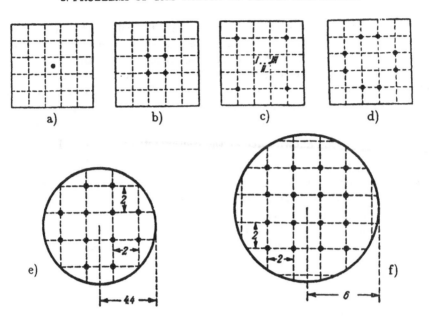

Figure 62

One can see in the table above that the deflection at the center of a girderless floor construction in all three cases are very close, as well as the deflections at the other points of the middle panel which were considered.

The description of the calculation implies that we can extend these scheme of the calculation onto the solution of the problem on forced oscillations of a three-dimensional frame with pillars of the circular transversal cross-section and a cross-plate in the construction of a girderless floor.

For this purpose, we should write the solution of the problem on the loading of a plate on the circular domains in closed form, using the formulae given in the first chapter. Then, we should construct the equations concerning linear as well as angular displacements of the capitals. In this case we should: a) set the average value of the deflection in the domain of support and the deflection at some place of the support equal to zero and b) set the slope of the pillar and the slope of the plate equal to each other.

Obviously, the same scheme of the calculation can be applied to the solution of the problem on the eigenoscillations of a girderless floor construction.

36. The problem on equilibrium of an unbounded plate which lies on a uniform elastic foundation whose model has a circular symmetry

Consider a plate which lies on a solid foundation. We assume that the upper bound of the foundation is the horizontal plane. If the vertical unit force acts on this plane, then, as a result of deforming the upper bound of the foundation, a surface will be formed, whose z-coordinate can be called the *kernel of the model of the elastic foundation*. Suppose that the equation of this surface, written in a system of coordinates with the origin at the point of application of the force, remains the same while varying the position of the force. Such a foundation will be called *uniform*.

In order to pass to the study of the bending of beams and plates which lie on a linearly deformable uniform foundation, we first consider an important auxiliary problem on the dependence between the vertical forces acting on the foundation and the deflections of the foundation.

We will consider the kernels which can be graphically represented as a surface of rotation here, i.e., the kernel of the form

$$K(x - \xi, y - \eta) = K\left(\sqrt{(x - \xi)^2 + (y - \eta)^2}\right).$$

These are the most important in applications.

Many important properties of elastic uniform foundations with circular symmetry are related to a more general type of kernel.

$$K(x - \xi, y - \eta) = K\left(\sqrt{(x - \xi)^2 + \mu^2(y - \eta)^2}\right),$$

where μ is a constant. The appropriate models of the foundation can play an important role for anisotropy in the plane oxy as well as for the solution of some dynamical problems. With the help of the kernel of the elastic foundation we can easily write the integro-differential system of equations of the problem under consideration, which has a simple solution based on using the Fourier transformation in the case of an unbounded domain. However, we present this solution in another form which is more descriptive from the mechanical viewpoint. As a basis, we use the problem on the loading of the foundation by the pressure which is varied according to the rule $q = a \sin \alpha x \sin \beta y$. Let us show that the settling for such a loading is given, for any uniform elastic foundation, by the expression

$$w = bc \sin \alpha x \sin \beta y, \tag{36.1}$$

where $c(\alpha, \beta)$ is a function whose form depends on the chosen model of the foundation.

Let us denote $\alpha x = x_1$, $\beta y = y_1$, $\alpha \xi = \xi_1$ and $\beta \eta = \eta_1$. After the change of the variable x, y, ξ, η by x_1, y_1, ξ_1, and η_1, the kernel $K(x - \xi, y - \eta)$ takes the form $K_1(x_1 - \xi_1, y_1 - \eta_1)$.

Applying the principle of addition of the actions, we obtain

$$w = \int\limits_{-\infty}^{\infty} \int\limits_{-\infty}^{\infty} b \sin(\alpha x) \sin(\beta y) K(x - \xi, y - \eta)\, dx\, dy$$

$$= \frac{b}{\alpha\beta} \int\limits_{-\infty}^{\infty} \int\limits_{-\infty}^{\infty} \sin x_1 \sin y_1 K_1(x_1 - \xi_1, y_1 - \eta_1)\, dx_1\, dy_1. \tag{36.2}$$

Introducing the variables $x_1 - \xi_1 = z_1$ and $y_1 - \eta_1 = \zeta_1$, we rewrite expression (36.2) in the following form:

$$w = \frac{b}{\alpha\beta} \int\limits_{-\infty}^{\infty} \int\limits_{-\infty}^{\infty} \sin(\xi_1 + z_1) \sin(\eta_1 + \zeta_1) K(z_1/\alpha, \zeta_1/\beta)\, dz_1\, d\zeta_1$$

$$= \frac{b}{\alpha\beta} \sin \xi_1 \sin \eta_1 \int\limits_{-\infty}^{\infty} \int\limits_{-\infty}^{\infty} \cos z_1 \cos \zeta_1 K(z_1/\alpha, \zeta_1/\beta)\, dz_1\, d\zeta_1.$$

Hence, $w = bc \sin \xi_1 \sin \eta_1$, where

$$c = \frac{1}{\alpha\beta} \int\limits_{-\infty}^{\infty} \int\limits_{-\infty}^{\infty} \cos z_1 \cos \zeta_1 K(z_1/\alpha, \zeta_1/\beta) \, dz_1 \, d\zeta_1, \qquad (36.3)$$

QED.

In order to solve the problem concerning the motion for the periodic loading considered here, we should determine c with the help of formula (36.3). Let us pass to these computations for different types of the foundation.

Firstly, we consider the most widespread model, namely, we consider an elastic uniform isotropic half-space. From the theory of elasticity it is known that in this case

$$K(x - \xi, y - \eta) = \frac{1 - \sigma^2}{\pi E} \frac{1}{\sqrt{(x - \xi)^2 + (y - \eta)^2}}, \qquad (36.4)$$

where E is the modulus of elasticity, σ is Poisson's ratio.

Let us denote

$$\frac{E}{2(1 - \sigma^2)} = k.$$

Computing c by formula (36.3), one can obtain [19]

$$c = \frac{1}{k\sqrt{\alpha^2 + \beta^2}},$$

hence

$$w = \frac{b}{k\sqrt{\alpha^2 + \beta^2}} \sin \alpha x \sin \beta y.$$

This result is well known in the theory of bending a plate which lies on an elastic half-space (Woinowsky-Krieger [85]).

For different applications, it is convenient to transform formula (36.4), by passing from Cartesian coordinates to polar ones. We shall first do this for the special case of a kernel with an axial symmetry.

Let us write the formula which is a consequence of the Bessel integral:

$$\int\limits_{-\infty}^{\infty} \int\limits_{-\infty}^{\infty} \cos(\alpha x) \cos(\beta y) K\left(\sqrt{x^2 + y^2}\right) dx \, dy = 2\pi \int\limits_{0}^{\infty} rK(r)J_0(\gamma r) \, dr, \qquad (36.5)$$

where $\gamma^2 = \alpha^2 + \beta^2$, $r^2 = x^2 + y^2$. Thus, formula (36.3) can be written in the following form:

$$c = 2\pi \int\limits_{0}^{\infty} rK(r)J_0(\gamma r) \, dr; \qquad (36.6)$$

i.e. as the Hankel transform.

In the cases, when the kernel has no axial symmetry, it can nevertheless be represented in the form

$$K = K\left(\sqrt{(x - \xi)^2 + \mu^2(y - \eta)^2}\right).$$

We obtain

$$c(\gamma) = \frac{2\pi}{\alpha} \int\limits_0^\infty r_1 K(r_1) J_0(\gamma_1 r_1)\, dr_1,$$

where

$$r_1 = \sqrt{x^2 + y^2},$$
$$\gamma_1 = \sqrt{\alpha^2 + \mu^2 \beta^2}.$$

With the help of (36.6) we can find the function $c(\gamma)$ for the problems in which the kernel K is known.

Let us give a table containing several types of the kernels K, for which one can obtain rather simple expressions of the function $c(\gamma)$, by using well-known improper integrals.

$K(r)$	$c(\gamma)$
$\dfrac{B}{\sqrt{r^2 + \delta^2}}$	$\dfrac{2\pi B}{\gamma} \exp(-\delta\gamma)$
$\dfrac{B}{2\delta^2} \exp\left(-\dfrac{r^2}{4\delta^2}\right)$	$2\pi B \exp(-\delta^2\gamma^2)$
$B K_0(\delta r)$	$\dfrac{2\pi B}{\gamma^2 + \delta^2}$
$\dfrac{B}{2\pi r} \exp(-\delta r)$	$\dfrac{B}{\sqrt{r^2 + \delta^2}}$
$\dfrac{B}{2\pi(r^2 + \delta^2)}$	$B K_0(\gamma\delta)$

Thus, we have obtained several kernels, which can, in particular, be used in order to approximate the surfaces of the settlings obtained experimentally. In this table, B and δ are the parameters which can be varied, K_0 is the Macdonald function.

If the kernel K has the form $B K_0(\delta r)$, then it corresponds to a foundation with two elastic characteristics.

For the half-space, whose modulus of elasticity is varied with the depth according to the exponential rule, G.K. Krein has obtained [17] the kernel

$$K(x - \xi, y - \eta) = \frac{1}{\pi D_n r^{n+1}},$$

where D_n is a constant; n is a number which characterizes the rule of the variation of the modulus of elasticity with respect to the depth,

$$r = \sqrt{(x - \xi)^2 + (y - \eta)^2}.$$

In this case, the theory of improper integrals with Bessel functions implies (see page 72) that for $0.5 < n < 1$ we have

$$c(\alpha, \beta) = \frac{2^{1-n}}{D_n} \sqrt{(\alpha^2 + \beta^2)^{n-1}} \frac{\Gamma((1 - n)/2)}{\Gamma((1 + n)/2)},$$

where Γ is the gamma function.

Let us pass directly to the problem of the calculation of an unbounded plate.

First we consider an auxiliary problem on the bending of an unbounded plate, on which the loading

$$q = a \cos \alpha x \cos \beta y$$

acts.

The differential equation for the elastic surface of the plate has the form

$$D\nabla^2\nabla^2 w = q - p,$$

where $\nabla^2 = \frac{\partial}{\partial x^2} + \frac{\partial}{\partial y^2}$. We suppose that the reactive pressure is

$$p = b \cos \alpha x \cos \beta y, \qquad (36.7)$$

and represent the deflection of the plate in the form

$$w = bc^* \cos \alpha x \cos \beta y,$$
$$c^* = c + 1/k_0. \qquad (36.8)$$

Introducing (36.7) and (36.8) into the differential equation for the elastic surface of the plate, we obtain the equation

$$bc^* D(\alpha^2 + \beta^2)^2 = a - b;$$

therefore, $b = a/[1 + c^* D(\alpha^2 + \beta^2)^2]$.

If we consider the problem on a plate which is subjected to the action of the loading

$$q = \sum_{m=0}^{\infty} \sum_{n=0}^{\infty} a_{mn} \cos \alpha_m x \cos \beta_n y,$$

then, using the formulae given above, we obtain

$$p = \sum_{m=0}^{\infty} \sum_{n=0}^{\infty} \frac{a_{mn}}{1 + c^*_{mn} D(\alpha_m^2 + \beta_m^2)^2} \cos \alpha_m x \cos \beta_n y,$$

$$w = \sum_{m=0}^{\infty} \sum_{n=0}^{\infty} \frac{a_{mn} c^*_{mn}}{1 + c^*_{mn} D(\alpha_m^2 + \beta_m^2)^2} \cos \alpha_m x \cos \beta_n y.$$

If we pass to the problem on a plate uniformly loaded over rectanglar surfaces with the sides $2a$ and $2b$ which are located so that their axes of symmetry form a rectangular frame, and if we assume that the distance between the rectangles becomes infinite, and moreover, we assume that $4qab = P$ as $a \to 0$ and $b \to 0$, then after the limiting passage[9], we obtain the required solution in the following form:

$$w = \frac{P}{\pi^2} \int_0^\infty \int_0^\infty \frac{c^* \cos \alpha x \cos \beta y \, d\alpha \, d\beta}{1 + p_0 c^* (\alpha^2 + \beta^2) + c^* D(\alpha^2 + \beta^2)^2}.$$

This formula can easily be transformed, if we introduce the notation

$$\alpha^2 + \beta^2 = \gamma^2, \qquad x^2 + y^2 = r^2, \qquad \alpha/\beta = \tan \varphi.$$

Then we have

$$\alpha = \gamma \sin \varphi, \qquad \beta = \gamma \cos \varphi, \qquad d\alpha \, d\beta = \gamma \, d\gamma \, d\varphi.$$

---

[9] One can use the integral representation of the delta function instead of the limiting passage.

Applying once again the formula

$$J_0\left(\sqrt{x^2 + y^2}\right) = \frac{1}{2\pi}\int_0^{2\pi} \cos(x\cos\varphi)\cos(y\sin\varphi)\,d\varphi,$$

we obtain

$$w = \frac{P}{2\pi}\int_0^\infty \frac{c^*\gamma J_0(\gamma r)\,d\gamma}{1 + p_0 c^*\gamma^2 + c^* D\gamma^4}.$$

Consider some special cases, assuming that $p_0 = 0$ (p_0 is the stretching force in the median plane).

1. **The foundation is an elastic half-space.** In this case $c^8 = c = 1/k\gamma$, where $k = E/(2(1 - \sigma^2))$. We introduce the following notation: $l = \sqrt[3]{D/k}$, $\lambda = \gamma\sqrt[3]{D/k}$, $r/l = \xi$. The reactive pressures and deflections are given by the expressions

$$p(\xi) = \frac{P}{2\pi l^2}\int_0^\infty \frac{\lambda J_0(\lambda\xi)\,d\lambda}{1 + \lambda^3}, \qquad (36.9)$$

$$w(\xi) = \frac{P l^2}{2\pi D}\int_0^\infty \frac{J_0(\lambda\xi)\,d\lambda}{1 + \lambda^3}, \qquad (36.10)$$

where l has the same dimension as the length.

Formulae (36.9) and (36.10) were obtained in 1932 in another way by S. Woinow-sky-Krieger, who, however, considered only the problem on an elastic half-space. This problem as well as a series of important and more complicated problems on a plate on an elastic layer and so on have been solved by O.Ya. Shekhter [59], who has also composed the tables necessary for the calculation.

2. **The Winkler-type foundation (the model of the coefficient of subgrade reaction).** In this case we have $c^* = 1/k_0$. Denoting $l_1 = \sqrt[4]{D/k_0}$, $\lambda_1 = \gamma l_1$, $r/l_1 = \xi$, we obtain

$$w(\xi) = \frac{P}{2\pi k_0 l_1^2}\int_0^\infty \frac{\lambda_1 J_0(\lambda_1\xi)\,d\lambda_1}{1 + \lambda_1^4}.$$

One can easily verify that this solution coincides with the one given in Section 32 of Part II, if $p_0 = 0$, i.e.

$$w(\xi) = \frac{P l^2}{4D} f_0(\xi).$$

For this purpose we should refer to Section 21 (page 74), where we considered improper integrals.

Using the addition theorem (page 26), one can easily show that if the solution of the problem on an unbounded plate which is loaded by a concentrated force has the form

$$w(\xi) = A\int_0^\infty \frac{J_0(\lambda\xi)\,d\lambda}{F(\lambda)},$$

then for the loading which is distributed along the circumference (of the reduced radius ρ) according the rule $q \cos n\theta$, we have

$$w = 2\pi q l A \rho \cos n\theta \int_0^\infty \frac{J_n(\lambda\rho)J_n(\lambda\xi)\,d\lambda}{F(\lambda)}. \qquad (36.11)$$

Let us pass to the solution of the problem on the bending of a plate which is loaded by a distributed loading which is varied according to the rule $q = B\rho^n \cos n\theta_1$ and acts on the area of the circular ring whose radii are equal to ρ_2 and ρ_1, where $\rho_2 > \rho_1$. Using the principle of addition of the actions, we integrate (36.11)

$$w = 2ABl^2\pi \cos n\theta \int_{\rho_1}^{\rho_2} \rho^{n+1}\,d\rho \int_0^\infty \frac{J_n(\lambda\rho)J_n(\lambda\xi)\,d\lambda}{F(\lambda)}.$$

Changing the order of integration, we rewrite the integral obtained in the following form:

$$w = 2ABl^2\pi \cos n\theta \int_0^\infty \frac{J_n(\lambda\xi)\,d\lambda}{F(\lambda)} \int_{\rho_1}^{\rho_2} \rho^{n+1} J_n(\lambda\rho)\,d\rho.$$

With the help of (15.1) we obtain

$$w = 2ABl^2\pi \cos n\theta \int_0^\infty \frac{J_n(\lambda\xi)[\rho_2^{n+1} J_{n+1}(\lambda\rho_2) - \rho_1^{n+1} J_{n+1}(\lambda\rho_1)]\,d\lambda}{\lambda F(\lambda)}.$$

For $n = 0$ we have the solution for the case, when the loading is uniformly distributed on the area of a circular ring

$$w = 2ABl^2\pi \int_0^\infty \frac{J_0(\lambda\xi)[\rho_2 J_1(\lambda\rho_2) - \rho_1 J_1(\lambda\rho_1)]\,d\lambda}{\lambda F(\lambda)}.$$

If $\rho_1 = 0$, then the ring becomes the circle of the radius $\rho_2 = \rho$ and the solution takes the form

$$w = 2ABl^2\rho\pi \int_0^\infty \frac{J_0(\lambda\xi)J_1(\lambda\rho)\,d\lambda}{\lambda F(\lambda)}.$$

CHAPTER 4

Problems of the theory of oscillations, hydrodynamics and heat transfer

37. On the oscillations of a thread

In this section we consider the Bernoulli problem on free oscillations of a suspended thread, as well as the problem on the forced oscillations of the thread. The case, when the thread has a variable density is also considered.

Consider free oscillations of a flexible heavy uniform thread. Suppose that the lower end of the thread is free; the x-axis will be directed up along the vertical straight line passing through the axis of the thread in the equilibrium state; the origin will be taken at the end of the thread. Let ρ denote the mass of the thread per unit length, g be the acceleration of gravity. The movement of a point of the thread in the direction perpendicular to the x-axis will be denoted by y. Only small oscillations are considered.

Consider the element of the length dx. The transversal force in the cross-section is $x Q_x = N\, dy/dx$, where N is the normal force. From the equilibrium condition for the element of the length of the thread, we obtain by projecting the forces applied to the element onto the y-axis,

$$\frac{d}{dx}\left(N\frac{dy}{dx}\right) = -q,$$

where q is the loading per unit length which is directed parallel to the y-axis.

In the case, when the thread performs free oscillations, we have

$$q = -\rho\frac{\partial^2}{\partial t^2} \quad \text{and} \quad \frac{\partial}{\partial x}\left(N\frac{\partial y}{\partial x}\right) - \rho\frac{\partial^2 y}{\partial t^2} = 0.$$

If we set $y(x,y) = w(x)\sin(\omega t + \varphi_0)$, then

$$\frac{d}{dx}\left(N\frac{dw}{dx}\right) + \rho\omega^2 w = 0.$$

If the density ρ is constant, then the thread is stretched by the force $N = \rho g x$. In this case we have

$$\frac{d}{dx}\left(x\frac{dw}{dx}\right) + \frac{\omega^2}{g}w = 0.$$

Setting $x = gs^2/(4\omega^2)$, we reduce the equation obtained to the Bessel equation of zero index with the argument s

$$\frac{d^2 w}{ds^2} + \frac{1}{s}\frac{dw}{ds} + w = 0.$$

Hence

$$w(s) = A J_0(s) + B Y_0(s)$$
$$= A J_0 \left(2\omega \sqrt{\frac{x}{g}} \right) + B Y_0 \left(2\omega \sqrt{\frac{x}{g}} \right).$$

The constant B is equal to zero because the function w cannot become infinite at $x = 0$. Thus, we obtain the solution

$$y(x,t) = A J_0 \left(2\omega \sqrt{\frac{x}{g}} \right) \sin(\omega t + \varphi_0).$$

The boundary condition at the place of fixation implies that for $x = l$ we have $y(l,t) = 0$.

Hence, $J_0 \left[2\omega \sqrt{l/g} \right] = 0$.

Denoting the roots of the Bessel function of zero index by $a_1, a_2, \ldots, a_k$, we immediately obtain the frequencies of the eigenoscillations $\omega_k = \dfrac{a_k \sqrt{g/l}}{2}$; the coefficients A_k and the initial phases φ_{0k} can be determined from the initial conditions.

If the initial form of the curving of the thread and the initial velocities are equal to $f_1(x)$ and $f_2(x)$, respectively, then, seeking the general solution in the form

$$y(x,t) = \sum_{k=1}^{\infty} A_k J_0 \left(2\omega_k \sqrt{\frac{x}{g}} \right) \sin(\omega_k t + \varphi_{0k}).$$

we obtain for $t = 0$

$$\sum_{k=1}^{\infty} A_k J_0 \left(2\omega_k \sqrt{\frac{x}{g}} \right) \sin \varphi_{0k} = f_1(x),$$

$$\sum_{k=1}^{\infty} A_k \omega_k J_0 \left(2\omega_k \sqrt{\frac{x}{g}} \right) \cos \varphi_{0k} = f_2(x).$$

From these equalities we can easily determine A_k and φ_{0k}, using the orthogonality of the Bessel functions.

If the fixing of the thread is elastic, then the boundary condition at $x = l$ has the form

$$N_{x=l} \frac{\partial y(l,t)}{\partial x} = c y(l,t),$$

where c is the so-called quasielastic coefficient of fixing, and if $N_{x=l} = g\rho l$, then

$$-\frac{g\rho l J_1 \left(2\omega_k \sqrt{l/g} \right) \omega_k}{\sqrt{lg}} = c J_0 (2\omega_k \sqrt{l/g}).$$

In this case the problem can be reduced to the Dini series.

It should be noted that in problems of dynamics and seismic stability of buildings at the present time the so-called shear vibrations of the buildings are considered, for which the differential equation is similar to the equation of the oscillations of a thread. One often assumes that ρ depends on x and is usually an exponential function.

Consider the oscillations of the thread, assuming that N is variable and denoting it by N_x. Suppose that the density ρ is a given function of the coordinate x.

Then the differential equation of the form of the free oscillations takes the form

$$\frac{d}{dx}\left(N_x \frac{dw}{dx}\right) + \rho(x)\omega^2 w = 0,$$

where the normal force, which is equal to the sum of external forces applied to the rod on the interval $(0, x)$, is equal to

$$N_x = g \int_0^x \rho(z)\, dz.$$

Hence,

$$N_x \frac{d^2 w}{dx^2} + g\rho(x)\frac{dw}{dx} + \omega^2 \rho(x)w = 0.$$

Consider the simplest rule of varying of $\rho(x)$: $\rho(x) = \rho_0(\alpha x)^\mu$. Then

$$N_x = g \int_0^x \rho_0(\alpha x)^\mu\, dx = \frac{g\rho_0}{\alpha}\frac{(\alpha x)^{\mu+1}}{\mu + 1}.$$

Hence,

$$\frac{d^2 w}{dx^2} + (\mu + 1)\frac{1}{x}\frac{dw}{dx} + \omega^2 \frac{\mu + 1}{gx} w = 0. \qquad (37.1)$$

The last equation is a special case of the equation

$$x^2 \frac{d^2 w}{dx^2} + (2\alpha_1 - 2\beta\nu + 1)x\frac{dw}{dx} + \left[\beta^2\gamma^2 x^{2\beta} + (\alpha_1 - 2\beta\nu)\alpha_1\right] w = 0,$$

whose solution has the form $w = x^{\beta\nu - \alpha_1} Z_\nu(\gamma x^\beta)$ (as can be seen in Section 13, page 32). Setting

$$2\alpha_1 - 2\beta\nu + 1 = \mu + 1, \qquad \alpha_1 - 2\beta\nu = 0,$$

$$2\beta = 1, \qquad \beta^2\gamma^2 = \frac{\omega^2(\mu + 1)}{g},$$

we obtain the solution of the problem considered in the form

$$w = C_1 J_\mu\left(2\omega\sqrt{(\mu + 1)/g}x^{1/2}\right) x^{-\mu/2} + C_2 Y_\mu\left(2\omega\sqrt{(\mu + 1)/g}x^{1/2}\right) x^{-\mu/2}.$$

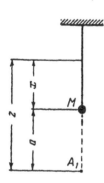

Figure 63

Assuming, as before, the origin to be at the lower end of the rod, we immediately obtain $C_2 = 0$, and for determination of the frequency of the free oscillations we have the transcendental equation

$$J_\mu\left(2\omega\sqrt{(\mu + 1)/g}\sqrt{l}\right) = 0.$$

Consider in more detail the question concerning the oscillations of a thread of constant density, which has a concentrated mass M at the end. The limiting cases lead to the problem on the mathematical pendulum and to the problem on the thread in the classical formulation which was considered at the beginning of this section. If we take, as before, the origin at the end of the thread, then $N_x = Mg + \rho gx$. Let us carry over the origin into the point A_1 (Fig. 63), where $a = M/\rho$.

In this case

$$y(z) = AJ_0\left(2\omega\sqrt{\frac{z}{g}}\right) + BY_0\left(2\omega\sqrt{\frac{z}{g}}\right).$$

The first boundary condition of the problem is: at $z = a$ we have $Q = -M\omega^2 y$. Hence,

$$N\frac{dy}{dz}\bigg|_{z=a} = -M\omega^2 y|_{z=a}$$

or

$$Mg\frac{dy}{dz} + M\omega^2 y|_{z=a} = 0. \tag{37.2}$$

The second boundary condition is not changed compared to the problems considered before, and if the thread has a stiff support, then the second boundary condition has the form

$$y_{z=a+l} = 0. \tag{37.3}$$

Both the boundary conditions are homogeneous. For the existence of a non-trivial solution it is necessary that the following transcendental equation holds

$$J_0\left(2\omega\sqrt{\frac{a+l}{g}}\right)\left[-Y_1\left(2\omega\sqrt{\frac{a}{g}}\right) + \omega\sqrt{\frac{a}{g}}Y_0\left(2\omega\sqrt{\frac{a}{g}}\right)\right]$$

$$- Y_0\left(2\omega\sqrt{\frac{a+l}{g}}\right)\left[-J_1\left(2\omega\sqrt{\frac{a}{g}}\right) + \omega\sqrt{\frac{a}{g}}J_0\left(2\omega\sqrt{\frac{a}{g}}\right)\right] = 0,$$

which can be obtained from (37.2) and (37.3).

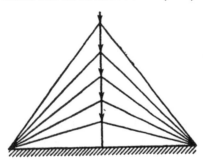

Figure 64

One cannot always assume that the normal force is caused only by the self-weight of the thread. Obviously, a problem can arise in which an additional compression or stretching of the rod is caused by other causes, for instance, by the influence of the tension of the cables (Fig. 64). Thus, the necessity arises to consider the influence of a normal force which does not depend on the mass per unit length. The simplest example is a pillar of a constant cross-section working on the shift which is compressed at the end by cables; we neglect the mass of the cables in the oscillations. In this case we have $N = -N_0 - g\rho x$.

By analogy with the previous example, we should move the origin to the point A which is located at the distance $a = N_0/g\rho$ from the end of the pillar, and introduce the coordinate $z = x + a$.

In conclusion, we consider the problem on the forced harmonic oscillations of a thread in more detail, which are caused by the action of the loading $q = \varphi(t)\sin pt$. In this case we assume that the frequency of the disturbing loading p does not coincide with any of the frequencies of the free oscillations of the thread.

Then the differential equation takes the form

$$\frac{\partial}{\partial x}\left(N\frac{\partial y}{\partial x}\right) - \rho\frac{\partial^2 y}{\partial t^2} = p(x,t) = \varphi(x)\sin \rho t.$$

Setting $y(x,t) = w(x)\sin pt$, we obtain an inhomogeneous ordinary differential equation

$$\frac{d}{dx}\left(N\frac{dw}{dx}\right) + \rho p^2 w = \varphi(x).$$

First, we consider in detail the forced oscillations of a thread with a constant density. Then, passing to the independent variable $s = 2p\sqrt{x/g}$, we obtain the following inhomogeneous equation:

$$\frac{d^2 w}{ds^2} + \frac{1}{s}\frac{dw}{ds} + w = \frac{g}{p^2}\varphi\left[\frac{g}{4p^2}s^2\right].$$

If the right-hand side represents a power function $\varphi = A^* x^\mu$, then we have

$$\frac{d^2 w}{ds^2} + \frac{1}{s}\frac{dw}{ds} + w = \frac{g}{p^2}A^*\left(\frac{g}{4p^2}s^2\right)^\mu = 2^{-2\mu}\left(\frac{g}{p^2}\right)^{1+\mu}A^* s^{2\mu}.$$

A partial solution of the problem can be expressed in this case via the Lommel functions $s_{\mu,\nu}$.

The general solution of the problem under the boundary conditions considered above has the form

$$w = AJ_0(s) + 2^{-2\mu}(g/p^2)^{1+\mu}A^* s_{2\mu+1,0}(s).$$

Let us note that the given frequency p is present here in the variable s and the constant A is determined from the boundary condition at $x = l$. Thus, denoting the value of s at $x = l$ by β, we obtain

$$A = -\left(\frac{g}{p^2}\right)^{1+\mu}\frac{A^* s_{2\mu+1,0}(\beta)}{2^{2\mu}J_0(\beta)}.$$

Note that if ρ is constant and $N(x)$ is varying according to a rule which is different from a linear one, then we can take $N(x)\,dw/dx = Q_x$ as an independent variable. Then the equation with respect to the function Q_x takes the form

$$\frac{d^2 Q_x}{dx^2} + \rho p^2 N^{-1}(x)Q_x = 0.$$

In the conclusion of this section we note that a lot of problems which are described by second order differential equations which can be reduced as a result of a change of variables to the Bessel equation, arises in very different applications. The problems of the stability of rods of variable cross-sections which can be reduced to the Bessel equation, have been considered in greaat detail by A.N. Dinnik [12].

38. Stability and transverse-longitudinal bending of a rectilinear rod; stability of the plane form of the bending of a strip

It is known that some problems of the theory of stability of rods have an exact solution in special, mostly Bessel, functions. The works devoted to the search for exact solutions are rather numerous. The expedience of continuing the work in this area has not diminished hitherto. The progress in studying and tabulating special functions, including Bessel, reveals a series of new opportunities, unused up to now, which allow one to bring to the number of exact solutions the exact solution of many problems concerning longitudinal bending which had not been considered before. In this section we consider some problems of the theory of stability of rods, which allow one to obtain an exact solution in the Bessel functions.

Stability of a rod compressed by forces applied at the ends. Consider the problem on stability of rods of a variable cross-section which are compressed by a longitudinal force P which is constant along the length of the rod. For a rod simply supported at the end as well as for a cantilever rod, these problems allow solutions in Bessel functions in a number of cases.

In particular, one can most simply obtain the solution, when the stiffness varies according to the exponential law. These problems have been considered in detail by A.N. Dinnik. Dinnik has also considered the longitudinal bending of a rod whose stiffness varies according to the sinusoidal law.

One can extend the circle of the problem on transverse bending which can be solved in Bessel functions. On the one hand, one can obtain the solution for some laws of the variation of the stiffness, which are not considered hitherto. Moreover, even for the laws investigated by other authors, the appearance of new tables of Bessel functions and functions close to them allow one to obtain numerical solutions for much larger domains of the variation of the principal parameters which characterize the change of the stiffness of the rod.

On the other hand, one can consider boundary conditions which result in a more complicated statement of the problem: the fixing of both ends, the fixing of one end and the simply supporting of the other end, the elastic supporting. In these cases the solution contains the Bessel functions as well as the contiguous functions, in particular, the Lommel functions.

The problem on the longitudinal bending of a rod of a variable cross-section which is compressed by a longitudinal force P and is simply supported at the ends, can be reduced to the integration of the differential equation

$$EJ\frac{d^2y}{dx^2} + Py = 0 \tag{38.1}$$

under the boundary conditions

$$x = 0, \qquad\qquad\qquad y = 0;$$
$$x = l, \qquad\qquad\qquad y = 0,$$

where l is the length of the rod.

Suppose that $EJ = EJ_0(1 + \beta x)^m$. Denoting $1 + \beta x = z$, we rewrite (38.1) in the following form:

$$\frac{d^2y}{dz^2} + \frac{P}{\beta^2 z^m EJ_0}y = 0. \tag{38.2}$$

The boundary conditions take the form

$$\text{at } z = 1, \qquad\qquad\qquad y = 0,$$
$$\text{at } z = 1 + \beta l, \qquad\qquad\qquad y = 0.$$

Denoting $\dfrac{P}{\beta^2 EJ_0} = b_1$, we can rewrite (38.2) in the following form:

$$\frac{d^2y}{dz^2} + b_1 z^{-m}y = 0. \tag{38.3}$$

In order to reduce (38.3) to the Bessel equation, we should use a transformation of the variables. For this purpose we should introduce the new variable

$$v = z^{-1/2}y,$$

and pass from the independent variable z to the variable

$$t = \frac{2}{2-m}\sqrt{b_1}\,z^{1-\frac{m}{2}}. \tag{38.4}$$

Then, we obtain the Bessel equation of the index $1/(2-m)$ with respect to the function v from equation (38.3).

Thus, for an integer index we have

$$v = A_1 J_{1/(2-m)}(t) + A_2 Y_{1/(2-m)}(t). \tag{38.5}$$

If the index is fractional, then we have

$$v = A_1 J_{1/(2-m)}(t) + A_2 Y_{-1/(2-m)}(t). \tag{38.6}$$

The final result for an integer index is:

$$y = \sqrt{z}\left[A_1 J_{1/(2-m)}\left(\frac{2}{2-m}\sqrt{b_1}\,z^{1-\frac{m}{2}}\right) + A_2 Y_{1/(2-m)}\left(\frac{2}{2-m}\sqrt{b_1}\,z^{1-\frac{m}{2}}\right)\right]. \tag{38.7}$$

If the index is non-integer, then the term Y_ν in the second term should be replaced by $J_{-\nu}$. A.N. Dinnik has made the calculation for the cases $m = 1, 2, 3, 4$.

Let us show how the critical force can be computed.

The general approach consists in the fact that, substituting the solution (38.7) which was found into the boundary conditions, we obtain two linear homogeneous algebraic equations with respect to the coefficients A_1 and A_2. The case $A_1 = 0$, $A_2 = 0$ represents the trivial solution, which is of no interest for us.

The necessary condition for at least one of the two quantities A_1 and A_2 to be non-zero, is the vanishing of the determinant of the system obtained. This gives a transcendental equation. By solving this equation, we determine the eigenvalues of the problem under consideration. Then we easily compute the critical force.

As an example, we find the critical force for a rod, whose stiffness is varying according the linear law.

Suppose that

$$EJ = EJ_0(1 + \beta x).$$

Then $\dfrac{1}{m-2} = -1$ and the solution of equation (38.1) takes the form

$$v = A_1 J_1(t) + A_2 Y_1(t), \tag{38.8}$$

where

$$z = 1 + \beta x, \qquad t = \frac{2}{2-m}\sqrt{b_1}\,z^{1-\frac{m}{2}} = 2\sqrt{b_1}\,z^{1/2}.$$

Thus,

$$y = v\sqrt{z} = \sqrt{z}\left[A_1 J_1\left(2\sqrt{b_1 z}\right) + A_2 Y_1\left(2\sqrt{b_1 z}\right)\right]. \tag{38.9}$$

The boundary conditions imply that

$$y = 0 \text{ at } z = 1.$$

Therefore, we obtain the equation

$$0 = A_1 J_1 \left(2\sqrt{b_1}\right) + A_2 Y_1 \left(2\sqrt{b_1}\right). \tag{38.10}$$

For $z = 1 + \beta l$ we have $y = 0$, and we obtain the second equation

$$0 = A_1 J_1 \left(2\sqrt{b_1(1 + \beta l)}\right) + A_2 Y_1 \left(2\sqrt{b_1(1 + \beta l)}\right). \tag{38.11}$$

Equations (38.10) and (38.11) imply the equation

$$J_1(\xi) Y_1(\xi_1) - J_1(\xi_1) Y_1(\xi) = 0, \tag{38.12}$$

where

$$2\sqrt{b_1} = \xi, \quad \sqrt{1 + \beta l} = \delta, \quad \xi \delta = \xi_1.$$

From this equation we find ξ, and then we compute b_1. The critical force P_{cr} is defined by the formula

$$P_{cr} = \beta^2 E J_0 b_1 = \beta^2 E J_0 \frac{\xi^2}{4}. \tag{38.13}$$

Note that the solution can easily be obtained for the values $m = \dfrac{1}{4}, \dfrac{1}{3}, \dfrac{1}{2}, \dfrac{2}{3}, \dfrac{3}{4},$ $\dfrac{-1}{4}, \dfrac{-1}{3}, \dfrac{-1}{2}, \dfrac{-2}{3}, \dfrac{-3}{4}, \dfrac{-5}{4}, \dfrac{-4}{3}, \dfrac{-3}{2}, \dfrac{-5}{3}, \dfrac{-7}{4}, -2,$ using the tables of generalized Airy functions. The calculation of the critical forces for these cases of varying the stiffness was not hitherto performed.

According to the notation used in the tables mentioned, the generalized Airy functions satisfy the equation

$$U''(s) + s^a U(s) = 0, \tag{38.14}$$

whose solution has the form

$$U = A_1 U_1(s) + A_2 U_2(s). \tag{38.15}$$

The functions $U_1(s)$, $U_2(s)$ and their first derivatives have been tabulated. Comparing (38.2) and (38.14), we obtain

$$m = -\sigma, \qquad z = s^{a+2}\sqrt{\frac{\beta^2 E J_0}{P}}.$$

When the ends of the rod are hinged, equations (38.10) and (38.11) take the following form:

$$A_1 U_1 \left(^{a+2}\sqrt{\frac{P}{\beta^2 E J_0}} \right) + A_2 U_2 \left(^{a+2}\sqrt{\frac{P}{\beta^2 E J_0}} \right) = 0, \tag{38.16}$$

$$A_1 U_1 \left[^{a+2}\sqrt{\frac{P}{\beta^2 E J_0}} (1 + \beta l) \right] + A_2 U_2 \left[^{a+2}\sqrt{\frac{P}{\beta^2 E J_0}} (1 + \beta l) \right] = 0. \tag{38.17}$$

From these equations we obtain the following transcendental equation:

$$U_1 \left(^{a+2}\sqrt{\frac{P}{\beta^2 E J_0}} \right) U_2 \left[^{a+2}\sqrt{\frac{P}{\beta^2 E J_0}} (1 + \beta l) \right]$$

$$- U_2 \left(^{a+2}\sqrt{\frac{P}{\beta^2 E J_0}} \right) U_1 \left[^{a+2}\sqrt{\frac{P}{\beta^2 E J_0}} (1 + \beta l) \right] = 0, \tag{38.18}$$

which can be solved with the help of the tables mentioned above.

Let us make one more remark concerning the use of the tables mentioned. The derivatives of the functions U_1 and U_2 can also be expressed via the Bessel functions of a fractional order with the help of the formulae

$$U_1'(s,\alpha) = -(\alpha+2)^{-1/(\alpha+2)}\Gamma\left(\frac{\alpha+1}{\alpha+2}\right)s^{(\alpha+1)/2}J_{\frac{\alpha+1}{\alpha+2}}\left(\frac{2}{\alpha+2}\cdot s^{(\alpha+2)/2}\right), \quad (38.19)$$

$$U_2'(s,\alpha) = (\alpha+2)^{1/(\alpha+2)}\Gamma\left(\frac{\alpha+3}{\alpha+2}\right)s^{(\alpha+1)/2}J_{-\frac{\alpha+1}{\alpha+2}}\left(\frac{2}{\alpha+2}\cdot s^{(\alpha+2)/2}\right). \quad (38.20)$$

Therefore, using these tables, we can obtain the solutions for all cases, when $-\frac{1}{m-2} = \frac{\alpha+1}{\alpha+2}$, i.e. for $m = 2 - \frac{\alpha+2}{\alpha+1}$; for instance, for $\alpha = 1/4$ we have $m = 0.2$; for $\alpha = -3/4$ we have $m = -3$ and so on.

The question can arise concerning the practical importance of the solution of problems for fractional values of the number m. In our opinion, the importance of these solutions as well as other new solutions is connected with consideration of the elasto-plastic problems of the transverse bending, which can lead to very different laws of the variation of the stiffness along the length of the rod.

Let us pass to consideration of other laws of variation of the stiffness. If in the equation

$$EJy'' + Py = 0$$

the stiffness is varying so that

$$\frac{EJ}{P} = \frac{x^2}{b^2(c^2 x^{2b} - \nu^2) + \frac{1}{4}}, \quad (38.21)$$

then we obtain, using an appropriate change of variables, the solution for non-integer ν (for $b \neq 0$, $c \neq 0$) in the form

$$y = \sqrt{x}\left[A_1 J_\nu(cx^b) + A_2 J_{-\nu}(cx^b)\right]. \quad (38.22)$$

Now consider some more complicated laws of variation of the stiffness of the rod, which require more complicated changes of variables with the help of the Lommel method. It is shown that the solution of the equation

$$\frac{d^2 y}{dz^2} - \left[\frac{\psi''(z)}{\psi'(z)} + (2\mu-1)\frac{\psi'(z)}{\psi(z)}\right]\frac{dy}{dz} + [\mu^2 - \nu^2 + \psi^2(z)]\left[\frac{\psi'(z)}{\psi(z)}\right]^2 y = 0 \quad (38.23)$$

has the form

$$y = [\psi(z)]^\mu Z_\nu[\psi(z)]. \quad (38.24)$$

Therefore, we find that the equation

$$\frac{d^2 y}{dx^2} + (e^{2x} - \nu^2)y = 0 \quad (38.25)$$

can be transformed with the help of the change of the variable $t = e^x$ into the Bessel equation

$$\frac{d^2 y}{dt^2} + \frac{1}{t}\frac{dy}{dt} + \left(1 - \frac{\nu^2}{t^2}\right)y = 0. \quad (38.26)$$

Therefore,

$$y = Z_\nu(e^x). \quad (38.27)$$

We use this solution in order to consider the problem on the longitudinal bending of a rod, whose stiffness is varying according to the exponential law or a law close to this one.

Suppose that in equation (38.1) we have

$$\frac{P}{EJ} = B_1 e^{kx} - B_2. \tag{38.28}$$

Let us introduce a new variable $z = ax + b$, where a and b are unknown values which should be determined. Introducing (38.28) into equation (38.1), we obtain

$$\frac{d^2y}{dx^2} + (B_1 e^{kx} - B_2)y = 0. \tag{38.29}$$

Introducing the new variable $z = ax + b$, we obtain

$$a^2 \cdot \frac{d^2y}{dz^2} + \left[B_1 e^{k(z-b)/a} - B_2\right] y = 0 \tag{38.30}$$

In order that (38.30) coincides with (38.25) we should set

$$\frac{B_2}{a^2} = \nu^2, \qquad \frac{B_1}{a^2 \cdot e^{kb/a}} = 1, \qquad \frac{k}{a} = 2. \tag{38.31}$$

Therefore, we obtain

$$k = 2a,$$

$$b = \frac{a}{k}(\ln B_1 - 2\ln a),$$

$$\nu = \frac{\sqrt{B_2}}{a} = \frac{2\sqrt{B_2}}{k}.$$

Thus, in this case the index ν depends on the relation between the physical constants B_2 and k. Taking into account what has been said about the tables of the generalized Airy functions, we see that the solutions can be obtained in a very large range of the ratios B_2/k. This allows one to investigate a much larger number of the problems than the exponential law of variation of the stiffness gives, when $B_2 = 0$ and the index $\nu = 0$. However, it should be noted that the solution of the transcendental equations for $B \neq 0$ is, in fact, more difficult than in the cases considered before, because the unknown quantity P is present in the argument as well as in the index of the Bessel function. This requires interpolation with respect to the argument as well as with respect to the index.

We are now going to the investigation of the problems with other boundary conditions. Consider the case, when the stiffness is varied according to the power law. Suppose that both the ends of the rod are fixed. The rod is symmetric with respect to its middle which will be taken as the origin. Consider the symmetric forms of the loss of stability. The form which gives the calculated value of the critical force will be among them.

Let us write the differential equation and the boundary conditions for the right-hand half of the rod

$$EJy'' + Py = -M_0, \tag{38.32}$$

where M_0 is the supporting moment,

$$\left.\begin{array}{l} \text{at } x = 0, \quad \dfrac{dy}{dx} = 0, \\[3mm] \text{at } x = \dfrac{l}{2}, \quad y = 0, \quad \dfrac{dy}{dx} = 0. \end{array}\right\} \tag{38.33}$$

From these three conditions we find the two constants of the integration and the supporting moment M_0.

The right-hand side of equation (38.32) is constant and its particular solution also represents a constant. Since the equation has variable coefficients, the determination of the particular solution in the case, when the right-hand side represents a power function whose exponent is different from one, is connected with additional calculations.

We proceed from the previous scheme of the solution, which consists in the fact that after an appropriate change of variable the equation without the right-hand side becomes the Bessel equation. In this case the right-hand side is transformed into a power function. Therefore, we will seek a particular solution for the equation

$$x^2 u'' + x u' + (x^2 - \nu^2) u = f(x), \tag{38.34}$$

where $f(x)$ is a power function.

We can use two methods in order to solve the equation with the right-hand side.

The first one consists in the fact that the particular solution for the right-hand side which represents a power function, is expressed by the Lommel formula. The second method consists in the representation of the particular solution in the form of an integral, whose integrand contains the Bessel function multiplied by a power function. However, it is known that these integrals can be expressed via the Lommel functions. This question will be considered below in more detail. Now we are going to problems on rods which are compressed by a longitudinal force whose value varies along the length of the rod.

Stability of a rod compressed by a distributed loading. First, consider the Greenhill problem on the stability of a rod with a constant cross-section under the action of a longitudinal force, which is a linear function of the coordinates of the cross-section.

Suppose that the rod is fixed at one end; the second end is free. We take the origin at the fixed end, the x-axis will be directed along the axis of the rod (Fig. 65). We investigate the case of the longitudinal bending for a constant uniformly distributed loading q. In this case the longitudinal force is $N = q(l - x)$, where l is the length of the rod. Suppose that when losing the stability, the rod bends along a curve $y(x)$. The bending moment in the cross-section $x = x_1$ can be expressed in the form of the following definite integral:

$$M = -\int_{x_1}^{l} q(\eta - y)\, dx, \tag{38.35}$$

Figure 65

where $\eta = y(x_1)$.

It is known that the equation of the elastic line of the bent rod has the following form:

$$EJ\frac{d^2y}{dx^2} = -M; \tag{38.36}$$

therefore,

$$EJ\frac{d^3y}{dx^2} = \frac{-dM}{dx} = -Q. \tag{38.37}$$

In our case, the forces q remain vertical under the bending. The value of the main vector of the forces located above the cross-section considered is equal to $R = q(l - x)$ and this vector is directed vertically. In order to find the transverse force, we should project the main vector onto the normal to the elastic line. Thus, we have

$$Q = R \cdot \frac{dy}{dx} = q(l - x)\frac{dy}{dx}. \tag{38.38}$$

Since we are considering small deflections, we replace the sine by the tangent.

Hence, the differential equation can be written in the following form:

$$EJ\frac{d^3y}{dx^3} = -q(l - x)\frac{dy}{dx}. \tag{38.39}$$

Denoting $\frac{dy}{dx} = u, l - x = \xi$, we have

$$EJ\frac{d^2u}{d\xi^2} + q\xi_1 u = 0. \tag{38.40}$$

Let us denote $\kappa = \sqrt[3]{\frac{q}{EJ}}$ and set $\kappa\xi_1 = \xi$. Then the differential equation (38.40) takes the form

$$\frac{d^2u}{d\xi^2} + \xi u = 0. \tag{38.41}$$

As is known, the Airy integral which can be reduced to the Bessel function of the order 1/3 satisfies this equation.

The integral of equation (38.41) has the following form:

$$u = A\xi^{1/2}J_{1/3}\left(\frac{2}{3}\xi^{3/2}\right) + B\xi^{1/2}J_{-1/3}\left(\frac{2}{3}\xi^{3/2}\right). \tag{38.42}$$

In order to determine the constants of integration, we should consider the boundary conditions.

From the condition of sealing the lower end, we have:

$$\frac{dy}{dx} = 0 \quad \text{at } x = 0.$$

Hence $u = 0$ for $\xi = \kappa l$. At the upper end we have

$$\frac{d^2y}{dx^2} = 0 \quad \text{at } x = l,$$

or, what is the same

$$\frac{du}{d\xi} = 0 \quad \text{for } \xi = 0.$$

Expanding the function (38.42) into a power series and keeping the first terms of this series, we obtain

$$u = A\frac{(1/3)^{\frac{1}{3}\xi}}{\Gamma(\frac{4}{3})} + B\frac{(1/3)^{-\frac{1}{3}\xi}}{\Gamma(\frac{2}{3})}.$$

(38.43)

After differentiating this expression and introducing u' into the second boundary condition, we obtain $A = 0$. Then the first boundary condition can be written in the following form:

$$BJ_{-1/3}\left(\frac{2}{3}\sqrt{\frac{ql^3}{EJ}}\right) = 0.$$

(38.44)

Hence, the critical value of q can be expressed via the roots of the Bessel function $J_{-1/3}(x)$. The smallest root of this function is $x \approx 1.87$. This corresponds to

$$q_{cr} \approx \frac{7.87EJ}{l^3}.$$

(38.45)

Let us generalize this known solution and consider more complicated problems, whose solution will be sought in Bessel functions and some other functions contiguous to these functions.

Suppose that the stiffness of the rod B and the weight of a unit of its length q are varied along the rod according the following law:

$$B = b(l - x)^m,$$
$$q = c(l - x)^n.$$

(38.46)

Moreover, let a transverse loading q_1 act on the rod, by causing a transverse force in its cross-sections which is equal to

$$-Q_x^0 = f(x).$$

The differential equation of the elastic line can be written in the form

$$\frac{d}{dx}\left(B\frac{d^2y}{dx^2}\right) = -Q$$

(38.47)

or

$$B\frac{d^3y}{dx^3} + \frac{dB}{dx}\frac{d^2y}{dx^2} = -Q_1^{(x)} + Q_x^0,$$

(38.48)

where $Q_1^{(x)} = \frac{dy}{dx} \cdot \int_x^l q\,dx$. Substituting the expressions for B, q, $Q_1^{(x)}$ into the differential equation, we obtain

$$y''' - \frac{m}{l-x}y'' + \frac{c}{b(n+1)}(l-x)^{n-m+1}y' - \frac{Q_x^0}{b(l-x)^m} = 0.$$

(38.49)

After denoting $y' = z$, equation (38.49) takes the form

$$z'' - \frac{m}{l-x}z' + \frac{c}{b(n+1)}(l-x)^{n-m+1}z - \frac{Q_x^0}{b(l-x)^m} = 0.$$

(38.50)

Let us reduce equation (38.50) to the Bessel equation. Applying the substitutions

$$l - x = A\xi^k, \qquad \Phi = z\xi^p,$$

(38.51)

we obtain

$$\frac{dz}{dx} = \frac{dz}{d\xi} \cdot \frac{d\xi}{dx} = \frac{d}{d\xi}(\xi^{-p}\Phi) \cdot \frac{d\xi}{dx}. \tag{38.52}$$

By virtue of (38.51) we have

$$\frac{d\xi}{dx} = -\frac{1}{Ak}\xi^{1-k}, \tag{38.53}$$

$$\frac{dz}{dx} = -\frac{\xi^{1-p-k}}{Ak}\left(-\frac{p}{\xi}\Phi + \frac{d\Phi}{d\xi}\right); \tag{38.54}$$

therefore,

$$\frac{d^2 z}{dx^2} = \frac{d}{d\xi}\left(\frac{dz}{dx}\right) \cdot \frac{d\xi}{dx}$$

$$= \frac{\xi^{-p-2k+2}}{A^2 k^2}\left[\frac{d^2\Phi}{d\xi^2} + \frac{1-2p-k}{\xi}\frac{d\Phi}{d\xi} + \frac{p(p+k)}{\xi^2}\Phi\right]. \tag{38.55}$$

Substituting the values obtained for z, $\dfrac{dz}{dx}$ and $\dfrac{d^2 z}{dx^2}$ into equation (38.50), we have

$$\frac{d^2\Phi}{d\xi^2} + \frac{1-2p-k+mk}{\xi}\frac{d\Phi}{d\xi}$$

$$+ \left[\frac{p(p+k)-mkP}{\xi^2} + \frac{ck^2}{b(n+1)}A^{n-m+3}\xi^{k(n-m+3)-2}\right]\Phi$$

$$= \frac{Q_x^0}{b(l-x)^m} \cdot A^2 k^2 \cdot \xi^{p+2k-2}. \tag{38.56}$$

In order that equation (38.56) becomes the Bessel equation, the coefficients A, k and p should be chosen in such a way that the following equalities hold

$$\left.\begin{array}{r} 1 - 2p - k + mk = 1, \\ k(n - m + 3) - 2 = 0, \\ \dfrac{ck^2}{b(n+1)}A^{n-m+3} = 1. \end{array}\right\} \tag{38.57}$$

Therefore,

$$k = \frac{2}{n-m+3},$$

$$p = \frac{m-1}{n-m+3}, \tag{38.58}$$

$$A = \left[\frac{4c}{b(n+1)(n-m+3)^2}\right]^{-1/(n-m+3)}.$$

As a result, equation (38.56) takes the form

$$\frac{d^2\Phi}{d\xi^2} + \frac{1}{\xi}\frac{d\Phi}{d\xi} + \left(1 - \frac{p^2}{\xi^2}\right)\Phi = \frac{k^2 A^{2-m}}{b}Q_\xi^0 \cdot \xi^{\frac{m-2n-3}{n-m+3}}, \tag{38.59}$$

where Q_ξ^0 is the value of Q_x^0 after replacing $l-x$ by ξ in accordance with expressions (38.51).

Suppose that Q_x^0 varies according to the rule

$$Q_x^0 = D(l - x)^r.$$
(38.60)

Then

$$Q_\xi^0 = D(A\xi^k)^r = DA^r \xi^{\frac{-2r}{n-m+3}},$$
(38.61)

and the right-hand side of equation (38.59) is equal to

$$\frac{k^2 DA^{2-m+r}}{b} \xi^{\frac{m-2n-3+2r}{n-m+3}} = R\xi^t$$
(38.62)

where

$$R = \frac{k^2 DA^{2-m+r}}{b},$$
(38.63)

$$t = \frac{m - 2n - 3 + 2r}{n - m + 3}.$$
(38.64)

If $Q_x^0 = 0$, then the solution of the corresponding homogeneous equation has the form

$$\Phi = C_1 J_p(\xi) + C_2 J_{-p}(\xi),$$
(38.65)

where p is given by formula (38.58). Then by virtue of expressions (38.51), we obtain

$$z = \xi^{-p} \Phi = \xi^{-p} [C_1 J_p(\xi) + C_2 J_{-p}(\xi)].$$
(38.66)

Expression (38.66) can be used for the determination of the critical parameters of the longitudinal loading for different boundary conditions.

One can show that the expression (38.66) can be reduced to the form (38.42) for $m = 0$.

If the right-hand side of equation (38.56) is not equal to zero, which corresponds to the presence of a transversal loading or to the case when there is a transversal force which varies according the linear law, then a question arises concerning how to obtain a particular solution of the inhomogeneous equation

$$\frac{d^2\Phi}{d\xi^2} + \frac{1}{\xi}\frac{d\Phi}{d\xi} + \left(1 - \frac{p^2}{\xi^2}\right)\Phi = R\xi^t.$$
(38.67)

If we introduce the operator

$$\nabla_\xi^{(p)} = \left[\frac{d^2}{d\xi^2} + \frac{1}{\xi}\frac{d}{d\xi} + \left(1 - \frac{p^2}{\xi^2}\right)\right] \cdot \xi^2,$$
(38.68)

then equation (38.67) can be rewritten in the form

$$\nabla_\xi^{(p)} = R\xi^{t+2}.$$
(38.69)

It is known from the theory of the Bessel functions that the solution of this equation can be obtained in terms of Lommel functions.

In the special case, when $p = t+1$, the solution can be expressed via the Struve functions $\mathbf{H}_\nu(x)$, which are particular solutions of the inhomogeneous equation

$$\nabla_x^{(\nu)}\mathbf{H}_\nu(x) = \frac{4\left(\frac{1}{2}x\right)^{\nu+1}}{\Gamma\left(\frac{1}{2}\right)\cdot\Gamma\left(\nu + \frac{1}{2}\right)}.$$
(38.70)

Comparing (38.9) and (38.70), we see that, by virtue of linearity of the operator $\nabla_x^{(\nu)}$ the particular solution of equation (38.69) for $t+1 = p$ has the form

$$\Phi(\xi) = \frac{\Gamma\left(\frac{1}{2}\right) \cdot \Gamma\left(p + \frac{1}{2}\right)}{4\left(\frac{1}{2}\right)^{p+1}} R \cdot \mathbf{H}_p(\xi). \tag{38.71}$$

Adding (38.71) to the solution of the homogeneous equation, we find the general solution up to the constants which are determined by the boundary conditions. If $p \neq t+1$, then the integral of the homogeneous equation (38.69) can be expressed via the Lommel function of one variable, which is a particular solution of the equation

$$\nabla_x^{(\nu)} S_{\mu,\nu}(x) = x^{\mu+1}. \tag{38.72}$$

Thus, the particular solution of equation (38.69) has the form

$$\Phi(\xi) = R S_{\mu,p}(\xi), \tag{38.73}$$

where $S_{\mu,p}(\xi)$ is the Lommel function, and the index

$$\mu = t+1. \tag{38.74}$$

Let us study several problems of the theory of the longitudinal bending, which can be reduced to these functions. Assume that $n = m = 0$. In this case we have the problem on a rod of constant cross-section, which is compressed by an uniformly distributed longitudinal loading q. The transversal loading is assumed up to now to be arbitrary.

From expressions (38.51), (38.58), (38.64) and (38.74) we can see that in this case

$$\xi^2 = \frac{4c}{9b}(l - x)^3,$$

$$p = -\frac{1}{3},$$

$$t = -1 + \frac{2}{3}r,$$

$$\mu = \frac{2}{3}r.$$

The solution in the Struve functions will be obtained for

$$p = t+1 = \mu,$$

i.e. for

$$r = -\frac{1}{2}.$$

Hence, a solution in the Struve functions can only be obtained in the rather unnatural case, when

$$Q_x^0 = D(l - x)^{-1/2} = \frac{D}{\sqrt{l - x}}. \tag{38.75}$$

One can easily see that the intensity of the corresponding transversal loading is equal to

$$q_1 = \frac{dQ_x}{dx} = -\frac{1}{2}\frac{D}{\sqrt{(l - x)^3}}.$$

Now consider the inverse problem, namely, we assume the law of the variation of the transversal loading to be given. First, we assume that $Q_x^0 = $ const. This problem can be met, for instance, when considering the transverse-longitudinal bending of a rod, which is fixed at one end and is loaded at the other end by a horizontal force Q_x^0. Moreover, let the rod be loaded by an uniformly distributed longitudinal loading q. In this case $r = 0$. Hence, $t = -1$, $\mu = t + 1 = 0$. Thus, the particular solution of equation (38.69) has the form

$$\Phi(\xi) = z\xi^{-1/3} = RS_{0,-1/3}(\xi),$$

or

$$z = R\xi^{1/3}S_{0,-1/3}(\xi). \tag{38.76}$$

The constants C_1 and C_2 should be determined from the boundary conditions:

$$\text{at } x = 0 \qquad \frac{dy}{dx} = z = 0,$$

$$\text{at } x = l \qquad \frac{d^2y}{dx^2} = \frac{dz}{dx} = 0.$$

The solution of this problem has been obtained by Engelhardt [92]. This author has obtained numerical results by expanding the Lommel and Bessel functions into power series. Now suppose that Q_x^0 is a linear function

$$Q_x^0 = D_0 + D_1(l - x). \tag{38.77}$$

The particular solution, which corresponds to the term D_0 has been considered above. Thus, it remains to consider the case, when the right-hand side of the differential equation has the following form:

$$Q_x^0 = Q_1 = D_1(l - x). \tag{38.78}$$

Using, as before, the general solution, we have in this case

$$r = 1; \qquad \mu = \frac{2}{3}; \tag{38.79}$$

and the particular solution of equation (38.69) is

$$\Phi(\xi) = R \cdot S_{2/3,-1/3}(\xi). \tag{38.80}$$

This problem can also be solved in terms of functions which are integrals of the Airy integrals. Consider in detail the special case, which has been investigated by Rothman [96]. Suppose that the rod is simply supported at the ends, has a constant cross-section and is loaded by a loading q uniformly distributed along its length, whose vector forms a constant angle θ with the direction of the axis of the rod (Fig. 66). In this case the normal component of the loading as well as its tangent component are constant. Hence, in this case, the integral of the homogeneous equation has, as before, the form

$$z = \left[C_1J_{1/3}(\xi) + C_2J_{-1/3}\right]\xi^{1/3}. \tag{38.81}$$

The transversal force in this case is equal to

$$Q_x^0 = q\left(\frac{1}{2} - x\right)\sin\theta. \tag{38.82}$$

Hence, the corresponding particular solution can be written as a linear combination of the Lommel functions written above. However, this solution can be written in another form, given by Rothman.

Before passing to the construction of the solution due to Rothman, we give some data concerning Airy integrals.

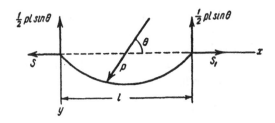

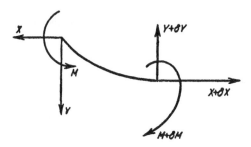

Figure 66

Let us denote the Airy integral

$$\int_0^\infty \cos\left(tz + \frac{1}{3}t^3\right)\, dt = \text{Ai}(z) \tag{38.83}$$

and the adjoint function

$$\frac{1}{\pi}\int_0^\infty \left[e^{tz - \frac{1}{3}t^3} + \sin\left(tz + \frac{1}{3}t^3\right)\right]\, dt = \text{Bi}(z). \tag{38.84}$$

Moreover, we introduce the notation

$$\frac{1}{\pi}\int_0^\infty \sin\left(tz + \frac{1}{3}t^3\right)\, dt = \text{Gi}(z), \tag{38.85}$$

$$\frac{1}{\pi}\int_0^\infty \exp\left(tz - \frac{1}{3}t^3\right)\, dt = \text{Hi}(z). \tag{38.86}$$

In the book of G.N. Watson there are formulae concerning these functions.

Let us now give the solution of Rothman.

As we can see in Fig. 66, the differential equation of the elastic line of the rod has the form

$$EJ\frac{d^3y}{dx^3} + X \cdot \frac{dy}{dx} = Y, \tag{38.87}$$

where

$$X = qx \cos\theta + S,$$
$$\left. Y = qx \sin\theta - \frac{1}{2}ql \sin\theta = -Q. \right\}$$

(38.88)

Setting

$$x = \xi,$$

(38.89)

we can rewrite equation (38.87) in the form

$$\frac{d^3y}{d\xi^3} + A\xi \frac{dy}{d\xi} = B\xi + C,$$

(38.90)

where

$$A = (EJq^2 \cos^2\theta)^{-1},$$
$$B = (EJq^3 \cos^3\theta)^{-1} \tan\theta,$$
$$\left. C = -\left(\frac{1}{2}ql \sin\theta + S \tan\theta\right)(EJq^3 \cos^3\theta)^{-1}. \right\}$$

(38.91)

Now, denoting

$$\frac{dy}{d\xi} = W,$$

(38.92)

we reduce (38.90) to the form

$$\frac{d^2W}{d\xi^2} + A\xi \cdot W = B\xi + C.$$

(38.93)

The solution of the corresponding homogeneous equation has the form

$$W = \xi^{1/2}\left[C_1 J_{1/3}\left(\frac{2}{3}A^{1/2}\xi^{3/2}\right) + C_2 J_{-1/3}\left(\frac{2}{3}A^{1/2}\xi^{3/2}\right)\right].$$

(38.94)

This solution can also be expressed via the functions Ai and Bi, which are given by expressions (38.83) and (38.84)

$$W = C_1^* \operatorname{Ai}(-\xi) + C_2^* \operatorname{Bi}(-\xi).$$

(38.95)

The particular solution of equation (38.90) has the form

$$y = \frac{B\xi}{A} + C\sum_{r=0}^{\infty}(-3A)^r \cdot r! \cdot \frac{\xi^{3r+3}}{(3r+3)!}.$$

(38.96)

One can show that the series present in this expression is equal to

$$-\frac{1}{\pi}\int_0^\xi \operatorname{Gi}(t)\, dt,$$

(38.97)

where the function Gi(t) is given by expression (38.85). Thus, we obtain

$$y = -C_1^*\int_0^{-\xi} \operatorname{Ai}(t)\, dt - C_2^*\int_0^{-\xi} \operatorname{Bi}(t)\, dt - \frac{C^*}{\pi}\int_0^{-\xi} \operatorname{Bi}(t)\, dt + \frac{B}{A}\xi + F,$$

(38.98)

where C_1^*, C_2^*, and F are arbitrary constants. The corresponding boundary condition have the form

$$y = \frac{d^2y}{dx^2} = 0 \quad \text{for } x = 0 \quad \text{and for } x = t$$

or, what is the same

$$y = \frac{d^2y}{dx^2} = 0 \quad \text{for } \xi = S \quad \text{and for } \xi = ql\cos\theta + S.$$

(38.99)

Rothman gives tables of the integrals of the functions Ai($\pm x$) and Bi($\pm x$) which allows one to solve the problem formulated numerically.

One can see from what is said above that the particular solution which can be represented in the form of the Lommel function, can also, as was done by Rothman, be represented in the form of the integral of the functions, which give the solution of the homogeneous equation, multiplied by a polynomial. In order to clarify the connection between the particular solution obtained by Rothman and the Lommel functions $S_{\mu,\nu}(z)$ and $s_{\mu,\nu}(z)$, we give the known representation of the Lommel function $s_{\mu,\nu}(z)$ via the integrals of Bessel functions, namely,

$$s_{\mu,\nu}(z) = \frac{2}{\sin\nu\pi} \int_0^z x^\mu [J_\nu(z)J_{-\nu}(x) - J_{-\nu}(z)J_\nu(x)]\, dx,$$

(38.100)

for non-integer ν, or

$$s_{\mu,\nu}(z) = \frac{1}{2}\pi \int_0^z x^\mu [Y_\nu(z)J_\nu(x) - Y_\nu(x)J_\nu(z)]\, dx$$

(38.101)

if ν is an integer.

One can easily see that the expressions in the square brackets differ only by a constant factor from the functions which have the property of the unit matrix and which are given in different chapters of this book. This implies that in all the cases, when the solution can be expressed as the integral of the product of a power function and an appropriate fundamental function, it can also be represented directly via the Lommel functions.

Stability of the plane form of the bending of a strip. The problem of the stability of the plane form of the bending of a strip of a rectangular cross-section was investigated by many authors, including L. Prandtl, A. Michell, A.N. Dinnik, A.P. Korobov. S.P. Tomoshenko and others.

We shall consider several such problems, where the solution can be obtained in closed form in the cylindrical functions.

First, consider the relatively simple problem on the stability of a strip of a rectangular cross-section, which is simply supported at the ends and loaded at the end cross-sections by unequal moments M_1 and M_2, which lie in the plane passing through the axis of the strip and is parallel to the bigger side of the rectangle h. The smaller side will be denoted by b. Let us introduce the following notation (Fig. 67): u, v are the displacements of the axis of the rod in the direction of the x and y axes; β is the twist angle, B_1 and B_2 are the stiffness in the direction of the x and y axes, C is the twist stiffness, ξ, η, ζ are the moving axes.

We give the deduction of the main equations obtained by L. Prandtl.

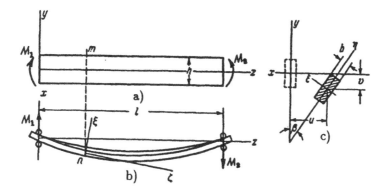

Figure 67

Let us construct the differential equation of the bending

$$B_1 u'' = M_\eta,\qquad\qquad (38.102)$$
$$B_2 v'' = -M_\xi,\qquad\qquad (38.103)$$

and of the twisting

$$C\beta' = M_\zeta,\qquad\qquad (38.104)$$

where the prime denotes differentiation with respect to z and M_η, M_ξ M_ζ are the projections on the ξ, η, ζ axes of the main moment of the forces which act from one and the same side of the cross-section. Let

$$M_x = M_x^0,$$
$$M_y = 0,\qquad\qquad (38.105)$$
$$M_z = -Qu,$$

where Q is the transversal force, and M_x^0 is the bending moment in the direction of the greater stiffness. We give the values of the direction cosines of the angles formed by the x, y, z and ξ, η, ζ axes, under the assumption that the deformations are small

	x	y	z
ξ	1	β	$-u'$
η	$-\beta$	1	$-v'$
ζ	u'	v'	1

We replace M_ξ, M_η and M_ζ in the differential equations by their values, negleccting the small quantities of higher orders everywhere:

$$M_\xi = M_x = M_x^0, \tag{38.106}$$

$$M_\eta = -\beta M_x^0, \tag{38.107}$$

$$M_\zeta = M_x^0 u' - Qu. \tag{38.108}$$

Thus, we have

$$B_1 u'' = -\beta M_x^0, \tag{38.109}$$

$$B_1 v'' = -M_x^0, \tag{38.110}$$

$$C\beta' = M_x^0 u' - Qu. \tag{38.111}$$

Equations (38.109) and (38.111) form a system, which is used in the theory of stability of the plane form of the bending in order to determine the critical loading. In our case

$$\left.\begin{array}{l} Q = -\dfrac{M_1 - M_2}{l} = -\dfrac{M_2}{l}(r-1), \\[2mm] M_x^0 = M_1 + Qz = M_2\left[r + (1-r)\dfrac{z}{l}\right], \\[2mm] r = \dfrac{M_1}{M_2}, \end{array}\right\} \tag{38.112}$$

and the system of equations, which allows one to determine the critical loadings, takes the form

$$B_1 u'' = -\beta M_2\left[r + (1-r)\frac{z}{l}\right], \tag{38.113}$$

$$C\beta' = M_2\left[r + (1-r)\frac{z}{l}\right]u' + \frac{(r-1)M_2}{l}u. \tag{38.114}$$

Let us differentiate the second equation with respect to z:

$$\begin{aligned} C\beta'' &= M_2\left[r + (1-r)\frac{z}{l}\right]u'' + \frac{M_2(r-1)}{l}u' + \frac{(r-1)M_2}{l}u' \\[2mm] &= M_2\left[r + (1-r)\frac{z}{l}\right]u''. \end{aligned} \tag{38.115}$$

Eliminating u'' from equations (38.113) and (38.115), we obtain

$$\beta'' + \frac{M_2^2}{B_1 C}\left[r + (1-r)\frac{z}{l}\right]^2 \beta = 0, \tag{38.116}$$

under the following boundary conditions

$$\beta(0) = 0, \qquad \beta(l) = 0. \tag{38.117}$$

Let us now introduce a new independent variable

$$z_1 = r + (1-r)\frac{z}{l} \tag{38.118}$$

and denote

$$k^2 = \frac{M_2^2 l^2}{(1-r)^2 B_1 C}. \tag{38.119}$$

Then equation (38.116) and the boundary conditions (38.117) can be written in the following form

$$\frac{d^2\beta}{dz_1^2} + k^2 z_1^2 \beta = 0,$$ (38.120)

$$\beta(r) = 0, \qquad \beta(l) = 0.$$ (38.121)

The general solution of this equation has the form

$$\beta(z_1) = z_1^{1/2} \left[C_1 J_{1/4}\left(\frac{kz_1^2}{2}\right) + C_2 J_{-1/4}\left(\frac{kz_1^2}{2}\right) \right].$$ (38.122)

Substituting this solution into the boundary conditions, we obtain the transcendental equation with respect to k:

$$J_{1/4}\left(\frac{kr^2}{2}\right) J_{-1/4}\left(\frac{k}{2}\right) - J_{1/4}\left(\frac{k}{2}\right) J_{-1/4}\left(\frac{kr^2}{2}\right) = 0, \qquad \text{for } r \geq 0,$$ (38.123)

$$J_{1/4}\left(\frac{kr^2}{2}\right) J_{-1/4}\left(\frac{k}{2}\right) + J_{1/4}\left(\frac{k}{2}\right) J_{-1/4}\left(\frac{kr^2}{2}\right) = 0, \qquad \text{for } r \leq 0.$$ (38.124)

A.N. Dinnik [25] has shown that the problem of stability of the plane form of the bending in a more general case can be reduced to the integration of the differential equation

$$\beta'' + \frac{M^2}{B_1 C}\beta = 0,$$ (38.125)

where β is the bending momentum.

A.N. Dinnik, mainly, considered problems, whose solution can be obtained in Bessel functions of the order 1/4. Later, the investigation of such problems was continued by Salvadory [97].

Let us note that Dinnik also considered problems of stability of a strip under loadings which are uniformly distributed and vary over a triangle.

Let us pass to consideration of the problem of the stability of a cantilever of a variable cross-section, which is loaded by a concentrated force at the end. Suppose that the transversal cross-section represents a narrow rectangle, whose height is varied according to a power law. This problem has been considered by K. Federhofer [93]. In this case

$$B_1 = \frac{b^3 h}{12}, \qquad C \approx \frac{2B_1}{1+\sigma}$$

where σ is the Poisson coefficient. Let the height of the rectangle be

$$h = h_0 \left(1 - \frac{z}{l}\right)^n,$$

where $0 \leq n \leq 1$. Then

$$B_1 = \frac{b^2}{12}h_0 \left(1 - \frac{z}{l}\right)^n = B_0 \left(1 - \frac{z}{l}\right)^n,$$

$$C = C_0 \left(1 - \frac{z}{l}\right)^n.$$

Replacing B_1 and C in equations (38.109) and (38.111) by their expressions, eliminating u and introducing the new variable $\dfrac{l-z}{l} = z_2$, we obtain the equation

$$\frac{d^2\beta}{dz_2^2} + \frac{n}{z_2}\frac{d\beta}{dz_2} + \left(\frac{P^2 l^4}{B_0 C_0}\right) z_2^{2(1-n)}\beta = 0. \tag{38.126}$$

Introducing the new variables

$$z_3 = \frac{k}{2-n} z_2^{2-n}, \tag{38.127}$$

$$\Phi = \beta z_2^{(n-1)/2}, \tag{38.128}$$

where

$$k = \sqrt{\frac{P^2 l^4}{B_0 C_0}}, \tag{38.129}$$

we reduce equation (38.126) to the Bessel equation

$$\frac{d^2\Phi}{dz_3^2} + \frac{1}{z_3}\frac{d\Phi}{dz_3} + \left(1 - \frac{p^2}{z_3^2}\right)\Phi = 0, \tag{38.130}$$

where

$$p = \frac{1-n}{2(2-n)}. \tag{38.131}$$

If p is not an integer, then

$$\Phi = C_1 J_p(z_3) + C_2 J_{-p}(z_3), \tag{38.132}$$

and, passing to the main variables, we obtain

$$\beta = z_2^{(1-n)/2}\left[C_1 J_p\left(\frac{k}{2-n}z_2^{2-n}\right) + C_2 J_{-p}\left(\frac{k}{2-n}z_2^{2-n}\right)\right]. \tag{38.133}$$

The boundary conditions of the problem have the following form:

$$\left.\begin{array}{l} \beta = 0 \quad \text{at } z_2 = 1, \\[2mm] \dfrac{d\beta}{dz_2} = 0 \quad \text{at } z_2 = 0. \end{array}\right\} \tag{38.134}$$

Consider in more detail the case, when the value n satisfies the condition $0 < n < 1$ which also corresponds to the condition $1/4 > p > 0$. In order to use conditions (38.134), we represent the function $\dfrac{d\beta}{dz_2}$ in the form

$$\frac{d\beta}{dz_2} = \frac{d\beta}{dz_3}\frac{dz_3}{dz_2} = \frac{d\beta}{dz_3} k z_2^{1-n}. \tag{38.135}$$

Replacing z_2 in (38.133) by its expression via z_3 in accordance with formula (38.127) and taking into account (38.131), we obtain

$$\frac{d\beta}{dz_3} = \left(\frac{2-n}{k}z_3\right)^p [C_1 J_{p-1}(z_3) - C_2 J_{-(p-1)}(z_3)]. \tag{38.136}$$

At the same time

$$z_2^{1-n} = \left(\frac{2-n}{k}z_3\right)^{\frac{1-n}{2-n}} = \left(\frac{2-n}{k}z_3\right)^{2p}. \tag{38.137}$$

Thus,

$$\frac{d\beta}{dz_2} = k\left(\frac{2-n}{k}z_3\right)^{3p}[C_1 J_{p-1}(z_3) - C_2 J_{-(p-1)}(z_3)].\tag{38.138}$$

Note that z_3 also tends to zero as $z_2 \to 0$. Therefore, it is sufficient to study the behaviour of the function $\dfrac{d\beta}{dz_3}$ as $z_3 \to 0$.

After expanding the functions $J_{p-1}(z_3)$ and $J_{-(p-1)}(z_3)$ in expression (38.138) into a power series in z_3, we obtain

$$\frac{d\beta}{dz_2} = k\left(\frac{2-n}{k}\right)^{3p} z_3^{3p}\left\{C_1 \frac{z_3^{p-1}}{2^{p-1}\Gamma(p)}\left(1 - \frac{z_3^2}{2\cdot 2p} + \dots\right)\right.$$

$$\left. - C_2 \frac{z_3^{1-p}}{2^{1-p}\Gamma(2-p)}\left[1 - \frac{z_3^2}{2(4-2p)} + \dots\right]\right\}.\tag{38.139}$$

Thus, up to the higher powers of the arguments z we have

$$\frac{d\beta}{dz_2} = k\left(\frac{2-n}{k}\right)^{3p}\left[C_1 \frac{z_3^{4p-1}}{2^{p-1}\Gamma(p)} - C_2 \frac{z_3^{2p+1}}{2^{1-p}\Gamma(2-p)}\right].\tag{38.140}$$

The second term in this expression also tends to zero as $z_3 \to 0$. Since $4p - 1 < 0$, the condition

$$\frac{d\beta}{dz_2} \to 0 \quad \text{as } z_3 \to 0$$

can only be satisfied if we set

$$C_1 = 0.$$

The critical loading can now be determined, using the first boundary condition (38.134), which leads to the transcendental equation

$$J_{-p}\left(\frac{k}{2-n}\right) = 0.\tag{38.141}$$

If $n = 1$, then $p = 0$ and

$$\beta = C_1 J_0 + C_2 N_0.\tag{38.142}$$

In this case the transcendental equation (38.141) takes the form

$$J_0(k) = 0.$$

The table below contains the two first roots of equation (38.139) for different values of n:

n	p	k_1	k_2
0	1/4	4.013	10.246
1/4	3/14	3.614	9.066
1/2	1/6	3.214	7.885
3/4	1/10	2.811	6.703
1	0	2.405	5.520

Briefly consider the problem of stability of the plane form of pure bending of a cantilever by the moment M in the case, when the stiffnesses B_1 and C vary according to the exponential law

$$B_1 = B_0 e^{\kappa z}, \qquad C = C_0 e^{\kappa z}.$$

This problem can be met when considering the bending of the strip whose transversal cross-section is a narrow rectangle, whose height is varying according to the exponential law.

The differential equations of the problem have the form

$$B_1 \frac{d^2 u}{dz^2} = -\beta M, \tag{38.143}$$

$$C \frac{d\beta}{dz} = \frac{du}{dz} M. \tag{38.144}$$

After differentiating the second equation with respect to z and replacing $\frac{d^2 u}{dz^2}$ by its value from the first equation, we obtain as a result

$$C \frac{d^2 \beta}{dz^2} + \frac{dC}{dz} \frac{d\beta}{dz} = -\frac{M^2}{B_1} \beta. \tag{38.145}$$

Replacing C and B_1 in this equation by their representations, we obtain

$$C_0 e^{\kappa z} \frac{d^2 \beta}{dz^2} + \kappa C_0 e^{\kappa z} \frac{d\beta}{dz} = -\frac{M^2}{B_0} e^{-\kappa z} \beta, \tag{38.146}$$

or

$$\frac{d^2 \beta}{dz^2} + \kappa \frac{d\beta}{dz} + \frac{M^2}{B_0 C_0} e^{-2\kappa z} \beta = 0. \tag{38.147}$$

Let us reduce this equation to the Bessel equation, by applying the method of Lommel.

As a basis, we take the fact that the general solution of the equation

$$\frac{d^2 \beta}{dz^2} - \left[\frac{\psi''(z)}{\psi'(z)} + (2\mu - 1) \frac{\psi'(z)}{\psi(z)} \right] \frac{d\beta}{dz} + \left[\mu^2 - \nu^2 + \psi^2(z) \right] \left[\frac{\psi'(z)}{\psi(z)} \right]^2 \beta = 0 \tag{38.148}$$

has the form

$$\beta = [\psi(z)]^\mu Z_\nu [\psi(z)], \tag{38.149}$$

where Z is a cylindrical function. In our case we have

$$\left. \begin{array}{l} \dfrac{\psi''(z)}{\psi'(z)} + (2\mu - 1) \dfrac{\psi'(z)}{\psi(z)} = -\kappa, \\[3mm] [\mu^2 - \nu^2 + \psi^2(z)] \left[\dfrac{\psi'(z)}{\psi(z)} \right]^2 = \dfrac{M^2}{B_0 C_0} e^{-2\kappa z}. \end{array} \right\} \tag{38.150}$$

Let us set

$$\psi(z) = A e^{kz} \tag{38.151}$$

and check, whether the formulated conditions can hold in this case. For this purpose we substitute expression (38.151) into equations (38.150). Then we obtain

$$k + (2\mu - 1)k = -\kappa, \tag{38.152}$$

$$[\mu^2 - \nu^2 + A^2 e^{2kz}] k^2 = \frac{M^2}{B_0 C_0} e^{-2\kappa z}. \tag{38.153}$$

The first condition implies that

$$2\mu k = -\kappa.$$

The second condition can be satisfied, by setting

$$\mu^2 - \nu^2 = 0;$$

$$A^2 k^2 = \frac{M^2}{B_0 C_0};$$

$$k = -\kappa.$$

Thus, we have

$$k = -\kappa, \qquad \mu = \frac{1}{2}, \qquad \nu = \pm\frac{1}{2}, \qquad A = \frac{M}{\kappa}\frac{1}{\sqrt{B_0 C_0}}, \tag{38.154}$$

and the solution of equation (38.147 has the form

$$\beta = C_1\sqrt{e^{-\kappa z}}\,J_{1/2}(Ae^{-\kappa z}) + C_2\sqrt{e^{-\kappa z}}\,J_{-1/2}(Ae^{-\kappa z}). \tag{38.155}$$

It is known that Bessel functions whose index is equal to a half integer, can be expressed via elementary functions. Thus the solution obtained can be written using the symbols of elementary functions.

39. Plane heat waves in a half-space and a layer; heat waves in a rod

In this section we consider some one-dimensional problems of the theory of heat waves, which are described by differential equations with variable coefficients, which can be reduced to the Bessel equations.

Let us place the plane oxy in one of the two parallel planes which bound the layer. The z-axis is directed inside the layer whose thickness is equal to h. We only consider the problem on plane waves, i.e., we suppose that the temperature of a point of the layer is $T(x, y, z, t) = f(z, t)$.

The problems concerning plane heat waves in a layer are considered in detail in heat engineering in connection with the study of the influence of periodic variations of external temperatures onto the heat fields in the cladding — walls and coverings of buildings. These problems are also of interest in other technical questions as well as in problems of geophysics. For instance, when studying the heat fields in the upper layer of soil which arise as a result of the diurnal variations of the temperature, we have to solve a problem on heat waves.

In problems on heat waves in the soil, one usually assumes that the heat capacity per unit volume is constant, and the coefficient of thermal conductivity is assumed to be a given function of the z-coordinate, i.e., one assumes that $\lambda = \lambda(z)$ depends on the depth. The case, when $\lambda(z)$ is a linear function of the depth, has been considered by A.F. Chudnovski [58]. The solution of this problem is obtained in terms of Bessel functions of zero index. A more difficult problem on heat waves in a layer with variable thermal physics characteristics has been investigated by A.A. Dorodnitsyn [14].

We will consider several problems here on heat waves for different laws of the variation of the coefficient of thermal conductivity $\lambda(z)$ with respect to the depth of the layer. The differential equation of the heat conduction in the case considered has the form

$$c\frac{\partial T}{\partial t} = \frac{\partial}{\partial z}\left[\lambda(z)\frac{\partial T}{\partial z}\right]. \tag{39.1}$$

The boundary conditions of problems on heat waves must contain the given functions of the time of the type $D_n \sin \kappa t$ and $D_n^{(1)} \cos \kappa t$ in the right-hand side, where κ is the angular frequency of the variation of the heat effect. As an example, we write the boundary conditions of the first and the second kind for the problem on the layer considered below.

In the first case the temperature should be given

$$T = D_1 \sin \kappa t + D_2 \cos \kappa t, \qquad\qquad \text{for } z = 0, \qquad\qquad (39.2)$$

$$T = D_3 \sin \kappa t + D_4 \cos \kappa t, \qquad\qquad \text{for } z = h. \qquad\qquad (39.3)$$

In the second case the quantity of the heat flow through the boundaries of the layer is given by:

$$q = D_1^* \sin \kappa t + D_2^* \cos \kappa t, \qquad\qquad \text{for } z = 0,$$

$$q = D_3^* \sin \kappa t + D_4^* \cos \kappa t, \qquad\qquad \text{for } z = h$$

where q is the heat flow.

When solving problems on heat waves, we assume that the temperature in the layer is equal to

$$T(z,t) = \varphi_1(z) \sin \kappa t + \varphi_2(z) \cos \kappa t, \qquad\qquad (39.4)$$

and the heat flow in the direction of the z-axis is equal to

$$q(z,t) = \psi_1(z) \sin \kappa t + \psi_2(z) \cos \kappa t.$$

Obviously,

$$\psi_1(z) = -\lambda(z)\frac{d\varphi_1(z)}{dz},$$

$$\psi_2(z) = -\lambda(z)\frac{d\varphi_2(z)}{dz}.$$

Substituting expression (39.4) into (39.1) and making the coefficients of $\sin \kappa t$ and $\cos \kappa t$ equal, we obtain the following system of equations

$$\frac{d}{dz}\left[\lambda(z)\frac{d\varphi_2(z)}{dz}\right] = c\kappa\varphi_1(z), \qquad\qquad (39.5)$$

$$\frac{d}{dz}\left[\lambda(z)\frac{d\varphi_1(z)}{dz}\right] = -c\kappa\varphi_2(z). \qquad\qquad (39.6)$$

Multiplying equation (39.5) by i and adding equation (39.6) to it, we obtain

$$\frac{d}{dz}\left\{\lambda(z)\left[\frac{d\varphi_1(z)}{dz} + i\frac{d\varphi_2(z)}{dz}\right]\right\} = c\kappa(i\varphi_1 - \varphi_2). \qquad\qquad (39.7)$$

Using the notation $\Phi(z) = \varphi_1(z) + i\varphi_2(z)$, we can rewrite (39.7) in the following form:

$$\frac{d}{dz}\left[\lambda(z)\frac{d\Phi(z)}{dz}\right] = ic\kappa\Phi(z). \qquad\qquad (39.8)$$

In some cases it is convenient to consider the function

$$\Psi(z) = \psi_1(z) + i\psi_2(z) = -\lambda(z)\frac{d\Phi}{dz}. \qquad\qquad (39.9)$$

Heat flow $q(z)$ for a constant c satisfies the equation

$$\frac{\partial^2 q}{\partial z^2} = \frac{c}{\lambda(z)}\frac{\partial q}{\partial t}.$$

From the dependences given above we can easily obtain the following equations for the functions ψ_1, ψ_2:

$$c\kappa\psi_2 = \lambda(z)\frac{\partial^2\psi_1}{\partial z^2}, \tag{39.10}$$

$$c\kappa\psi_1 = -\lambda(z)\frac{\partial^2\psi_2}{\partial z^2}. \tag{39.11}$$

Using the notation of (39.9), we find from (39.10) and (39.11)

$$\frac{d^2\Psi(z)}{dz^2} = \frac{c\kappa}{\lambda(z)}i\Psi(z).$$

Let us pass to the consideration of various laws of variation of the coefficient of heat conductivity.

First, consider the problem on heat waves, when the coefficient of heat conductivity is varying according to a power law: let $\lambda(z) = \lambda_0(1 + d_0z)^{-m}$.

In this case the differential equation (39.8) takes the form $\frac{d}{d\zeta}\left(\zeta^{-m}\frac{d\Phi}{d\zeta}\right) = i\delta^2\Phi$, where $\zeta = 1 + d_0z$, $\delta^2 = c\kappa/\lambda_0 d_0^2$. Note that δ and ζ are dimensionless quantities.

We rewrite the equation obtained in the form

$$\frac{d^2\Phi}{d\zeta^2} - \frac{m}{\zeta}\frac{d\Phi}{d\zeta} - i\zeta^m\delta^2\Phi = 0, \tag{39.12}$$

which is a special case of the equation

$$y'' + \frac{1-2\alpha}{\kappa}y' + \left[(\beta\gamma x^{\gamma-1})^2 + \frac{\alpha^2 - p^2\gamma^2}{x^2}\right]y = 0. \tag{39.13}$$

With the help of the change of the variables $y = x^\alpha v$, $z = \beta x^\gamma$ (see page 32), equation (39.13) can be reduced to the Bessel equation with respect to the function v. Hence, it has solutions of the form $y = x^\alpha Z_p(\beta x^\gamma)$. In our case we have $1 - 2\alpha = -m$, $\alpha^2 = p^2\gamma^2$, $2(\gamma - 1) = m$, $i\delta^2 = \beta^2\gamma^2$; therefore, for non-integer $(m + 1)/)m + 2)$ the solution of equation (39.12) can be written in the following form:

$$\Phi = \zeta^{(1+m)/2}\left\{A_1 J_{(m+1)/(m+2)}\left(\frac{2\delta}{m+2}\zeta^{(m+2)/2}\sqrt{-i}\right)\right.$$

$$\left. + A_2 J_{-(m+1)/(m+2)}\left(\frac{2\delta}{m+2}\zeta^{(m+2)/2}\sqrt{-i}\right)\right\}. \tag{39.14}$$

If $(m + 1)/(m + 2)$ is an integer, then

$$\Phi = \zeta^{(1+m)/2}\left\{A_1 J_{(m+1)/(m+2)}\left(\frac{2\delta}{m+2}\zeta^{(m+2)/2}\sqrt{-i}\right)\right.$$

$$\left. + A_2 H^{(2)}_{(m+1)/(m+2)}\left(\frac{2\delta}{m+2}\zeta^{(m+2)/2}\sqrt{-i}\right)\right\}. \tag{39.15}$$

Let us note that taking Ψ as the function to be determined, we obtain the differential equation

$$\frac{d^2\Psi}{d\zeta^2} - i\delta^2\zeta^m\Psi = 0.$$

If $1/(m+2)$ is non-integer, then the solution of this equation has the following form:

$$\Psi = \sqrt{\zeta}\left[A_1 J_{1/(m+2)}\left(\frac{\delta\sqrt{-i}\zeta s^{m/2+1}}{1+m/2}\right) + A_2 J_{-1/(m+2)}\left(\frac{\delta\sqrt{-i}\zeta s^{m/2+1}}{1+m/2}\right)\right],$$
(39.16)

if $1/(m+2)$ is an integer, then we have

$$\Psi = \sqrt{\zeta}\left[A_1 J_{1/(m+2)}\left(\frac{\delta\sqrt{-i}\zeta s^{m/2+1}}{1+m/2}\right) + A_2 H^{(2)}_{1/(m+2)}\left(\frac{\delta\sqrt{-i}\zeta s^{m/2+1}}{1+m/2}\right)\right]. \quad (39.17)$$

Let us show how one can find the real functions φ_1, φ_2, ψ_1, and ψ_2, which are present in the solution of the problem on heat waves.

As an example we consider the determination of φ_1 and φ_2 in the case, when Φ is defined by formula (39.15). Let us denote $A_1 = B_1 + iB_2$, $A_2 = B_3 + iB_4$, where B_1, B_2, B_3 and B_4 are real numbers.

Let us denote

$$J_\nu\left(z\sqrt{-i}\right) = u_\nu(z) - iv_\nu(z),$$

$$H^{(2)}_\nu\left(z\sqrt{-i}\right) = f_\nu(z) - ig_\nu(z).$$

After replacing A_1 and A_2 in expression (39.15) by their values, and substituting the right-hand sides of equalities (39.11) and (39.12) instead of $J_\nu(z)$ and $H^{(2)}_\nu(z)$, we obtain by separating the real and imaginary parts

$$\varphi_1 = \zeta^{(1+m)/2}\left\{B_1 u_{(m+1)/(m+2)}\left(\frac{2\delta}{m+2}\zeta^{(m+2)/2}\right)\right.$$

$$+ B_2 v_{(m+1)/(m+2)}\left(\frac{2\delta}{m+2}\zeta^{(m+2)/2}\right) + B_3 f_{(m+1)/(m+2)}\left(\frac{2\delta}{m+2}\zeta^{(m+2)/2}\right)$$

$$\left. + B_4 g_{(m+1)/(m+2)}\left(\frac{2\delta}{m+2}\zeta^{(m+2)/2}\right)\right\}, \quad (39.18)$$

$$\varphi_2 = \zeta^{(1+m)/2}\left\{-B_1 v_{(m+1)/(m+2)}\left(\frac{2\delta}{m+2}\zeta^{(m+2)/2}\right)\right.$$

$$+ B_2 u_{(m+1)/(m+2)}\left(\frac{2\delta}{m+2}\zeta^{(m+2)/2}\right) - B_3 g_{(m+1)/(m+2)}\left(\frac{2\delta}{m+2}\zeta^{(m+2)/2}\right)$$

$$\left. + B_4 f_{(m+1)/(m+2)}\left(\frac{2\delta}{m+2}\zeta^{(m+2)/2}\right)\right\}. \quad (39.19)$$

The solution of problems on heat waves in the layer is expressed here via Bessel functions of a complex argument. Obviously, from the practical point of view we should first of all use solutions which can be expressed via tabulated functions.

It is known that the functions of the complex argument $x\sqrt{i}$ are tabulated for three values of the index: 0; 1 and $1/3$.

Consider three cases in detail.

1. In the first case we have $\nu = (m+1)/(m+2) = 0$. Thus, $m = -1$. This means that the coefficient of heat conductivity varies according to a linear law. We have the problem here which was mentioned above and which was investigated by A.F. Chudnovsky [58].

In this case

$$\Phi = A_1 J_0 \left(2\delta\zeta^{1/2}\sqrt{-i}\right) + A_2 H_0^{(2)} \left(2\delta\zeta^{1/2}\sqrt{-i}\right).$$ (39.20)

Denoting $2\delta\sqrt{\zeta} = \xi$, for the sake of brevity, we obtain

$$\varphi_1 = B_1 u_0(\xi) + B_2 v_0(\xi) + B_3 f_0(\xi) + B_4 g_0(\xi).$$ (39.21)

The function φ_2 can be expressed with the help of formula (39.6) via φ_1; therefore,

$$\varphi_2 = -B_1 v_0(\xi) + B_2 u_0(\xi) - B_3 g_0(\xi) + B_4 f_0(\xi).$$ (39.22)

The functions u_0, v_0, f_0, g_0 and the constants B_1, B_2, B_3, B_4 are real.

In order to determine B_1, B_2, B_3, B_4 we should take the boundary conditions into account. Let us consider the case, when the boundary conditions of the first kind are given. By substituting φ_1 and φ_2 from expressions (39.21) and (39.22) into (39.4) and afterwards into (39.3) and making the coefficients of sines and cosines equal, we obtain

$$\left.\begin{array}{l} D_1 = \varphi_1(\xi_0), \\ D_2 = \varphi_2(\xi_0), \\ D_3 = \varphi_1(\xi_h), \\ D_4 = \varphi_2(\xi_h). \end{array}\right\}$$

At $z = 0$ we have $\zeta = 1$, hence, $\xi_0 = 2\delta$; at $z = h$ we have $\zeta = 1 + d_0 h$, hence, $\xi_h = 2\delta\sqrt{1 + d_0 h}$. Let us rewrite the system obtained in the expanded form:

$$B_1 u_0(\xi_0) + B_2 v_0(\xi_0) + B_3 f_0(\xi_0) + B_4 g_0(\xi_0) = D_1,$$
$$B_1 v_0(\xi_0) - B_2 u_0(\xi_0) + B_3 g_0(\xi_0) - B_4 f_0(\xi_0) = -D_2,$$
$$B_1 u_0(\xi_h) + B_2 v_0(\xi_h) + B_3 f_0(\xi_h) + B_4 g_0(\xi_h) = D_3,$$
$$B_1 v_0(\xi_h) - B_2 u_0(\xi_h) + B_3 g_0(\xi_h) - B_4 f_0(\xi_h) = -D_4.$$

When we have determined B_1, B_2, B_3, and B_4 from this system of equations, the problem can be treated as solved.

In the special case when $h \to \infty$, we should set $B_1 = 0$, $B_2 = 0$, because the functions u_0 and v_0 increase infinitely as $\xi \to \infty$.

In this case we have

$$\varphi_1 = B_3 f_0(\xi) + B_4 g_0(\xi),$$ (39.23)
$$\varphi_2 = -B_3 g_0(\xi) + B_4 f_0(\xi).$$ (39.24)

2. Let us pass to consideration of a law of variation of the coefficient of heat conductivity, when the solution can be expressed via the functions of the index 1. Let $(m + 1)/(m + 2) = -1$; then $m = -3/2$ and $\lambda = \lambda_0(1 + d_0 z)^{1/2}$. The solution of the problem which is given by formula (39.15) for $m = -3/2$, has the form

$$\Phi = \zeta^{-1/4} \left\{ A_1 J_{-1} \left(4\delta\sqrt{-i}\zeta^{1/4}\right) + A_2 H_{-1}^{(2)} \left(4\delta\sqrt{-i}\zeta^{1/4}\right) \right\}.$$ (39.25)

Let us pass to the functions of a positive index with the help of the formulae

$$u_{-1}(z) = -u_1(z),$$
$$v_{-1}(z) = -v_1(z),$$
$$f_{-1}(z) = -f_1(z),$$
$$g_{-1}(z) = -g_1(z).$$

Thus, the functions φ_1 and φ_2 can be written in the following form:

$$\varphi_1 = \zeta^{-1/4}\{B_1 u_1(\eta) + B_2 v_1(\eta) + B_3 f_1(\eta) + B_4 g_1(\eta)\}, \tag{39.26}$$

$$\varphi_2 = \zeta^{-1/4}\{-B_1 v_1(\eta) + B_2 u_1(\eta) - B_3 g_1(\eta) + B_4 f_1(\eta)\}, \tag{39.27}$$

where $\eta = 4\delta^{1/4}$.

3. Consider the case of the variation of the coefficient of heat conductivity, when

$$\frac{m+1}{m+2} = \frac{1}{3}.$$

This implies that $m = -1/2$; hence, in this case we have

$$\lambda = \lambda_0 \sqrt{1 + d_0 z}.$$

It is convenient for us to write the solution in the form

$$\Phi = \zeta^{1/4}\left\{ A_1 H_{1/3}^{(1)}\left(\frac{4\delta}{3}\sqrt{-i}\zeta^{3/4}\right) + A_2 H_{1/3}^{(2)}\left(\frac{4\delta}{3}\sqrt{-i}\zeta^{3/4}\right)\right\}. \tag{39.28}$$

The solution of the formulated problem on heat waves in a layer with a co-efficient of the heat conductivity which varies according to a power law, can be expressed via elementary functions in the case, when Bessel functions of a half-integer index are present in (39.19). Let us find values of m, for which the index of the Bessel function, which is equal to $\nu = (m+1)/(m+2)$ is a half-integer.

Suppose that $(m+1)/(m+2) = n+1/2$, where n is an integer. We successively assign to n the following values: $n = 0$, 1, 2, 3. In this case we respectively obtain $m = 0$, -4, $-8/3$, $-12/5$. The case $m = 0$ is of no interest, because it corresponds to the problem on the layer with a constant coefficient of heat conductivity.

As an example we consider the case $n = 1$ roughly. In this case $m = -4$, hence, $\lambda = \lambda_0(1 + d_0 z)^4$. The function $\Phi(z)$ can be written in the following form:

$$\Phi = \zeta^{-3/2}\left\{ A_1 J_{3/2}\left(\delta\zeta^{-1}\sqrt{-i}\right) + A_2 J_{-3/2}\left(\delta\zeta^{-1}\sqrt{-i}\right)\right\}. \tag{39.29}$$

Let us denote

$$J_{-3/2}\left(z\sqrt{\pm i}\right) = u_{-3/2}(z) \pm i v_{-3/2}(z),$$

$$J_{3/2}\left(z\sqrt{\pm i}\right) = u_{3/2}(z) \pm i v_{3/2}(z).$$

Moreover, denoting $\delta\zeta^{-1} = \xi_1$, we can rewrite the solution of the problem in the form $\Phi = \varphi_1 + i\varphi_2$, where

$$\varphi_1 = \zeta^{-3/2}\{B_1 u_{-3/2}(\xi_1) + B_2 v_{-3/2}(\xi_1) + B_3 u_{3/2}(\xi_1) + B_4 v_{3/2}(\xi_1)\},$$

$$\varphi_2 = \zeta^{-3/2}\{-B_1 v_{-3/2}(\xi_1) + B_2 u_{-3/2}(\xi_1) - B_3 v_{3/2}(\xi_1) + B_4 u_{3/2}(\xi_1)\}.$$

The temperature in the layer is, as before, expressed by the formula $T = \varphi_1 \sin \kappa t + \varphi_2 \cos \kappa t$.

In order to compute the functions $u_{3/2}$, $u_{-3/2}$, $v_{3/2}$, $v_{-3/2}$ we can use (see (7.4)) the formulae

$$J_{3/2}(z) = \left(\frac{2}{\pi z}\right)^{1/2}\left(\frac{\sin z}{z} - \cos z\right),$$

$$J_{-3/2}(z) = \left(\frac{2}{\pi z}\right)^{1/2}\left(\frac{-\cos z}{z} - \sin z\right).$$

We should substitute $z = x\sqrt{i}$ in these formulae and then separate the imaginary and real parts of the expression obtained. We have

$$\sin\left(x\sqrt{i}\right) = \sin\left[\frac{x}{\sqrt{2}} + \frac{xi}{\sqrt{2}}\right]$$

$$= \sin\left(\frac{x}{\sqrt{2}}\right)\cosh\left(\frac{x}{\sqrt{2}}\right) + i\cos\left(\frac{x}{\sqrt{2}}\right)\sinh\left(\frac{x}{\sqrt{2}}\right),$$

$$\cos\left(x\sqrt{i}\right) = \cos\left[\frac{x}{\sqrt{2}} + \frac{xi}{\sqrt{2}}\right]$$

$$= \cos\left(\frac{x}{\sqrt{2}}\right)\cosh\left(\frac{x}{\sqrt{2}}\right) - i\sin\left(\frac{x}{\sqrt{2}}\right)\sinh\left(\frac{x}{\sqrt{2}}\right);$$

therefore,

$$J_{\pm 3/2}\left(x\sqrt{i}\right) = \left(\frac{2}{\pi x\sqrt{i}}\right)^{1/2}\left\{\frac{-\frac{\sin}{\cos}\left(x\sqrt{i}\right)}{x\sqrt{i}} - \frac{\sin}{\cos}\left(x\sqrt{i}\right)\right\}.$$

One can easily obtain the solution of a similar problem, if $\lambda(z)$ varies according to the exponential law (see [20]).

40. A die which lies on an elastic half-space whose modulus of elasticity is a power function of the depth

The problem on the distribution of the pressure on the foot of the die which is pressed in an elastic half-space, has attracted the attention of numerous researchers. A series of important problems concerning the dies has been solved by N.I. Muskhelishvili, A.I. Lurier and others. This problem has been studied in detail in the monographs by I.Ya. Staerman [60] and L.A. Galin [9]. In most cases these works are devoted to a die which lies on an elastic uniform half-space.

In this section we formulate the problem on a die which lies on a foundation with a homogeneous axially symmetric kernel, i.e., we assume that the settling of a point (x, y) of the foundation, caused by the unit vertical force applied at the point (ξ, η), is a function only of the distance between these points $K(r) = K\left(\sqrt{(x - \xi)^2 + (y - \eta)^2}\right)$.

This problem is solved for the case, when the modulus of elasticity of the foundation varies with the depth according to a power law.

If the surface of the foundation is exposed to the loading $p(x, y)$ (for the sake of brevity of the calculation, we assume that p is an even function with respect to each of the coordinates), then

$$w(x, y) = \int_0^\infty \int_0^\infty c\left(\sqrt{z_1^2 + z_2^2}\right) a(z_1, z_2)\cos z_1 x \cos z_2 y\, dz_1\, dz_2,$$

$$p(x, y) = \int_0^\infty \int_0^\infty a(z_1, z_2)\cos z_1 x \cos z_2 y\, dz_1\, dz_2,$$

where

$$a(z_1, z_2) = \frac{4}{\pi^2} \int_0^\infty \int_0^\infty p(x, y) \cos z_1 x \cos z_2 y \, dx \, dy,$$

$$c(z_1, z_2) = \int_{-\infty}^\infty \int_{-\infty}^\infty K(x, y) \cos z_1 x \cos z_2 y \, dy \, dx.$$

If we suppose that for the points $B(x, y)$ which lie inside the domain F, the settling is $w = \varphi(x, y)$, and for the points outside this domain we have $p(x, y)$, where $\varphi(x, y)$ is a known function and $p(x, y)$ is an unknown function, then the problem on the die can be formulated in the following form:
inside the domain occupied by the die we have

$$\int_0^\infty \int_0^\infty c(z_1, z_2) a(z_1, z_2) \cos z_1 x \cos z_2 y \, dz_1 \, dz_2 = \varphi(x, y),$$

and outside this domain we have

$$\int_0^\infty \int_0^\infty a(z_1, z_2) \cos z_1 x \cos z_2 y \, dz_1 \, dz_2 = 0.$$

By solving these equations with respect to the function $a(z_1, z_2)$, we can easily find $p(x, y)$.

In the axially symmetric case we obtain, using the Parseval integral,

$$\frac{1}{2\pi} \int_{-\infty}^\infty \int_{-\infty}^\infty p\left(\sqrt{x^2 + y^2}\right) \cos z_1 x \cos z_2 y \, dx \, dy = \int_0^\infty r p(r) J_0(\gamma r) \, dr = \tilde{p}(\gamma), \quad (40.1)$$

where

$$\gamma = \sqrt{z_1^2 + z_2^2},$$
$$r = \sqrt{x^2 + y^2}.$$

Thus, we have

$$a(\gamma) = 2\tilde{p}(\gamma)/\pi, \tag{40.2}$$

and the problem on the die of the radius R, which has the settling $w_0(r)$, will be formulated in the following way:

$$\int_0^\infty \gamma c(\gamma) a(\gamma) J_0(\gamma r) \, d\gamma = \frac{2w_0}{\pi}, \qquad r < R,$$

$$\int_0^\infty \gamma a(\gamma) J_0(\gamma r) \, d\gamma = 0, \qquad r > R,$$

where

$$c(\gamma) = 2\pi \int_0^\infty r K(r) J_0(\gamma r) \, dr.$$

Denoting $r/R = \rho$ and $\gamma R = \beta$, we have $a(\gamma) = a(\beta/R) = a_1(\beta)$, $c(\gamma) = c(\beta/R) = c_1(\beta)$, $w_0(r) = w_0(\rho)$ and

$$\int_0^\infty \beta c_1(\beta) a_1(\beta) J_0(\beta \rho)\, d\beta = \frac{2Rw_0^*(\rho)}{\pi}, \qquad \rho < 1, \qquad (40.3)$$

$$\int_0^\infty \beta a_1(\beta) J_0(\beta \rho)\, d\beta = 0, \qquad \rho > 1. \qquad (40.4)$$

The solution of this system of equations for the case, when $c_1(\beta)$ is a power function, is well known (see Section 22). It will be used below when considering the die which lies on a foundation whose modulus of elasticity is varying according to the power law.

In this case the kernel has the form

$$K(r) = \frac{1}{\pi E_n r^{n+1}}.$$

Suppose that the modulus of elasticity increases with the depth, i.e., $n > 0$.

In this case

$$c(\gamma) = 2\pi \int_0^\infty r K(r) J_0(\gamma r)\, dr = \frac{2^{1-n}}{E_n} \frac{\Gamma\left(\frac{1-n}{2}\right)}{\Gamma\left(\frac{1+n}{2}\right)} \gamma^{n-1}.$$

For the sake of brevity, we denote

$$\frac{2^{1-n}}{E_n} \frac{\Gamma\left(\frac{1-n}{2}\right)}{\Gamma\left(\frac{1+n}{2}\right)} = A_n, \qquad n - 1 = \alpha,$$

$$\frac{2}{\pi} w_0^* \frac{R^{n+1}}{A_n} = g(\rho), \qquad \beta a_1(\beta) = f(\beta).$$

Then equations (40.3) and (40.4) take the form

$$\int_0^\infty \beta^\alpha f(\beta) J_0(\beta \rho)\, d\beta = g(\rho), \qquad 0 < \rho < 1,$$

$$\int_0^\infty f(\beta) J_0(\beta \rho)\, d\beta = 0, \qquad 1 < \rho < \infty.$$

If we set $1 > n > 0$, then $\alpha < 0$ and the solution of the problem can be obtained, using the result formulated on page 90. Taking into account the fact that the right-hand side of the second integral equation vanishes, and the index of the Bessel function is equal to zero, we obtain the formula

$$f(\beta) = \frac{2^{-\alpha/2}\beta^{-\alpha}}{\Gamma(1+\alpha/2)} \left[\beta^{1+\alpha/2} J_{\alpha/2}(\beta) \int_0^1 y(1-y^2)^{\alpha/2} g(y)\, dy \right.$$

$$\left. + \int_0^1 u(1-u^2)^{\alpha/2}\, du \int_0^1 g(yu)(\beta y)^{2+\alpha/2} J_{1+\alpha/2}(\beta y)\, dy \right].$$

If we set $g = g_0 = \text{const}$, i.e., we assume that the die has a plane lower boundary, then as a result of the calculation, which can be reduced to the determination of

$f(\beta)$ and hence, $a_1(\beta)$, with the help of relations (40.2) and (40.1) we find that the reactive pressure is

$$p(\rho) = \frac{\pi}{2}g_0\frac{2^{(1-n)/2}}{\Gamma\left(\frac{n+1}{2}\right)}\int_0^\infty \beta^{(1-n)/2}J_{(n+1)/2}(\beta)J_0(\beta\rho)\,d\beta.$$

The problem is now reduced to the computation of the improper Weber–Sonin–Schafheitlin integral, which in our case gives the following result:

$$p(\rho) = \frac{\pi g_0}{2^n\Gamma^2\left(\frac{n+1}{2}\right)}{}_2F_1\left(1,\frac{1-n}{2};1;\rho^2\right)$$

$$= \frac{\pi g_0}{2^n\Gamma^2\left(\frac{n+1}{2}\right)}\frac{1}{(1-\rho^2)^{(1-n)/2}}.$$

41. Application of integral equations to the solution of some problems on membranes and plates

In this section we consider several problems which can be reduced to integral Fredholm equations of the second kind, whose kernels can be expressed via Bessel functions.

First, consider the problem on the forced oscillations of a membrane, which can be reduced to the solution of the internal problem for the equation

$$\nabla^2 w + \lambda^2 w = 0$$

under the condition that on the contour C (simple, piecewise smooth), which bounds the domain considered (finite, simply connected), we have $w = w(\sigma) = f$, where σ is the arc coordinate of the point of the contour.

We represent the solution of the problem in terms of a source of the form

$$w(\sigma) = \int_C \mu(s)K_1^*(\sigma, s)\,ds,$$

where $\mu(s)$ is a density of the moments which will be defined, whose vectors are directed along the tangent to the contour; s is the arc coordinate of the point of the contour considered. The equation of the elastic surface of the membrane, to which the concentrated moment is applied, can be represented in the form

$$K_1^* = Y_1(r)\cos(n_\sigma, r), \tag{41.1}$$

where n_σ is the normal, $r = \sqrt{(x-\xi)^2 + (y-\eta)^2}$, x, y are the coordinates of the point of observation which belongs to the domain considered; ξ, η are Cartesian coordinates of the point of the contour with the arc coordinate σ.

It follows from the expansion of cylindrical functions of the second kind with an integer index into power series that the kernel K_1^* can be represented in the form

$$Y_1(r)\cos(n_\sigma, r) = \left(-\frac{2}{\pi r} + \psi\right)\cos(n_s, r), \tag{41.2}$$

where ψ is a regular function. Therefore, using the results of the classical theory of the logarithmic potential, we immediately obtain an integral equation for the determination of $\mu(s)$

$$2\mu(\sigma) + \int_C \mu(s)K_1^*(s,\sigma,\lambda)\,ds = f(\sigma).$$

The uniqueness of the solution of this equation follows from the fact that the kernel (41.1) is the imaginary part of the potential which satisfies the Sommerfeld principle of radiation. The investigation of the existence and uniqueness of the solution can be performed in the usual way.

By introducing the kernel $K_2^* = Y_0(r)$, we can obtain integral equations for the Neumann problem.

The problem of obtaining, investigating and solving equations of this kind in problems of the propagation of plane waves is discussed in the literature in detail and repeatedly. First of all, we should mention the work by V.D. Kupradze [32] here. The kernels of the integral equations for an unbounded domain contain Hankel functions of the second kind, which allow one to satisfy the conditions at infinity which represent the Sommerfeld principle of radiation.

Let us pass to the problem in which the kernel of the integral equation is a modified Bessel function.

Consider the equilibrium of a membrane which lies on an elastic Winkler-type foundation. This problem can be reduced to the integration of the equation

$$\nabla^2 w - \lambda^2 w = 0. \tag{41.3}$$

Consider the internal problem, assuming that $w = f$ on the contour. We represent the solution of the problem in the form

$$w = \int_C \mu(s)K_1^*(x,y,\xi,\eta)\,ds, \tag{41.4}$$

where $K_1^* = K_1(r)\cos(n_s,r)2/\pi$; $K_1(r)$ is the Macdonald function of the first kind. Noting that a formula similar to (41.2) also holds for the kernel K_1^*, we obtain a Fredholm equation for the determination of μ:

$$-2\mu(\sigma) + \int_C \mu(s)K_1^*(s,\sigma)\,ds = f(\sigma).$$

Obviously, the solution of equation (41.3), which is represented by integral the (41.4) for an infinite domain under the condition that $w = 0$ or $dw/dn = 0$ on C, is identically zero. This fact immediately follows from the asymptotic expansion of the Macdonald function $K_n(r)$ for large values of the argument

$$K_n(r) \sim \sqrt{\frac{\pi}{2r}}e^{-r}\sum_{m=0}^{\infty}\frac{(p,m)}{2r^m},$$

where $(p,m) = \dfrac{\Gamma(p+m+1/2)}{m!\Gamma(p-m+1/2)}$. In our case we set $K_n(r) \sim \sqrt{\frac{\pi}{2r}}e^{-r}$.

One can now easily prove that the integral equation of this problem has a solution and this solution is unique.

In problems of the theory of heat conductivity, in particular, in the investigation of the plane problem of the theory of heat waves, problems arise which are similar to the considered ones, but which differ from those by the fact that the kernels are expressed via cylindrical functions of the complex argument; they have been considered in [20].

Let us pass to the consideration of plane problems which can be reduced to the system of two Helmholtz equations. Such problems arise, in particular, in the theory of diffraction of plane waves. They have been considered in detail by V.D. Kupradze [29]. These problems can be reduced to the solution of a system of two singular integral equations whose kernels can be expressed via cylindrical functions. As a result of regularization these equations are transformed into integral Fredholm equations of the second kind. We shall here consider problems on plates with a clamped edge; the boundary conditions of these problems are fundamentally different from those considered in the problems of diffraction. We will largely use the paper by Lauricella [70] here, who, with the help of introducing special potentials has reduced the first basic problem of the biharmonic equation to the integral Fredholm equations of the second kind. It is interesting to note that the potentials introduced by Lauricella are closely connected with the basic Green function for the biharmonic equation which, up to a constant factor, has the form $K(r) = r^2 \ln r$, $r = \sqrt{(x - \xi)^2 + (y - \eta)^2}$. The potentials introduced by him can easily be expressed via the function $K(r)$ in the following way

$$u_0' = \frac{\partial^2 K}{\partial x^2}; \qquad v_0' = u_0'' = \frac{\partial^2 K}{\partial x \, \partial y}; \qquad v_0' = \frac{\partial^2 K}{\partial y^2}.$$

Consider the problem on a plate which lies on a Winkler-type elastic foundation which occupies a finite simply connected domain and is bounded by a simple contour with a continuously rotating tangent. In has been noted in the previous examples that the problem of the equilibrium of a plate which lies on an elastic foundation can be reduced to the integration of the differential equation $\nabla^2 \nabla^2 w + \lambda^4 w = q/D$, where λ^4 is a positive constant.

In our case the boundary condition on the contour have the form $w = 0$, $\frac{\partial w}{\partial n} = 0$. Let us denote

$$u = \frac{\partial w}{\partial x}, \qquad v = \frac{\partial w}{\partial y}. \tag{41.5}$$

We will seek the solution, by reducing the problem to the solution of a homogeneous differential equation with inhomogeneous boundary conditions, which we write in the following form:

$$w = f_1(s), \qquad \frac{dw}{dn} = f_2(s).$$

Note that (41.5) implies that $\frac{\partial u}{\partial y} = \frac{\partial v}{\partial x}$.

Let $M(x, y)$ be a point of the domain; $P(\xi, \eta)$ a point of the contour,

$$r = \lambda \sqrt{(x - \xi)^2 + (y - \eta)^2}, \qquad \frac{1}{\lambda} \frac{\partial r}{\partial x} = \cos \varphi, \qquad \frac{1}{\lambda} \frac{\partial r}{\partial y} = \sin \varphi,$$

where φ is the angle between the radius-vector r and the x-axis.

Let us introduce the functions (the notation $g_0(r)$, $g_2(r)$ see on page 13):

$$u_0' = g_2(r) \cos 2\varphi - g_0(r),$$
$$u_0'' = g_2(r) \sin 2\varphi,$$
$$v_0' = g_2(r) \sin 2\varphi,$$
$$v_0'' = -g_2(r) \cos 2\varphi - g_0(r).$$

There are the following relations between these functions

$$\frac{\partial u_0'}{\partial y} = \frac{\partial v_0'}{\partial x} \quad \text{and} \quad \frac{\partial u_0''}{\partial y} = \frac{\partial v_0''}{\partial x}.$$

The functions u_0', v_0', u_0'', and v_0'' can be written in the form

$$u_0' = \left[-\frac{1}{\pi} + \psi_1^* \right] \cos 2\varphi - \frac{2}{\pi} \left[\ln r + \ln \frac{\gamma}{2} - \frac{\pi}{4} + \psi_2^* \right],$$

$$v_0' = \left[-\frac{1}{\pi} + \psi_1^* \right] \sin 2\varphi,$$

$$u_0'' = \left[-\frac{1}{\pi} + \psi_1^* \right] \sin 2\varphi,$$

$$v_0'' = -\left[-\frac{1}{\pi} + \psi_1^* \right] \cos 2\varphi - \frac{2}{\pi} \left[\ln r + \ln \frac{\gamma}{2} - \frac{\pi}{4} + \psi_2^* \right],$$

where

$$\psi_1^* = \sum_{m=1}^{\infty} A_m r^{2m} + \sum_{m=1}^{\infty} B_m r^{4m} \ln r,$$

$$\psi_2^* = \sum_{m=1}^{\infty} C_m r^{2m} + \sum_{m=1}^{\infty} D_m r^{2m} \ln r.$$

We form the functions $u = \frac{1}{\lambda} \frac{\partial w}{\partial x}$ and $v = \frac{1}{\lambda} \frac{\partial w}{\partial y}$ with the help of the expressions

$$u = \int_C \rho(s) \frac{du_0'}{dn_\sigma} \, ds + \int_C \mu(s) \frac{du_0''}{dn_\sigma} \, ds, \tag{41.6}$$

$$v = \int_C \rho(s) \frac{dv_0'}{dn_\sigma} \, ds + \int_C \mu(s) \frac{dv_0''}{dn_\sigma} \, ds, \tag{41.7}$$

which satisfy the homogeneous differential equation and relation (41.5); n_σ denotes the external normal here.

Let us rewrite the kernels of expressions (41.6) and (41.7) in the form

$$\frac{du_0'}{dn_\sigma} = \Phi_1 - \frac{2}{\pi r} \sin 2\varphi \sin(n, r) - \frac{2}{\pi} \frac{\cos(n, r)}{r},$$

$$\frac{dv_0'}{dn_\sigma} = \Phi_2 + \frac{2}{\pi r} \cos 2\varphi \sin(n, r),$$

$$\frac{du_0''}{dn_\sigma} = \Phi_2 + \frac{2}{\pi r} \cos 2\varphi \sin(n, r),$$

$$\frac{dv_0''}{dn_\sigma} = \Phi_3 + \frac{2}{\pi r} \sin 2\varphi \sin(n, r) - \frac{2}{\pi} \frac{\cos(n, r)}{r}.$$

Here, Φ_1, Φ_2, Φ_3 are regular functions.

Suppose that the point M of the domain tends from inside to the contour. Then, taking into account formulae of the theory of logarithmic potential, we obtain integral equations for the determination of μ and ρ

$$u(\sigma) = -2\rho(\sigma) + \text{v.p.} \int_C \rho(s) \left[-\frac{2}{\pi r} \sin 2\varphi \sin(n, r) + \Phi_1 \right] ds$$

$$+ \text{v.p.} \int_C \mu(s) \left[\frac{2}{\pi r} \cos 2\varphi \sin(n, r) + \Phi_2 \right] ds, \qquad (41.8)$$

$$v(\sigma) = -2\mu(\sigma) + \text{v.p.} \int_C \rho(s) \left[\frac{2}{\pi r} \cos 2\varphi \sin(n, r) + \Phi_2 \right] ds$$

$$+ \text{v.p.} \int_C \mu(s) \left[\frac{2}{\pi r} \sin 2\varphi \sin(n, r) + \Phi_3 \right] ds. \qquad (41.9)$$

The integrals in equations (41.8) and (41.9) are understood in the sense of their principal values.

These equations form a system of singular integral equations. Let us transform these systems into a regular one. We will perform the regularization process following S.G. Mikhlin who in [38] reduced the system of singular integral equations for the plane problem of elasticity theory of an anisotropic medium to a regular system.

For this purpose we first of all note that the expressions $\sin(n, r)/r$, which have a singularity of a simple pole type, can be written in the form

$$\frac{\sin(n, r)}{r} ds = \frac{1}{2} \cotan \frac{\tau - t}{2} d\tau + p(t, \tau),$$

where τ, t are parameters, which are varied from 0 to 2π; $p(t, \tau)$ is a continuous function. For the regularization we can use the Hilbert formula

$$\frac{1}{4\pi^2} \int_0^{2\pi} \cotan \frac{\tau - t}{2} \left\{ \int_0^{2\pi} \cotan \frac{s - \tau}{2} \rho(s)\, ds \right\} d\tau = -\rho(t) + \frac{1}{2\pi} \int_0^{2\pi} \rho(\tau)\, d\tau.$$

Let us rewrite the system, which was obtained before, in the form

$$L_1 = -2\rho(\sigma) + \text{v.p.} \int_C \rho(s) \left[-\frac{2}{\pi r} \sin 2\varphi \sin(n, r) + \Phi_1 \right] ds$$

$$+ \text{v.p.} \int_C \mu(s) \left[\frac{2}{\pi r} \cos 2\varphi \sin(n, r) + \Phi_2 \right] ds = u_1(\sigma),$$

$$L_2 = -2\mu(\sigma) + \text{v.p.} \int_C \rho(s) \left[\frac{2}{\pi r} \cos 2\varphi \sin(n, r) + \Phi_2 \right] ds$$

$$+ \text{v.p.} \int_C \mu(s) \left[\frac{2}{\pi r} \sin 2\varphi \sin(n, r) + \Phi_3 \right] ds = v_1(\sigma).$$

Let us set

$$L_1^*(\mu_1, \rho_1) = 2\rho_1(\sigma) + \text{v.p.} \int_C \rho_1(s) \left[-\frac{2}{\pi r} \sin 2\varphi \sin(n, r) + \Phi_1 \right] ds$$

$$+ \text{v.p.} \int_C \mu_1(s) \left[\frac{2}{\pi r} \cos 2\varphi \sin(n, r) + \Phi_2 \right] ds,$$

$$L_2^* = 2\mu_1(\sigma) + \text{v.p.} \int_C \rho_1(s) \left[\frac{2}{\pi r} \cos 2\varphi \sin(n, r) + \Phi_2 \right] ds$$

$$+ \text{v.p.} \int_C \mu_1(s) \left[\frac{2}{\pi r} \sin 2\varphi \sin(n, r) + \Phi_3 \right] ds.$$

After forming the system of integral equations

$$L_1^*(L_1, L_2) = L_1^*(u_1, v_1),$$
$$L_2^*(L_1, L_2) = L_2^*(u_1, v_1),$$

and using the Hilbert formula, we obtain the regular system of equations

$$2\rho(\sigma) + \ldots = L_1^*(u_1, v_1),$$
$$2\mu(\sigma) + \ldots = L_2^*(u_1, v_1).$$

The dots here denote regular integrals.

The problem on stabilized oscillations of a plate clamped along the contour can be reduced to the integration of the equation

$$\nabla^2 \nabla^2 w - \lambda^4 w = 0.$$

In this case the functions u_0', v_0', u_0'', v_0'' should be taken in the following form:

$$u_0' = -\left[Y_2(r) + \frac{2}{\pi} K_2(r) \right] \sin 2\varphi,$$

$$v_0' = \left[Y_2(r) + \frac{2}{\pi} K_2(r) \right] \cos 2\varphi + Y_0(r) - \frac{2}{\pi} K_0(r),$$

$$u_0'' = -\left[Y_2(r) + \frac{2}{\pi} K_2(r) \right] \cos 2\varphi + Y_0(r) - \frac{2}{\pi} K_0(r),$$

$$v_0'' = -\left[Y_2(r) + \frac{2}{\pi} K_2(r) \right] \sin 2\varphi.$$

One can easily verify that

$$\frac{\partial u_0'}{\partial y} = \frac{\partial v_0'}{\partial x}, \qquad \frac{\partial u_0''}{\partial y} = \frac{\partial v_0''}{\partial x}.$$

The problem on the bending of a plate compressed by a hydrostatical pressure p can be reduced to the integration of the differential equation $D\nabla^2\nabla^2 w + p\nabla^2 w = q(x, y)$.

For a plate clamped along the contour, the problem can be reduced to the integration of this equation without the right-hand side under the following conditions on the contour: $u(s) = f_1(s)$, $v(s) = f_2(s)$, where as before $u = \frac{\partial w}{\partial x}$, $v = \frac{\partial w}{\partial y}$.

In order to solve this problem we should introduce the functions

$$u_0' = -\left[Y_2(r) + \frac{4}{\pi r^2}\right]\sin 2\varphi,$$

$$v_0' = \left[Y_2(r) + \frac{4}{\pi r^2}\right]\cos 2\varphi + Y_0(r),$$

$$u_0'' = -\left[Y_2(r) + \frac{4}{\pi r^2}\right]\cos 2\varphi + Y_0(r),$$

$$v_0'' = -\left[Y_2(r) + \frac{4}{\pi r^2}\right]\sin 2\varphi.$$

Obtaining the integral equations in the last two problems can be performed just in the same way as in the previous problem on a plate on an elastic foundation.

42. On wave propagation in electric power lines

Consider the problem on wave propagation in electric power lines [55].

Let v and i, respectively, denote the voltage and the current in the single-wire system; moreover, let L, C, R and G, respectively, denote the inductance, the capacity, the resistance and the conductivity. It is known that the voltage and the current satisfy the following system of equations:

$$-\frac{\partial v}{\partial x} = L\frac{\partial i}{\partial t} + Ri, \tag{42.1}$$

$$-\frac{\partial i}{\partial x} = C\frac{\partial v}{\partial t} + Gv, \tag{42.2}$$

if $R = 0$, $G = 0$, then the line is called *non-corrupting*, and the system of equations (42.1), (42.2) leads to the well studied one-dimensional wave equation

$$\frac{\partial^2 v}{\partial x^2} = LC\frac{\partial^2 v}{\partial t^2}.$$

In the general case the system of equations (42.1), (42.2) leads to a differential equation of the second order, which is sometimes called the *telegraphers equation*

$$\frac{\partial^2 v}{\partial x^2} - LC\frac{\partial^2 v}{\partial t^2} - (RC + LG)\frac{\partial v}{\partial t} - RGv = 0.$$

A similar equation can also be written with respect to the function i.

Exactly the same equation can be obtained when considering the problem on the oscillations of a string which has a viscous resistance and is supported by an elastic foundation. In this case the deflection of the string w satisfies the equation

$$T\frac{\partial^2 w}{\partial x^2} - m\frac{\partial^2 w}{\partial t^2} - \gamma\frac{\partial w}{\partial t} - k_0 w = 0, \tag{42.3}$$

where k_0 is the coefficient of subgrade reaction, T is the tension, m is the mass per unit length, γ is a number which characterizes the dependence of the dissipative force on the velocity.

Obviously, in both cases we assume that all the physical characteristics are constant and the perturbations are applied only at the ends of the power line or the string.

Since the reader may wish to consider some analogies between the problem formulated and the problems of mechanics which were considered before, we will base our presentation on equation (42.3). Moreover, the physical interpretation will be presented for the problem on the longitudinal oscillations of a rod of a constant cross-section which is situated in an elastic medium. We should only assume that in equation (42.3) w is the longitudinal displacement and replace T by EF, where E is the modulus of elasticity, T is the area of the transversal cross-section and k_0 is the coefficient of subgrade reaction related to the whole perimeter of the cross-section. Note that the derivative of the required function w with respect to the coordinate x is connected with the normal force by the relation $N = EF\frac{\partial w}{\partial x}$. First, we make the necessary calculations in the case, when the attenuation is absent. Consider a semi-infinite rod. Let the origin be at its end and the x-axis be directed to the rod. Suppose that the end of the rod is exposed to the action of the force $1 \cdot e^{-i\omega t}$. Then for $\gamma = 0$ the solution of equation (42.3) takes the form

$$w = \frac{c}{T\sqrt{p^2 + \mu^2 c^2}} \exp\left[-\frac{x}{c}\sqrt{p^2 + \mu^2 c^2} + pt\right], \qquad (42.4)$$

where $p = -\omega i$, $\mu = \sqrt{k_0/T}$, $c = \sqrt{T/m}$.

Consider the problem on the action of a unit instantaneous impulse. From the physical point of view we should consider a superposition of harmonics whose sum gives the instantaneous impulse. If we use the terminology of the operational calculus and consider the factor of (42.4), which does not depend on the time

$$\Psi(x, p) = \frac{c}{T\sqrt{p^2 + \mu^2 c^2}} \exp\left[-\frac{x}{c}\sqrt{p^2 + \mu^2 c^2}\right]$$

as the representation of a function $\psi(x, t)$, i.e. if we assume that $\psi(x, t)$ satisfies the relation

$$\Psi(x, p) = \int\limits_0^\infty e^{-px}\psi(x, t)\, dt,$$

then $\psi(x, t)$ is the solution required.

The calculation of the signal gives

$$w(x, t) = \begin{cases} 0, & t < x/c, \\ \frac{c}{T} J_0\left(\mu c\sqrt{t^2 - (x/c)^2}\right), & t \geq x/c. \end{cases}$$

In the case, when equation (42.3) contains the first derivative with respect to the time, we can write the solution, setting $\gamma = b\kappa$, in the form

$$w(x, t) = \begin{cases} \frac{1}{mc} e^{-nc^2 t} J_0\left(\sqrt{\mu^2 - n^2 c^2}\sqrt{t^2 - x^2/c^2}\right), & t \geq x/c, \\ 0, & t < x/c, \end{cases}$$

where $n = b\kappa/(2T)$.

43. The action of an impulse on cylindrical and prismatic tanks filled with a fluid

In this section we consider the question on oscillations of the fluid in a vertical cylindrical or prismatic tank. We suppose that the fluid is ideal, and the oscillations are caused by the action of a horizontal impulse. The question is considered of what pressure is then acting on the walls, bottom and the internal columns of the tank.

The oscillations of the fluid in a vertical circular cylindrical tank have been investigated by Jacobsen [68]. The same author also considered the problem on oscillations of the fluid near a vertical round column. A.A. Okhotsimski [42] has solved the problem for a cylindrical tank with a round column placed vertically. The author of the paper [23] gave the general solution of a similar problem for a circular cylinder with a column placed eccentrically.

First consider the circular cylindrical tank with one round column placed eccentrically.

Suppose that at zero time, i.e., at $t = 0$ the fluid is at rest. The cylindrical tank, which is stationary at first, begins to move translationally at the instant $t = 0$ in the horizontal direction which we assume to coincide with the direction of the x-axis according to the law $x = f(t)$.

Consider the motion of the fluid with respect to the moving tank, assuming the upper surface of the fluid to be free. We shall seek the potential of the velocities in the form

$$\Phi(r, z, \theta; t) = \Phi_1(r, z, \theta) f'(t).$$

The function Φ_1 satisfies the equation $\nabla^2 \Phi_1 = 0$, where ∇^2 is the Laplace operator in cylindrical coordinates.

The boundary conditions on the lateral area of the cylinder can be obtained starting from the fact that the relative velocity of the movement of the fluid at the points belonging to the lateral area is directed along the tangent to this surface. This is the condition of impenetrability. Therefore, taking into account the known expression for the velocity v via the potential of the velocities Φ, we obtain the following boundary conditions on the lateral surfaces:

$$\left(\frac{\partial \Phi}{\partial r}\right)_{r=b} = f'(t) \cos \theta, \qquad (43.1)$$

$$\left(\frac{\partial \Phi}{\partial r_1}\right)_{r_1=a} = f'(t) \cos \beta, \qquad (43.2)$$

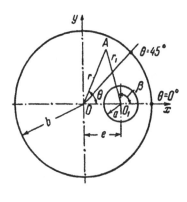

Figure 68

where r, r_1 are, respectively, the distances from the axis of the cylinder and the axis of the column to the point A; θ, β are the polar angles with the vertices on the axis of the cylinder and the axis of the column; b, a are the radii of the cylinder and the column; e is the distance from the axis of the column to the axis of the cylinder (Fig. 68).

The vertical component of the velocity on the lower surface of the cylinder should be equal to zero, i.e., $\left(\frac{\partial \Phi}{\partial z}\right)_{z=0} = 0$. On the free surface the pressure must be equal to zero. Therefore,

$$p_{z=h} = -\gamma \frac{\partial \Phi}{\partial t} + F(t), \qquad (43.3)$$

where γ is the density; the function $F(t)$ can be included in Φ, because its derivatives with respect to the coordinates are equal to zero, hence, it does not produce an effect on the distribution of the velocities. Thus, assuming $F(t) = 0$ in (43.3), we obtain $\gamma \frac{\partial \Phi}{\partial t} = 0$ for $z = h$, hence, $\Phi(r, h, \theta; t) = 0$. Since at zero time the fluid was at rest, $\Phi_1(r, \theta, h) = 0$.

We seek the solution of the problem in the form

$$\Phi(r, z, \theta; t) = \Phi_1(r, z, \theta) f'(t) = \sum_{n=0}^{\infty} \varphi_n(r, \theta) T_n(z) f'(t).$$

Assume that

$$T_n(z) = A_n \cos knz. \tag{43.4}$$

The conditions on the upper surface imply that $\cos knh = 0$. Hence, $k = \pi/2h$, and for even n we should set $A_n = 0$ in (43.4).

Obviously, the derivative of every term of the series with respect to z vanishes at $z = 0$, because

$$\sum_{n=1,3,5,\ldots}^{\infty} \varphi_n(r, \theta) f'(t) A_n kn \sin knz \Big|_{z=0} = 0.$$

The function φ_n satisfies the equation $\nabla^2 \varphi_n - k^2 n^2 \varphi_n = 0$. Thus, the problem considered can be reduced to the solution of the Helmholtz equation. Seeking a solution of this equation in the form

$$\varphi_n(r, \theta) = \sum R_{mn}(r) \Theta_{mn}(\theta),$$

we obtain

$$\varphi_n(r, \theta) = 2 \sum_{m=0}^{\infty}{}' \left[A_{mn}^0 I_m(knr) + B_{mn}^0 K_m(knr) \right] \left(A_{mn}' \cos m\theta + B_{mn}' \sin m\theta \right),$$

$$\tag{43.5}$$

where $I_m(knr)$ and $K_m(knr)$ are the modified Bessel functions, respectively, of the first and the second kind of the integer index m.

By virtue of the symmetry, we should set $B_{mn}' = 0$, and the expression for Φ_1 can be written in the following form:

$$\Phi_1(r, \theta, z) = 2 \sum_{m=0}^{\infty}{}' \sum_{n=1,3,5,\ldots}^{\infty} \{ A_{mn} I_m(knr) + B_{mn} K_m(knr) \} \cos knz \cos m\theta,$$

where $A_{mn} = A_{mn}^0 A_{mn}'$ and $B_{mn} = B_{mn}^0 A_{mn}'$.

In order to pass to the coordinate system whose axis coincides with the axis of the column, we should replace r by r_1 and θ by β in (43.5).

We shall seek the function Φ_1 as the sum of two series

$$\Phi_1 = 2 \sum_{m=0}^{\infty}{}' \sum_{n=1,3,5,\ldots}^{\infty} A_{mn} I_m(knr) \cos knz \cos m\theta$$

$$+ 2 \sum_{m=0}^{\infty}{}' \sum_{n=1,3,5,\ldots}^{\infty} B_{mn} K_m(knr_1) \cos knz \cos m\beta. \tag{43.6}$$

We suppose that the radius of the column is small with respect to the radius of the cylinder and only take the two first terms with respect to the index m which are present in (43.6). Thus, the expression for the potential of velocities takes the following form:

$$\Phi = \sum_{n=1,3,5,\ldots}^{\infty} \left\{ 2\sum_{m=0}^{\infty}{}' A_{mn} I_m(knr) \cos m\theta + B_{0n} K_0(knr_1) \right.$$

$$\left. + 2B_{1n} K_1(knr_1) \cos \beta \right\} \cos(knz) f'(t).$$

The symbol $\sum'$, here as everywhere, denotes that, when summing, we introduce the factor $1/2$ for $m = 0$. Let us go to the determination of the coefficients A_{mn}, B_{0n}, B_{1n}.

For this purpose we pass from the coordinates r_1, β to the coordinates r, θ and use the following addition formulae for Bessel functions:

$$K_0(knr_1) = 2\sum_{m=0}^{\infty}{}' K_m(knr) I_m(kne) \cos m\theta,$$

$$K_1(knr_1) \cos' \beta = 2\sum_{m=0}^{\infty}{}' K_{m+1}(knr) I_m(kne) \cos m\theta,$$

$$r_1 = \sqrt{r^2 + e^2 - 2re\cos\theta}, \qquad r \geq e.$$

Thus, condition (43.1) on the external contour implies, after the necessary calculations, that

$$A_{\nu n} = \frac{-2B_{0n} K'_\nu(knb) I_\nu(kne) - 4B_{1n} K'_\nu(knb) I'_\nu(kne) + \psi/(kn)}{2I'_\nu(knb)},$$

where

$$\nu = m; \qquad \psi = \begin{cases} (-1)^{(n-1)/2}\dfrac{4}{\pi n} & \text{for } \nu = 1, \\ 0 & \text{for } \nu \neq 1. \end{cases}$$

Consider condition (43.2) on the internal contour, by setting $r_1 = a$. In order to satisfy this condition, we should rewrite all the terms of (43.6) containing A_{mn} with the help of the appropriate addition formulae so as to pass to the coordinate system r_1, β. Then one can collect all the terms which contain the same factors $\cos \nu\beta$; however, this does not allow one to satisfy the conditions for all ν, since at the beginning we assumed that only the terms which contain B_{0n} and B_{1n} remain. Therefore, in order to determine B_{0n} and B_{1n} it is necessary to have only two equations. These equations can be obtained, by satisfying the boundary conditions for $\nu = 0$ and $\nu = 1$.

We only write the final result without the intermediate calculations. In order to determine B_{on} and B_{1n} we have two equations

$$C_{1n} B_{0n} + D_{1n} B_{1n} = F_{1n};$$

$$C_{2n} B_{0n} + D_{2n} B_{1n} = F_{2n},$$

where

$$C_{1n} = -2\sum_{m=0}^{\infty}{}' K_m'(knb)\frac{I_m(kne)}{I_m'(knb)}I_0'(kna)I_m(kne) + K_0'(kna),$$

$$D_{1n} = -4\sum_{m=0}^{\infty}{}' K_m'(knb)\frac{I_m'(kne)}{I_m'(knb)}I_0'(kna)I_m(kne),$$

$$F_{1n} = -\frac{\psi}{kn}I_0'(kna)I_1(kne)\frac{1}{I_1'(knb)},$$

$$C_{2n} = -2I_1'(kna)\sum_{m=0}^{\infty}{}' K_m'(knb)\frac{I_m(kne)}{I_m'(knb)}[I_{m-1}(kne) + I_{m+1}(kne)],$$

$$D_{2n} = -4I_1'(kna)\sum_{m=0}^{\infty}{}' \frac{K_m'(knb)I_m'(kne)}{I_m'(knb)}[I_{m-1}(kne) + I_{m+1}(kne)] + 2K_1'(kna),$$

$$F_{2n} = -\frac{\psi}{kn}\frac{I_1'(kna)}{I_1'(knb)}[I_0(kne) + I_2(kne)] + \frac{4}{\pi}\cdot\frac{1}{n}(-1)^{(n-1)/2}\frac{1}{kn}.$$

If we keep p terms containing B_{mn} in the solution (43.6) instead of two such terms, then A_{mn} can be expressed with the help of conditions on the external contour via p unknown B_{mn}. We could obtain p equations with the unknown $B_{\nu n}$, where $\nu = 1,\ 2,\ 3,\ \ldots,\ p$ on the internal contour.

The similar calculation has been performed in the paper [23].

The problem on a rectangular tank which moves in a direction parallel to one of the sides of the rectangle forming the bottom, is essentially more simple and well studied. Obviously, if the expression for the potential of the velocities contains, as before, the series $\sum\varphi_n(x, y)\cos knz$, then $\varphi_n(x, y) = \varphi_n(x)$, i.e., the corresponding plane problem on the determination of the function φ_n becomes a one-dimensional problem.

The problem on a rectangular tank in the case when it moves along a straight line which is not parallel to the sides of the bottom is more difficult. The main difficulty in this case arises when representing the potential of the velocities as a product $f_1'(t)\sum\varphi_n(x, y)\cos knz$, and consists in the determination of the functions $\varphi_n(x, y)$ which satisfy the equation $\nabla^2\varphi_n - k^2 n^2\varphi_n = 0$ and the boundary conditions resulting from the restriction that the wall of the tank is impenetrable. These conditions can be reduced to the condition that the normal derivative of the function φ_n on the contour is equal to a given function of the point of the contour. In order to solve this problem we apply one of versions of the method of compensating loadings.

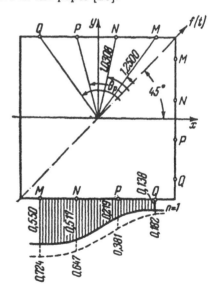

Figure 69

We seek a solution for a tank which has no internal columns in the following form:

$$\varphi_n(r, \varphi) = \sum_{k=1}^{\infty} A_k I_k(knr)[\cos k\theta + \alpha_k \sin k\theta],$$

and we select the coefficients A_k and α_k so that the boundary conditions on the contour hold in a sense.

As an example, we consider the problem of the determination of the pressure on the walls of a square tank, assuming that the direction of the impulse movement $f(t)$ is along its diagonal. We seek the functions $\varphi_n(r, \theta)$ in the form

$$\varphi_n(r, \theta) = \sum_{m=1,3}^{7} A_{mn} I_m(knr) \cos m\theta,$$

where θ and r are the polar coordinates of the point considered (Fig. 69). We require that the conditions of impenetrability hold at the points M, N, P, Q of one of the sides of the square, which are the mid-points of the segments obtained after dividing the side into four equal parts. Obviously, the same conditions holds at the corresponding points of all other sides. These conditions give the following system of equations

$$A_{1n} \left[\frac{dI_1(knr_i)}{dr} \cos\theta \sin(45° + \theta) - I_1(knr_i)\frac{1}{r_i} \sin\theta \cos(45° + \theta) \right]$$

$$+ A_{3n} \left[\frac{dI_3(knr_i)}{dr} \cos 3\theta \sin(45° + \theta) - I_3(knr_i)\frac{1}{r_i} 3\sin 3\theta \cos(45° + \theta) \right]$$

$$+ A_{5n} \left[\frac{dI_5(knr_i)}{dr} \cos 5\theta \sin(45° + \theta) - I_5(knr_i)\frac{1}{r_i} 5\sin 5\theta \cos(45° + \theta) \right]$$

$$+ A_{7n} \left[\frac{dI_7(knr_i)}{dr} \cos 7\theta \sin(45° + \theta) - I_7(knr_i)\frac{1}{r_i} 7\sin 7\theta \cos(45° + \theta) \right]$$

$$= \frac{4}{\pi\sqrt{2}} \frac{1}{n}(-1)^{(n-1)/2},$$

where $n = 1, 3, 5, 7$, $i = M, N, P, Q$ (in our case $r_M = r_Q$, $r_N = r_P$).

First, we should set $n = 1$, then $n = 3$ here. Taking the following numerical data: the length of the side of the square is $a = 1$, the height of the layer of the fluid is $h = \pi$, the density is $\gamma = 1$, and assuming in accordance with Fig. 69

$r_M = 1.25$,	$r_N = 1.03078$,	$r_P = 1.03078$,	$r_Q = 1.25$;
$\theta_M = 0.1419$,	$\theta_N = 0.5404$,	$\theta_P = 1.0304$,	$\theta_Q = 1.4289$,

we obtain the following system of equations for $n = 1$:

$$0.41321 A_{11} + 0.00616 A_{31} + 10^{-4} \cdot 0.209 A_{51} - 10^{-7} \cdot 0.58 A_{71} = 0.90032,$$

$$0.41177 A_{11} - 0.00246 A_{31} - 10^{-4} \cdot 0.479 A_{51} - 10^{-7} \cdot 0.84 A_{71} = 0.90032,$$

$$0.32008 A_{11} - 0.00815 A_{31} + 10^{-5} \cdot 0.56 A_{51} + 10^{-6} \cdot 0.117 A_{71} = 0.90032,$$

$$0.21809 A_{11} - 0.01104 A_{31} + 10^{-3} \cdot 0.1093 A_{51} - 10^{-6} \cdot 0.403 A_{71} = 0.90032.$$

Solving this system, we obtain

$$A_{11} = 2.33813,$$
$$A_{31} = -17.0810,$$
$$A_{51} = 2003.44,$$
$$A_{71} = 42569.9.$$

Similarly, for $n = 3$ we have

$$A_{13} = -0.14973;$$
$$A_{33} = 0.28253;$$
$$A_{53} = -1.45713;$$
$$A_{73} = -0.19584.$$

Now keeping only the terms with $n = 1$ and $n = 3$ in the expression for the potential of the velocities, we can easily find the pressure on the walls by the formula

$$p = -\gamma f''(t) \sum_{n=1,3} \sum_{m=1,3}^{7} A_{mn} I_m(knr_i) \cos m\theta_i.$$

APPENDIX A

Brief information on gamma functions

The gamma function of the argument z, which is also called the Euler integral of the second kind, is denoted by $\Gamma(z)$ and defined with the help of the equality

$$\Gamma(z) = \int_1^\infty e^{-x} x^{z-1} dx. \tag{1}$$

This integral converges, if $\Re z > 0$.

For $\Re z > 0$ the function $\Gamma(z)$ is regular. Using analytic continuation, one can consider $\Gamma(z)$ for $\Re z < 0$ and show that it is meromorphic and can be represented by the formula

$$\Gamma(z) = \sum_{n=0}^\infty \frac{(-1)^n}{n!} \frac{1}{z+n} + \int_1^\infty e^{-t} t^{z-1} dt, \qquad \Re z < 0. \tag{2}$$

From this equality we see that the function $\Gamma(z)$ has simple poles at the points $z = -n$, where $n = 0, 1, 2, \ldots$. The main property of the gamma function is expressed by the equality

$$z\Gamma(z) = \Gamma(z+1), \tag{3}$$

which can easily be obtained from (1).

Assuming that z lies inside the segment $[0,1]$ we can, by multiplying $\Gamma(z)$ and $\Gamma(1-z)$ represented by formula (1), obtain an important formula after easy computations

$$\Gamma(z)\Gamma(1-z) = \frac{\pi}{\sin \pi z}. \tag{4}$$

According to the principle of analytic continuation, formula (4) can be extended to the whole of the complex plane. Let us give two more functional relations

$$\Gamma(1/2+z)\Gamma(1/2-z) = \frac{\pi}{\cos \pi z}, \tag{5}$$

$$\frac{\Gamma(z)\Gamma(1-z)}{\Gamma(1/2+z)\Gamma(1/2-z)} = \frac{1}{\tan \pi z}; \tag{6}$$

the latter is a consequence of (4) and (5).

The duplication formula for the gamma functions has the form

$$\Gamma(2z) = 2^{2z-1} \frac{1}{\sqrt{\pi}} \Gamma\left(z + \frac{1}{2}\right) \Gamma(z). \tag{7}$$

Via gamma functions we can express the integral

$$\int_0^{\pi/2} \sin^{\mu-1} x \cos^{\mu_1-1} x \, dx = \frac{1}{2} \frac{\Gamma(\mu/2)\Gamma(\mu_1/2)}{\Gamma(\mu/2+\mu_1/2)}, \quad \Re\mu > 0, \quad \Re\mu_1 > 0. \tag{8}$$

For the logarithmic derivative of the gamma function

$$\psi(z) = \frac{d}{dz} \ln\Gamma(z) = \frac{\Gamma'(z)}{\Gamma(z)} \tag{9}$$

the following expansion holds

$$\psi(z) = -C - \frac{1}{z} - \sum_{k=1}^{\infty} \left[\frac{1}{z+k} - \frac{1}{k} \right], \tag{10}$$

where C is the Euler constant; its approximate value is equal to 0.5772157.

If n is a positive integer, then

$$\psi(n+1) = -C + 1 + 1/2 + \cdots + 1/n. \tag{11}$$

Let us give some particular values of the gamma function

$$\Gamma(1) = \Gamma(2) = \int_0^{\infty} e^{-x} dx = 1, \tag{12}$$

$$\Gamma(1/2) = \int_0^{\infty} e^{-x} x^{-1/2} dx = \sqrt{\pi}. \tag{13}$$

If n is a natural number, then

$$\Gamma(n) = (n-1)!, \tag{14}$$

$$\Gamma\left(n + \frac{1}{2}\right) = \frac{\sqrt{\pi}}{2^n} (2n-1)!!, \tag{15}$$

where $(2n-1)!! = 1 \cdot 3 \cdot 5 \ldots (2n-1)$.

More complete information concerning the gamma functions is given in [3], [31], [53], [72].

Bibliographical notes

The most detailed guidebook on the theory of the Bessel functions is the well-known book by G.N. Watson "A treatise on the theory of Bessel function" [7]. The account of Watson contains an extended bibliography.

The book by Watson contains a detailed and complete presentation of the theory of Bessel functions; it should be noted that among all the sources used by the author it is the most important.

The book by N. Nielsen [76], which is the most detailed account among those published before the appearance of the book by Watson and which contains a series of important original results, did not hitherto lose its significance. The book [49] in which a series of the classical investigations of N.Ya. Sonin concerning the cylindrical functions as well as some comments by N.I. Akhiezer are included, is of considerable interest.

A brief exposition of the theory of Bessel functions is given in the seventh chapter of the guidebook "Higher transcendental functions" [4]. This chapter also includes a very useful reference section which contains a great number of formulae obtained after 1922, which hence, were not mentioned in the book by Watson. Some topics of the theory of Bessel functions are considered from different points of view in courses of the equations of mathematical physics and the theory of special functions. It is not expedient to list these courses here.

The integrals of Bessel functions are considered in a very useful book [73]. It contains a very detailed list of indefinite integrals of Bessel functions and their product by a power and exponential functions. A special chapter is devoted to Airy functions. There is a series of formulae concerning definite, iterated and improper integrals. An emphasis is put on the integrals of Struve functions. The appendix contains some tables of Bessel functions and integrals of the Bessel functions. Extensive bibliographical notes are contained in this book.

A detailed list of integral transforms, whose kernels are the Bessel functions, is given in the known reference book in two volumes "Tables of integral transformations" [5, 6]. This edition also contains a lot of important improper integrals as well as tables of various integrals of Bessel functions.

A series of important improper integrals is contained in the very useful reference book "Integral transforms and operational calculus" [13].

Many formulae relating to different sections of the theory of Bessel functions, in particular, integrals, are contained in the reference book by I.S. Gradstein and I.M. Ryzhik [10].

Problems of the theory of incomplete cylindrical functions are considered in detail in the monograph by M.M. Agrest and M.Z. Maksimov [2]; tables of these functions are given in [1].

In the book by N.M. Sovetov and M.E. Averbukh [48] the functions are considered which in the present book are called the fundamental Cauchy functions of the Bessel equation, and in other papers of the author (for instance, [20]) are called functions having the property of the unit matrix.

Different generalizations of these functions are considered in the book [18], which also contains useful formulae and tables.

The well-known book by A. Gray and G.B. Matheus should be mentioned, which is one of the best guidebooks on the theory of Bessel functions. Many questions are considered in this book in more detail and withmore rigour than in the present book.

Among the works devoted to the presentation of the theory of Bessel functions we should mention the very useful book by R.O. Kuz'min [28], which is the first special monograph of an educational type on Bessel functions published in USSR and which played an important role in the popularization of the theory of Bessel functions.

A brief account of the tables of the Bessel functions is given in [20]. One can also find there references to more detailed reference books concerning tables; one should emphasize among them a very detailed reference book [33].

A detailed guidebook [78], which has mainly a reference character, is of interest.

The reader interested in applications of methods of the theory of functions of a complex variable, in particular, of contour integrals to the problems of the theory of Bessel functions, can use the book by M.A. Lavrentiev and B.V. Shabat [31]. This book contains a special chapter devoted to Bessel functions.

The book by E. Jahnke, F. Emde and F. Lösch "Special functions" [61] is very popular. This book contains formulae as well as graphs and tables of Bessel functions.

The applications of integral transformations, whose kernels are the cylindrical functions, are discussed in detail in the monographs by I. Sneddon [47] and Ya.S. Uflyand [54]. A series of applications of the theory of cylindrical functions to operational calculus can be found in the monograph by A.I. Lurie [35]. Applications of the theory of cylindrical functions to some problems of mathematical physics are given in numerous works. In particular, problems connected with the use of integral equations are widely used in the monograph by V.D. Kupradze [29].

The well-known monograph by A.N. Dinnik [12] is devoted to applications of Bessel functions to problems of the theory of elasticity. The theory of Bessel functions and their applications in radio engineering are presented in the book by T.A. Rozet [44]. Applications of Bessel functions are also considered in [20].

All the works listed above consider the theory of Bessel functions from the viewpoint of classical analysis. The applications of modern analysis, in particular, the theory of group representations in the theory of Bessel functions is very significant. The monograph by N.Ya. Vilenkin [8] is devoted to these questions.

After the book appeared in Russian, two other monographs were published which concern applications of Bessel functions and are included in the reference list. One of them is due to the author and is devoted to problems of the theory of heat conductivity and thermoelasticity; the other one has been written by the author with L.M. Reznikov and contains the topics relating to the theory of dynamic vibration absorbers; its third chapter is strongly associated with the use of Bessel functions. The reader of the English translation has the opportunity to use reference books on special functions, namely, the reference book by M. Abramowitz and

I. Stegun and the reference book by Y.L. Luke which are included in the reference list. Without doubt, it is very important for the reader to use the reference books on integrals and series. In this connection we included the reference book in three volumes by A.P. Prudnikov, Yu.A. Brychkov and O.I. Marichev which is published in English and is the most extensive edition of this type in the bibliography.

Bibliography

[1] Agrest M.M., Bekauri I.N., Maksimov M.S. *et al. Tables of Incomplete Cylindrical Functions*, Gos. kom. po ispol'z. atomnoj energii SSSR, 1966.

[2] Agrest M.M. and Maksimov M.S. *Theory of Incomplete Cylindrical Functions and their Application*, Springer-Verlag, Berlin–Heidelberg–New York, 1971.

[3] Bateman H. and Erdélyi A. *Higher Transcendental Functions. Vol. 1. Hypergeometric Function, Legendre Functions*. New York, McGraw-Hill 1953.

[4] ———. *Higher Transcendental Functions. Vol. 2. Bessel Functions, Functions of the Parabolic Cylinder, Orthogonal Polynomials*. New York, McGraw-Hill 1953.

[5] ———. *Tables of Integral Transforms. Vol. 1. Fourier, Laplace and Mellin Transforms*. New York, McGraw-Hill 1954.

[6] ———. *Tables of Integral Transforms. Vol. 2. Bessel Transforms. Integrals of Special Functions*. New York, McGraw-Hill 1954.

[7] Watson G.N. *A Treatise on the Theory of Bessel Functions*. 2. ed. Cambridge, The University Press 1959.

[8] Vilenkin N.Ya. *Special Functions and Theory of Group Representations*, AMS, Translations of Mathematical Monographs, 1968.

[9] Galin L.A. *Contact Problems of the Theory of Elasticity*, Gostehizdat, 1953.

[10] Gradstein I.S. and Ryzhik I.M. *Tables of Integrals, Sums, Series and Products*, 5th ed., Academic Press, Inc., 1994.

[11] Gray A. and Mathews G.B. *A Treatise on Bessel Functions and their Applications in Physics*. 3. ed. London, Macmillan 1948.

[12] Dinnik A.N. *Selected Works, Vol. II. Application of Bessel Functions to Problems of Elasticity Theory*. AN USSR, 1952.

[13] Ditkin V.A. and Prudnikov A.P. *Integral Transforms and Operational Calculus*, Pergamon Press, Oxford-Edinburgh-New York, 1965.

[14] Dorodnitsyn A.A. To the theory of diurnal movement of the temperature in the layer of intermixing, *Dokl. AN SSSR* **XXX** (1941), no. 5.

[15] Zabreiko P.P., Koshelev A.I., Krasnosel'skiy M.A., Mikhlin S.G., Rakovchshik L.S., and Stetsenko B.Ya. *Integral Equations*, Nauka, Moscow, 1968.

[16] Carslaw H.S. and Jaeger J.C. *Conduction of Heat in Solids*, 2. ed. Oxford, Clarendon Press 1959.

[17] Klein G.K. Calculation of beams on linearly-deformable half-space. *Sovetskii metropoliten*, (1940), no. 12.

[18] Korenev B.G. On a method of initial parameters in problems of circular plates and shells of rotation, *PMM* **10** (1946), no. 1.

[19] ———. *Problems of Calculation of Beams and Plates on an Elastic Foundation*, Gosstroyizdat, 1954.

[20] ———. *Some Problems of the Theory of Elasticity and Heat Conductivity Solvable in Bessel Functions*, Fizmatgiz, 1960.

[21] Korenev B.G. and Chernigovskaya E.I. *Calculation of Plates on an Elastic Foundation (manual for planners)*, Gosstroyizdat, 1962.

[22] ———. To the calculation of unbounded plates lying on an elastic foundation. *Stroitel'naya mehanika i raschet sooruzheniy* (1966), no. 2.

[23] ———. Action of an impulse onto cylindrical and prismatic tanks filled with a liquid. *Stroitel'naya mehanika*, Collect. of papers dedicated to I.M. Rabinovich. Gosstroyizdat, 1968.

[24] Korenev B.G. and Bukeyhanov S.R. On vibration of rods with a variable cross-section. *Stroitel'naya mehanika i raschet sooruzheniy* (1969), no. 6.

[25] Krylov A.N. *On Calculation of Beams Lying on an Elastic Foundation*, 3-d ed., AN SSSR, 1932

[26] ———. *Vibration of Vessels*, ONTI, 1936.

[27] Kuznetsov D.S. *Special Functions*, 2nd ed., Vysshaya shkola, 1965.

[28] Kuz'min R.O. *Bessel Functions*, 2nd ed., ONTI, 1935.

[29] Kupradze V.D. *Bound Problems of Vibration Theory and Integral Equations*, Gostekhizdat, 1950.

[30] Courant R. and Hilbert D. *Methods of Mathematical Physics. Vol. 1*. New York. Interscience Publ. 1962.

[31] Lavrentiev M.A. and Shabat B.V. *Methods for the Theory of Functions of a Complex Variable* 3d ed., Nauka, Moscow, 1965.

[32] Lardner T. Solution of transverse vibration problems of a certain class of rods with varying thickness expressed in terms of generalized hypergeometric functions. *Applied Mechanics* (Russian translation of Trans. of ASME). **35.E** (1968), no. 1.

[33] Lebedev A.V. and Fedorova R.M. *Reference Book on Mathematical Tables*, AN SSSR, 1956[1].

[34] Lebedev N.N. *Special Functions and their Applications*, Prentice Hall, Inc., Englewood Cliffs, N.J., 1965; Dover Publications, INc., New York, 1972.

[35] Lurieu A.I. *Operatianal Calculus and its Application to Problems of Mechanics*, 2nd ed., Gostekhizdat, 1951.

[36] Love A.E.H. *A Treatise on the Mathematical Theory of Elasticity*, 4. ed., Cambridge, The University Press, 1944.

[37] McLachlan N.W. *Theory and Applications of Mathieu Functions*, Oxford, Pergamon Press, 1947.

[38] Mikhlin S.G. *Plane Deformation in Anisotropic Medium*, AN SSSR, 1936.

[39] ———. *Applications of Integral Equations to some Problems of Mechanics, Mathematical Physics, and Technics*, Gostekhizdat, 1947.

[40] ———. *Direct Methods in Mathematical Physics*, Gostekhizdat, 1950.

[41] Morse P.M. and Feshbach H. *Methods of Theoretical Physics*, McGraw-Hill, New York, 1953.

[42] Okhotzimskiy A.A. To the theory of motion of a body with cavities partially filled by a liquid, *PMM*, (1956), no. 1.

[43] Privalov I.I. *Integral Equations* 2nd ed., ONTI, 1937.

[44] Rozet T.A. *Elements of the Theory of Cylindrical Functions with Applications to Radioengineering*, Sovetskoe Radio, 1956.

[45] Ruchimskiy M.N. *To Calculation of Conic and Sloping Spherical Shells under Axially Symmetrical Loading*, Gostoptehizdat, 1958.

[46] Smirnov V.I. *A Course of Higher Mathematics* v. III, p. 2, (International Series of Monographs in Pure and Applied Mathematics, Pergamon Press, Oxford, 1964.

[47] Sneddon I. *Fourier Transformations*, McGraw-Hill, New York, 1951.

[48] Sovetov N.M. and Averbukh M.E. *Difference Bessel Functions and their Applications in Technics*, Saratov Univ. Press, 1968.

[49] Sonin N.Ya. *Investigations on Cylindrical Functions and Special Polynomials*, Gostehizdat, 1954.

[50] Stratton J.A. *Electromagnetic theory*, McGraw-Hill, New York, London, 1941.

[51] Titchmarsh E. *Introduction to the Theory of Fourier Integrals*, Clarendon Press, Oxford, 1937.

[52] Tikhonov A.N. and Samarsky A.A. *Equations of Mathematical Physics*, Dover Publications Inc., New York, 1990.

[53] Whittaker E.T. and Watson G.N. *A Course of Modern Analysis. An Introduction to the General Theory of Infinite Processes and of Analytic Functions; with an Account of the Principal Trancendental Functions*, part II, Cambridge University Press, New York, 1996.

[54] Uflyand Ya.S. *Integral Transformations in Problems of the Theory of Elasticity*, 2nd ed., Nauka, Leningrad, 1967.

[55] Hayashi S. *Surges on transmission systems*, Denki-shoin, Kyoto, 1955.

[56] Tseitlin A.I. On bending of a circular plate resting on a linearly deformable foundation, *Izv. AN SSSR, Mekhanika Tverdogo Tela*, (1969), no. 1.

---

[1]see also N.M. Burunova. Addition no. 1 to [**33**], AN SSSR, 1959.

[57] _____ . On a method of dual integral equations and dual series and its applications to problems of mechanics, *PMM* (1966), no. 2.

[58] Chudnovskiy A.F. *Physics of Heat Exchange in the Soil*, Gostehizdat, 1948.

[59] Shekhter O.Ya. Calculation of unbounded plate resting on elastic foundation of finite and infinite power and loaded by a point force,. In: *NIS Fundamentstroya*, no. 10, 1939.

[60] Shtaerman I.Ya. *Contact Problem of the Theory of Elasticity*, Gostehizdat, 1949.

[61] Jahnke E., Emde F., and Lösch F. *Tafeln höherer Funktionen*. 6. Auflage. Stuttgart, B.G. Teubner, 1960. Russian transl. Nauka, 1968.

[62] Bowman F. *Introduction to Bessel Functions*, Dover books on mathematics, N.Y., 1958.

[63] Cooke R.G. On the theory of Schlömlich series, *Proc. of the London Math. Soc.*, second series **28** (1928), 207–241.

[64] Farrel O.J. and Ross B. *Solved Problems: Gamma and Beta Functions, Legendre Polynomials, Bessel Functions*, Macmillan, N.Y.; Collier Macmillan, London, 1963.

[65] Friedrichs K.O. *Ein Verfahren der Variationsrechnung*, Nachrichten der Ges. d. Wiss. zu Gottingen, 1929.

[66] Goudet G. *Les Fonctions de Bessel et leur Applications en Physique*, 2-ème ed., Masson, Paris, 1954.

[67] Hayes W. Neo-Schlömilch series, *J. of Math. and Physics*, XXXIV, no. 2 (1956).

[68] Jacobsen L.S. Impulsive hydrodynamics of fluid inside a cylindrical tank and fluid surrounding a cylindrical pier, *Bulletin of the Seismological Society of America* **39** (1949), no. 3.

[69] Jeffreys H. and Swirles B. (Lady Jeffreys). *Methods of Mathematical Physics*, 3-d ed., Univ. press, Cambridge, 1956.

[70] Lauricella G. Sur l'integration de l'equation relative à l'équilibre des plaques élastiques, *Acta Math.*, **32** (1909).

[71] Lommel E. *Studien über die Bessel'schen Funktionen*, Teubner, Leipzig, 1868.

[72] Lösch F. und Schoblik F. *Die Fakultät (Gammafunktion) und verwandte Funktionen mit besonderer Berücksichtung ihrer Anwendungen*, Teubner, Leipzig, 1951.

[73] Luke Y.L. *Integrals of Bessel Functions*, McGraw-Hill, N.Y.–Toronto–London, 1962.

[74] McLachlan N.W. *Bessel Functions for Engineers*, 2-d ed., Clarendon Press, Oxford, 1955.

[75] Neumann C.G. *Theorie der Bessel'schen Funktionen. Ein Analogen zur Theorie der Kugelfunktionen*, Teubner, Leipzig, 1867.

[76] Nielsen N. *Handbuch der Theorie der Cylinderfunktionen*, Teubner, Leipzig, 1904.

[77] Noble B. The solution of Bessel function dual integral equations by a multiplying — factor method, *Proc. Cambridge Philos. Soc.* **59** (1963), no. 2.

[78] Petiau G. *La Théorie des Fonctions de Bessel, Exposée en Vue de ses Applications à la Physique Mathematique*, Centre National de la Recherche Scientifique, Paris, 1955.

[79] Rehwald W. *Elementare Einführung in die Bessel, Neumann und Hankel Funktionen, Wesentliche Eigenschaften der Zylinderfunktionen mit zahlreichen Anwendungsbeispielen aus Physik und Technik*, Hirzel, Stuttgart, 1959.

[80] Pastor J. Rey y Brzezicki A. de Castro. *Funciones de Bessel Teoria Matematica y Applicationes a la Ciencia y a la Tecnica*, Dossat, Madrid, 1958.

[81] Rutgers J.G. Sur des séries et des intégrales définies contenantes les fonctions de Bessel, *Nederl. Akad. Wetensch. Proc.* **44** (1941), 464–474, 636–647, 744–753, 840–851, 978–988, 1092–1098.

[82] Schöbe W. Asymptotische Entwicklung für Zylinderfunktionen, *Acta Math.* **92** (1954), 1–2.

[83] Srivastav R.P. A pair of dual integral equations involving Bessel functions of the first and the second kind. *Proc. Edinburgh Math. Soc.* **14** (1964), no. 12.

[84] Weyrich R. *Die Zylinderfunktionen und ihre Anwendungen*, Teubner, Leipzig und Berlin, 1937.

[85] Woinowsky-Krieger S. Über die Biegung dünner rechteckiger Platten durch Kreislasten, *Ingenieur Archiv* III, H. 3 (1932).

[86] Korenev B.G. Problems of thermoelasticity and particular solutions associated with them of the inhomogeneous Bessel equation, *Dokl. AN SSSR*, **210** (1973), no. 4.

[87] _____ . *Problems of the Theory of Heat Conductivity and Thermoelasticity*, Nauka, Moscow, 1980.

[88] Prudnikov A.P., Brychkov Yu.A., and Marichev O.I. *Integrals and series*. Vol. 1. *Elementary functions*, Gordon and Breach Science Publishers, New York, 1986.

[89] _____. *Integrals and series*. Vol. 2. *Special functions*, Gordon and Breach Science Publishers, New York, 1988.

[90] _____. *Integrals and Series*. Vol. 3. *More special functions*, Gordon and Breach Science Publishers, New York, 1990.

[91] Abramowitz M. and Stegun I. (eds.). *Handbook of Mathematical Functions with Formulas, Graphs and Mathematical Tables*. Appl. Math. Ser. 55. U.S. Govt. Printing office. Washington, D.C., 1964.

[92] Engelhardt H. *Die einheitlichte Behandlung der Stabknickung mit Berücksichtigung des Stabelgegenwichts in des Eulerfällen I bis 4 als Eigenwertproblem*, Stahlbau, H. 4, 1954.

[93] Federhofer K. *Berechnung der Kipplasten gerader Stäbe mit veränderlicher Höhe*. Verhandlungen der 3. Inter. Kongress für Technische Mechanik. Stockholm, 1930.

[94] Korenev B.G. and Reznikov L.M. *Dynamic Vibration Absorbers. Theory and application*. John Wiley and sons. Chichester, New York, Brisbane, Toronto, Singapore. 1993.

[95] Luke Y.L. *Mathematical Functions and their Approximations*. Academic Press, New York. 1975.

[96] Rothman M. The problem of an infinite plate under an inclined loading with tables of the integrals of Ai(x) and Bi(x), *Quart. Journal Mech. and Applied Math.*, Oxford. B. VII, r.p., 1954.

[97] Salvadori M.J. *Lateral Buckling of Simply Supported Beams of Rectangular Csoss-Section under Bending and Shear*. Proc. of the First U.S. National Congress of Applied Mechanics, 1951.

Index

Printed and bound by CPI Group (UK) Ltd, Croydon, CR0 4YY

23/10/2024

01778248-0006